D1423421

Biopsychosocial Factors of Stress, and Mindfulness for Stress Reduction

Holly Hazlett-Stevens

Editor

Biopsychosocial Factors of Stress, and Mindfulness for Stress Reduction

 Springer

Editor
Holly Hazlett-Stevens
Department of Psychology
University of Nevada
Reno, NV, USA

ISBN 978-3-030-81244-7   ISBN 978-3-030-81245-4   (eBook)
https://doi.org/10.1007/978-3-030-81245-4

© Springer Nature Switzerland AG 2021
This work is subject to copyright. All rights are reserved by the Publisher, whether the whole or part of the material is concerned, specifically the rights of translation, reprinting, reuse of illustrations, recitation, broadcasting, reproduction on microfilms or in any other physical way, and transmission or information storage and retrieval, electronic adaptation, computer software, or by similar or dissimilar methodology now known or hereafter developed.
The use of general descriptive names, registered names, trademarks, service marks, etc. in this publication does not imply, even in the absence of a specific statement, that such names are exempt from the relevant protective laws and regulations and therefore free for general use.
The publisher, the authors, and the editors are safe to assume that the advice and information in this book are believed to be true and accurate at the date of publication. Neither the publisher nor the authors or the editors give a warranty, expressed or implied, with respect to the material contained herein or for any errors or omissions that may have been made. The publisher remains neutral with regard to jurisdictional claims in published maps and institutional affiliations.

This Springer imprint is published by the registered company Springer Nature Switzerland AG
The registered company address is: Gewerbestrasse 11, 6330 Cham, Switzerland

For Ian

Preface

Stress is an inevitable part of our shared human condition as we adapt and change with the constant flux of life. Stress involves the whole mind-body system, always functioning within a dynamic situational, psychological, biological, developmental, and social context. Decades of scientific theory and research have elegantly elucidated how perceptions and appraisals intersect with interconnected physiological systems – such as nervous, endocrine, and immune systems –in the course of stress reactivity. Although many adverse health consequences of chronic stress have been documented, newer research points to possibilities of resilience when adaptive appraisals lead to healthy responding. As early as 1979, people in the West have learned to practice an ancient Buddhist form of stress reduction in a secular format called *mindfulness meditation*. This approach promotes resilience and adaptive responding to the stress of living associated with medical conditions, psychological symptoms, and adverse life circumstances. The original protocol developed to teach mindfulness meditation in clinical settings, mindfulness-based stress reduction (MBSR), has become a widely accepted evidence-based intervention for many medical and stress-related conditions. Adaptations of this protocol for specific clinical problems, such as mindfulness-based cognitive therapy (MBCT) for depression, are efficacious as well.

This book brings together scientific knowledge on the intricate biopsychosocial nature of stress with proliferating research and theories demonstrating how mindfulness improves health and why mindfulness works. The chapters in Part I review historical and current perspectives on stress, explain underlying neurobiological, social, and immunological mechanisms, and explore the roles of perception and appraisal in stress responding. The chapters in Part II introduce the multifaceted topic of mindfulness and review research investigating both the effectiveness of interventions featuring mindfulness meditation (mindfulness-based interventions) and potential mechanisms of change. These chapters were written by experts across scientific disciplines to bridge the gap between the science of stress and the science of mindfulness. I am filled with gratitude and appreciation for each of these authors for their willingness to share their unique perspectives and expertise in collaboration with me on this project.

Holly Hazlett-Stevens
Department of Psychology
University of Nevada
Reno, NV, USA

Contents

About the Editor

Holly Hazlett-Stevens received her Ph.D. in clinical psychology from the Pennsylvania State University in 1999. Under the mentorship of Dr. Thomas D. Borkovec, she studied the nature of anxiety and worry as well as cognitive behavioral therapy for generalized anxiety disorder. From 1999 to 2001, Dr. Hazlett-Stevens was a post-doctoral fellow at the Anxiety Disorders Research Center in the Department of Psychology at the University of California, Los Angeles (UCLA) under the mentorship of Dr. Michelle Craske. There she received training in cognitive behavioral treatment for panic disorder, coordinated panic disorder intervention research projects, and continued her own program of anxiety and worry research.

In 2002, Dr. Hazlett-Stevens joined the faculty of the Department of Psychology at the University of Nevada, Reno, where she is currently an associate professor. She has published over 50 scholarly research articles and book chapters and authored two books, *Women Who Worry Too Much* and *Psychological Approaches to Generalized Anxiety Disorder*. She also co-authored *New Directions in Progressive Relaxation Training* with Douglas A. Bernstein and Thomas D. Borkovec, and the updated *Progressive Relaxation Training: A Guide for Practitioners, Students, and Researchers* with Douglas A. Bernstein.

Since 2010, Dr. Hazlett-Stevens attended a series of intensive professional training programs in Mindfulness-Based Stress Reduction (MBSR) instruction from the University of Massachusetts Medical School Center for Mindfulness in Medicine, Health Care, and Society. She received over 260 hours of professional education in MBSR instruction and is a certified MBSR instructor. Dr. Hazlett-Stevens currently conducts research examining the effects of MBSR, the nature of mindfulness, and how mindfulness training reduces stress and anxiety while improving health and well-being.

Contributors

Mohamad Baydoun Department of Oncology, Cumming School of Medicine, University of Calgary, Calgary, Alberta, Canada

Linda E. Carlson Department of Oncology, Cumming School of Medicine, University of Calgary, Calgary, Alberta, Canada

Laura Cohen Department of Psychological Science, University of Vermont, Burlington, VT, USA

Julie Deleemans Department of Oncology, Cumming School of Medicine, University of Calgary, Calgary, Alberta, Canada

Philip A. Desormeau Department of Psychological Clinical Science, University of Toronto Scarborough, Toronto, Ontario, Canada

Norman A. S. Farb Department of Psychology, University of Toronto Mississauga, Mississauga, Ontario, Canada

Michelle Flynn Department of Psychology, University of Calgary, Calgary, Alberta, Canada

Michael Gordon Department of Counseling Psychology, Santa Clara University, Santa Clara, CA, USA

Emily J. Hangen Department of Psychology, Harvard University, Cambridge, MA, USA

Holly Hazlett-Stevens Department of Psychology, University of Nevada, Reno, NV, USA

Tori Humiston Department of Psychological Science, University of Vermont, Burlington, VT, USA

Jeremy P. Jamieson Department of Psychology, University of Rochester, Rochester, NY, USA

Olena Kleshchova Department of Psychology, University of Nevada, Reno, NV, USA

Amy Hughes Lansing Department of Psychological Science, University of Vermont, Burlington, VT, USA

Andrew W. Manigault Department of Psychology, Ohio University, Athens, OH, USA

Devesh Oberoi Psychosocial Resources, Cumming School of Medicine, University of Calgary, Calgary, Alberta, Canada

Michaela Patton Department of Psychology, University of Calgary, Calgary, Alberta, Canada

Katherine-Anne Piedalue Division of Psychosocial Oncology, University of Calgary, Calgary, Alberta, Canada

Hassan Pirbhai Department of Oncology, Cumming School of Medicine, University of Calgary, Calgary, Alberta, Canada

Asha Putnam Department of Psychology, University of Nevada, Reno, NV, USA

Selma A. Quist-Møller Department of Psychology, University of Copenhagen, Copenhagen, Denmark

Shauna Shapiro Department of Counseling Psychology, Santa Clara University, Santa Clara, CA, USA

Utkarsh Subnis Department of Oncology, Cumming School of Medicine, University of Calgary, Calgary, Alberta, Canada

Daniel Szoke Department of Psychology, University of Nevada, Reno, NV, USA

Kirsti Toivonen Department of Psychology, University of Calgary, Calgary, Alberta, Canada

Mariann R. Weierich Department of Psychology, University of Nevada, Reno, NV, USA

Peggy M. Zoccola Department of Psychology, Ohio University, Athens, OH, USA

Part I
Stress Reactivity and Health

Chapter 1
Stress: Historical Approaches to Allostasis

Tori Humiston and Amy Hughes Lansing

Introduction

Stress research has evolved from animal models to human subject research, with early research focusing largely on the role of physiology and stress from an evolutionary perspective (Romero, 2004). Walter Cannon's fight-or-flight research and Hans Selye's General Adaptation Syndrome (GAS) research set the stage for later research to examine the role of stress on health. Richard Lazarus's research on human subjects provided researchers at the time with the cognitive appraisal of stress theory, paving the way for Thomas Holmes and Richard Rahe to examine the impact of stress on human health. Once the association between illness and stress was discovered, research focused on empirical studies aiming to find the mechanisms of why stress impacts health (McEwen et al., 2015; Taylor, 2010; Thoits, 2010). In the present chapter, a brief overview of stress research will be described. By describing the contributions of Cannon, Selye, and Holmes and Rahe, it will be clear how stress research has evolved from early to current day models such as allostatic load and research into specific mechanisms of stress.

Early Stress Research (1920s–1960s)

Early research on stress was led by Walter Cannon's fight-or-flight theory in animals, which outlines that animals will behave in ways that either aggress the stressor or escape the stressful situation. Fight or flight describes the physiological response of the animal that increases the chances that they survive emergent situations that provoke feelings of pain, fear, or rage (Cannon, 1929). The physiological response

T. Humiston · A. H. Lansing (✉)
University of Vermont, Burlington, VT, USA
e-mail: Amy.Hughes.Lansing@uvm.edu

© Springer Nature Switzerland AG 2021
H. Hazlett-Stevens (ed.), *Biopsychosocial Factors of Stress, and Mindfulness for Stress Reduction*, https://doi.org/10.1007/978-3-030-81245-4_1

of the animal is often referred to as the "active defense" response. Physiological changes during this response include an increase in activities of the sympathetic nervous system, increase in heart rate and respiration, increased blood pressure and blood flow to skeletal muscles, an increase in blood sugar and body temperature, and a decrease in blood flow to the skin and digestive organs. The aforementioned changes in the sympathetic nervous system are controlled by a release of epinephrine in the bloodstream and help the animal prepare to fight the stressor or to escape the stressor (Cannon, 1929).

Following Walter Cannon's cornerstone fight-or-flight theory, Hans Selye and colleagues extended research into physiological changes of animals, notably mice, during stressful events that they termed General Adaptation Syndrome (GAS). GAS, a term that indicates the interrelated adaptive reactions to general stress, posits that all organisms respond in a consistent reaction pattern, even when the cause of the stress varies (Selye, 1946). Selye and colleagues postulated that stress and its causes are always detrimental to an organism unless the organism has an adaptive response to the cause of stress. Selye and colleagues outlined three stages of GAS and their effect on an organism's physiology: alarm-reaction, resistance, and exhaustion. The three stages are an intricate balance between damage to the organism and defense of the organism. Physiological changes associated with the alarm-reaction stage of GAS are the breakdown of tissues, increase in blood pressure, release of adrenergic hormones such as epinephrine and norepinephrine, and a decrease in blood sugar. During the resistance stage, cortisol levels begin to lower, and many of the physiological changes that began in the alarm-reaction stage begin to regulate, including blood pressure and heart rate. In an ideal state, all physiological states return to baseline levels.

While these physiological changes may begin to downregulate, they may not reach pre-stress baseline levels. In the case of chronic stress, hormones, blood pressure, and heart rate may not lower to baseline in an effort to cope and adapt to the chronic stress. Chronic stress then leads to the third stage of Selye's model, which is the exhaustion stage. The exhaustion stage is categorized by many of the same physiological changes as the alarm-reaction stage (breakdown of tissue, low blood sugar, release of hormones from the adrenal cortex, etc.). One of Hans Selye's most important contributions to the stress research literature is that each organism's system balances a mix of passive damage to the organism and active defense, and that an organism has finite resources to adapt to stress based on its own genetic history (Selye, 1951).

As stress research evolved, Richard Lazarus and his colleagues were primarily focused on how stress impacts humans. Lazarus is prominently known for his work on cognitive appraisal of stress and his work with Susan Folkman on the transactional model of stress. Cognitive appraisals are the cognitive processes that lead to the magnitude and kind of emotional reaction. In turn, the emotional reaction is either reinforced or punished by the person's environment, therefore creating a continual interplay between the person and their environment. Lazarus believed that the cognitive process is an exchange between a person's personality traits and environmental stimuli (Lazarus & Monat, 1974). Emotions, as described by Lazarus, have three main components: subjective affect, physiological changes, and actions. Subjective affect includes cognitive appraisal and is the topography of the individual's emotional reaction. The physiological changes described by Lazarus are similar to that described by Selye and are meant to prepare the organism for active defense. The third component is that the actions exhibited by organisms are both pragmatic and expressive in nature. One point of contention between Selye's theory and Lazarus's theory of cognitive appraisal is that Lazarus believed that psychological processes may mediate GAS responses, rather than defense reactions being solely driven by biological processes. Lazarus (1966) and Mason (1971) began focusing on the effectiveness of the pituitary-adrenocortical system response's role in the stress response reaction, which encompasses the release and regulation of stress hormones such as adrenocorticotropic hormone (ACTH) and glucocorticoid hormones such as cortisol, and their impact on human health. Underlying Lazarus's work on cognitive appraisal theory is his work with Susan Folkman on the transactional model, a theory that posits that an individual and the environment are engaged in a bidirectional interchange, indicating that the person and the environment influence each other and that stress may result from the transaction between a person and their environment (Lazarus & Folkman, 1984).

Following early work on the association between stress and physiological functioning, Holmes and Rahe emphasized the measurement of stress and stress's impact on a person's health. Rahe, Holmes, and colleagues examined the hypothesis that many diseases have an onset or are exacerbated by an increase in social stress. Rahe et al. (1964) examined seven patient samples that encompassed five medical categories from the Seattle, Washington area. Patient samples included employees at a pulmonary institution who had contracted tuberculosis, tuberculosis outpatients, cardiac patients, patients with hernias, patients with skin diseases, pregnant women, women who had given birth out of wedlock, and two control groups. Participants in these samples were asked to complete the Schedule of Recent Experiences in addition to major social readjustments that may have occurred within the past 10 years (Rahe et al., 1964). The 1964 study served as a predecessor of the Holmes-Rahe Stress Inventory, which focused on the measurement of stress and its effect on health.

In 1967, Holmes and Rahe published analyzed data collected from medical records, from which they derived 43 life events, called Life Change Units (LCUs), which were assigned various level weights according to their associated impact on health. More stress equated to worse health prognoses and increased risk of illness. Holmes and Rahe found that the more stressful events someone experienced, taken into account with the events level of stress, was predictive of future illness. Examples of events include death of a spouse, detention in jail or other institution, pregnancy, and major holidays. Holmes and Rahe named this scale of events the Social Readjustment Rating Scale (SRRS; Holmes & Rahe, 1967); however, it has also been called the Holmes-Rahe Stress Inventory. Since the Social Readjustment Rating Scale's initial publication, new national normative data has been collected for individuals in the United States (Hobson & Delunas, 2001). Cross-cultural studies have also examined the use of the scale in various countries including Mexico, New Zealand, and Japan (Bruner et al., 1994; Isherwood & Adam, 1976; Yahiro et al., 1993). The findings from Mexico (Bruner, Acuña, & Gallardo) and New Zealand (Isherwood & Adam, 1976) found data congruent to Holmes and Rahe's initial findings regarding stressful life events in the SRRS. However, Yahiro et al. (1993) found that the SRRS should be modified for different cultures. Additionally, Blasco-Fontecilla et al. (2012) found that the SRRS may help predict an individual's suicide risk. Finally, Dekker and Webb (1974) found that individuals who reported more stressful life events were more likely to seek psychiatric care compared to those with less stressful life events and less anxiety.

Allostasis

Following Holmes and Rahe's research, studies examining the mechanisms of stress and health gained traction. In particular, the notion of allostasis began to make its mark in stress research. Allostasis can be conceptualized as a process by which the body attempts to obtain and maintain homeostasis. With regard to allostasis, three systems in particular have become the focus of human stress research: cardiovascular, neuroendocrine, and immune systems. As noted in many of the aforementioned theories, changes in blood pressure and blood flow are impacted by stress, which evolutionarily developed to help organisms fight or flee a stressful situation. Earlier models of the stress process provided the basic details into how physiology is affected by stress; however, the model of allostatic load provides a mechanistic approach to understanding the physiology of stress and predicting future illness. Based on the term, allostasis, the model is founded on the dynamic process of the body adapting to various environmental stressors by changing physiological functions to maintain homeostasis (Juster et al., 2010a, b). McEwen and Seeman (1999) note the importance of understanding the protective and damaging effects of allostasis and allostatic load, particularly in that allostasis and allostatic load may help the individual regulate their behavior and physiology in times of stress, but that while it can be perceived as protective in the short term, there may be damaging long-term effects.

The allostatic load model (ALM) encompasses the many changes that occur in real or imagined stress and impacts various organ systems associated with the sympathetic nervous system; it is the long-term cost of allostasis (Ellis & Del Guidice, 2014). ALM's long-term focus on survival by means of physiological adaptations to stress has implications for promoting illness. As the name suggests, changes in allostasis are the cornerstone to ALM. Changes in allostasis include increased inflammation, decreased cardiovascular response to stress (similar to resistance stage seen in Hans Selye's General Adaptation Syndrome), and increased levels of cortisol and hormones released by the adrenal gland (catecholamines such as dopamine, norepinephrine, and epinephrine). Activation of autonomic, neuroendocrine, metabolic, and immune systems is also central to ALM. Activation of these systems includes sympathetic nervous system and hypothalamic-pituitary-adrenal (HPA) axis responses, which help the organism in fight-or-flight responses and increase inflammation that evolutionarily served to heal wounds that may occur during fight or flight.

Additional mechanisms that comprise ALM include impairments in cognitive, behavioral, and emotional responses. Studies have shown reduced standardized scores on assessment of executive functions (Slavich & Shields, 2018). Stress in childhood has also been linked to increased psychopathology in adulthood (Raymond et al., 2018). A systematic review used ALM to conceptualize the impairments found in care partners, primarily women aged 60–75 years old, of older adults with neurocognitive disorders (Potier et al., 2018). Potier et al. (2018) found that increased allostatic load (measured by inflammatory biomarkers) was associated with increased reports of sleep disturbances, substance use, and increased incidence of chronic pain. As previously noted, ALM is targeted in in-the-moment survival for the organism; therefore, long-term outcomes from chronic stress adaptation can compromise quality of life and health (Juster et al., 2010a, b).

Although ALM is still widely held in stress research, other models such as the allostatic calibration model (ACM) have emerged to expand past the more limited scope of ALM. ACM examines the idiographic differences in how the stress response system functions and is based on evolutionary-developmental theory (Guidice et al., 2011). Guidice and colleagues (2011) state that ACM focuses on understanding an individual's development of stress responsivity and its ability to modify its development to adapt to its environment. One of the main differences between ALM and ACM is that ACM is a more individualistic approach. For example, ACM encompasses adaptive calibration of various system functions to best offset long-term detriments to the individual. One example of ACM is studies showing stunted growth of children adopted from orphanages, which is thought to occur as a balance in HPA axis functioning, wherein cortisol levels increase in response to stress and growth hormone secretion decreases reducing resource needs in a low-resource context (Gunnar, 2001; Hostinar & Gunnar, 2013; Johnson et al., 2011). Additionally, the scope of ACM also entails the study of self-regulation strategies that help an individual best adapt and match to their current environment (Ellis & Del Guidice, 2014). Acknowledging ALM in the history of stress research is important in understanding the development of ACM and the current state of stress research: examining individual system function in addition to the integration of systems in understanding an individual's predisposition to illness and health when adapting to stressful situations and environments.

Health Outcomes

As previously discussed, three main systems are impacted by chronic stress: cardiovascular, neuroendocrine, and immune systems. Early researchers like Cannon, Selye, and Holmes and Rahe broadly discussed the impact of stress on health, particularly regarding the cardiovascular, neuroendocrine, and immune systems. Further discussion of the role of each of these systems is presented alongside a summary of the current state of research in regard to these systems and how they are impacted by stress.

Cardiovascular System

Early research on the cardiovascular system was noted in Walter Cannon's fight-or-flight theory. While advanced mechanisms had not been previously studied, it was understood that blood flow and blood pressure changed during stressful situations which helped an organism escape or fight the stressor. Hans Selye expanded on the role of the cardiovascular mechanism in stress when he outlined the three stages of GAS where he discussed dynamic changes in blood flow and blood pressure change in various stages of stress, including the exhaustion stage, emphasizing the damage that stress creates within the organism. Current research on the interplay between the cardiovascular system and stress uses the reactivity hypothesis, stating that in the presence of increased or chronic stress, cardiovascular reactivity changes will lead to an increased risk of cardiovascular disease (Panaite et al., 2015; Treiber et al., 2003). Early research focused on increased cardiovascular reactivity as a problem for long-term health outcomes; however, recent research has found that low reactivity is problematic as well (Ginty & Conklin, 2011; Phillips et al., 2013). Lower cardiovascular reactivity has been conceptualized as a cardiovascular response pattern that is lower than that of baseline pattern. It has been hypothesized that lower cardiac reactivity in response to stress may be a response cost, in that lower reactions may indicate lower effort from the individual. Lower cardiac reactivity may be indicative of an individual's psychological coping strategy to stress and may be indicative of psychopathology such as depression in response to stress (Phillips et al., 2013).

In contrast to lower cardiac reactivity, extant studies have led to findings that indicate that stress reactivity and poststress recovery have important implications of the prognosis of cardiac health, by means of higher cardiac reactivity. Chida and Steptoe (2010) completed a meta-analysis examining the role of stress to predict future cardiac health and found that poor stress recovery and greater reactivity to in-the-moment stress were associated with poor cardiovascular functioning, as defined by elevated blood pressure, hypertension, atherosclerosis, and cardiac events. An additional meta-analysis examining how stress affects cardiovascular function has found that poorer stress recovery was predictive of poorer cardiovascular outcomes and that recovery from physical stress was associated with poorer cardiovascular outcomes when compared to psychological stress (Panaite et al., 2015). Carroll et al. (1995) examined the role of stress in middle-aged men living in London (aged 35–55) and found that a mental stressor was a significant predictor of higher blood pressure, even after a 5- and 10-year follow-up. However, in the present study, additional variance was accounted for by age, initial blood pressure screening, and baseline resting blood pressure (Carroll et al., 1995).

Additional research examined the role of discrimination in the association between stress and cardiovascular disease. Saban et al. (2018) noted the importance of examining the association of stress and cardiovascular disease in minority groups, as they experience a disproportionate rate of cardiovascular disease. In a laboratory study, Saban et al. (2018) studied the effects of a social stressor task called the Trier Social Stress Test (TSST) on salivary pro-inflammatory cytokine, interleukin-6, a cytokine that has been closely associated with cardiovascular disease. They found that women of color, who reported high levels of lifetime discrimination, had higher levels of interleukin-6 prior to the TSST and following the TSST. Future studies should parse out the complexities between salivary pro-inflammatory cytokines and cardiovascular disease on underrepresented populations (Saban et al., 2018).

Extant research regarding cardiovascular function and stress has examined mediators and moderators. For example, De Vogli et al. (2007) found significant associations between adverse close relationships and cardiac events over 12.2 years. The association was still significant after holding constant sociodemographic factors, biological factors, and other factors involved in social support (De Vogli et al., 2007). Additional studies have examined the role of social support in moderating cardiovascular reactivity. The findings indicated that when presented a stressful situation, subjects who received social support during the event had smaller increases in cardiovascular reactivity compared to the group who received no support (Gerin et al., 1992). Aside from social support in adult populations, research into childhood stress has also shown worse prognosis in adulthood. A retrospective cohort study found a significant association between data collected from the adverse childhood experiences (ACE) study and ischemic heart disease. Particularly, individuals in minoritized groups reported higher rates of ischemic heart disease. One finding from the study was that nine out of ten ACE categories predicted ischemic heart disease; the only ACE category that did not yield a significant association was parental marital discord (Dong, et al., 2004). The aforementioned research indicates that a number of factors mediate and moderate the impact of stress on the cardiovascular system; increased social support may reduce the cardiovascular reactivity to stress, while negative social relationships and adverse childhood experiences may increase cardiovascular reactivity to stress. Overall, chronic stress has been shown to negatively impact cardiac health, increasing risk of hypertension, cardiac events, and cardiovascular disease.

Neuroendocrine System

In addition to the cardiovascular system, neuroendocrine function plays an integral role in the body's management of stress and its health implications. Early researchers, such as Cannon and Selye, focused largely on the role of epinephrine in the organism's ability to engage in fight or flight and the increased levels of epinephrine in the alarm-reaction stage. As research evolved, Lazarus and his colleagues became interested in studying the role of other hormones in an individual's response to stress, such as adrenocorticotropic hormone (ACTH) that is responsible for regulating levels of the stress hormone cortisol in the blood. An emphasis on studying various hormones and brain functions led to research exploring the role of the brain and behavior in managing stress and health. One example is research into ACTH and the HPA axis. The HPA axis is responsible for elevations in glucocorticoids and cate-cholamines, which impacts not only baseline level of functioning but also the body's stress response (Elenkov et al., 1999). The HPA axis then became an integral part in understanding stress, health, and psychopathology, which has led to the HPA axis being labeled the physiological stress response system of the body (Kudielka et al., 2012).

As indicated in Cannon's fight-or-flight theory, reproductive and digestive systems function at a lower level in times of stress. Glucocorticoids play a part in decreasing the function of these systems in part by affecting levels of hormones responsible for their functioning. An excess in glucocorticoids can lead to a decreased level of functioning in reproductive system by decreasing luteinizing hormone, follicle-stimulating hormone, thyroid-stimulating hormone, and growth hormone. Additionally, excess glucocorticoid levels can lead to decreased functioning in digestive systems by creating a diabetogenic effect through increasing insulin secretion, increasing gluconeogenesis, and increasing fatty acid production (Kaltsas & Chrousos, 2007).

In the mid-1980s, a push toward research into the glucocorticoid cortisol led to cortisol being labeled "the stress hormone," and cortisol became a marker of HPA axis functioning (Hellhammer et al., 2009). HPA axis functioning and stress have been cited in association with a plethora of disorders and diseases such as depression (and other mood disorders) and coronary heart disease (Murray & Lopez, 1996). Cortisol has many functions within the body, and its ability to adapt to stress impacts metabolism, immune functioning, development, emotional processes, and sleep (Kudielka et al., 2012). Other hormones like corticotropin-releasing hormone (CRH) are also part and parcel with the body's stress response through the same autonomic activation pathway as cortisol (Habib et al., 2001). Two main CRH receptors are directly associated with an individual's response to stress: CRH-R1, found in the anterior pituitary, adrenal glands, and reproductive organs, and CRH-R2, found mainly in the cardiovascular system (Habib et al., 2001). Additionally, when an organism has higher levels of CRH, due to stress, the immune system is effected through activating mast cells leading to release of histamine, which in turn leads to inflammation. The effect of inflammation has been linked to autoimmune and inflammatory diseases, such as rheumatoid arthritis, autoimmune thyroid disease, and ulcerative colitis (Kaltsas & Chrousos, 2007).

Extant research has also examined the role of oxytocin in the body's response to stress, notably its role in reducing feelings of anxiety. Oxytocin is produced in the hypothalamus and transported to the pituitary and released into the bloodstream. Oxytocin has been largely studied within the context of maternal behavior, such as lactation in animal models (Carter, 1998; Hennessy et al., 2019). In human research, oxytocin has been found to play a role in caregiving as well as predicting quality of life following pregnancy. One study found that higher levels of oxytocin during pregnancy were associated with increased postpartum maternal caregiving behavior (Feldman et al., 2007). Another study noted that increased oxytocin levels were associated with participants feeling less anxious in stressful situations. For example, in an experiment using the Trier Social Stress Test, participants who were allowed to bring a friend along to a public speaking task reported feeling less anxious and also had increased levels of oxytocin compared to a control group who was not allowed to bring a friend to the task (Kirschbaum et al., 1993). Studies have indicated a clear connection between stress, oxytocin, and social support. Karelina et al. (2011) found that following an ischemic stroke, oxytocin mediated the association between neuroprotective factors and social support. Human subject research has also indicated that receiving social support may lead to more adaptive stress reactivity, as evidenced by increased levels of oxytocin, lower levels of cortisol, and lower blood pressure (Cosley et al., 2010). In all, neuroendocrine function is integral to the body's adaptation to stress through the regulation of neurotransmitters, HPA axis activity, and glucocorticoids. Poor health outcomes have been linked to heightened neuroendocrine system reactivity to chronic stress, which may lead to mental health concerns like anxiety and depression.

Immune System

It is clear that there is an interplay between the neuroendocrine system and the immune system. While the neuroendocrine system is key in adapting to stress, it is also important to examine the independent role of the immune system and how it is impacted by stress. Notably, research into the role of the immune system and stress began in the 1920s and gained traction again in the 1970s with Ader and Cohen's research into conditioned immunosuppression, which highlighted the bidirectional nature of the central nervous system and the immune system (Ader & Cohen, 1975). In the 1980s, experiments involved cloning cytokine interleukin to increase HPA function in sheep. Additional experiments showed the association between cytokine interleukin and its ability to induce fever (Besedovsky et al., 1986).

Studies up until the 1980s were primarily focused on long-range communication between the brain and the immune system. Long-range communication hinged on the idea that immune function was influenced by circulating hormones from the neuroendocrine system, such as cortisol, growth hormone, and other glucocorticoids, and through neurotransmitters such as epinephrine and norepinephrine. A shift to researching short-range communication pathways began when it was found that neuroendocrine factors and neuromediators could be produced by immune cells (Dantzer, 2018). In addition to these factors produced in immune cells, it also became clear that immune factors indirectly acted upon the brain via molecules such as prostaglandins, which could transport past the blood-brain barrier to create a fever within the individual in response to infectious pathogens (Dantzer, 2018).

As noted in early stress and immunology research, it is sometimes assumed that changes in stress levels and their change in immune function are directly related to health outcomes. However, there are underlying factors that should be considered when understanding the association of stress, immune function, and health outcomes. One underlying factor is the immune system functioning of the individual prior to stress (Kiecolt-Glaser & Glaser, 1995). An additional underlying factor is the evolutionary function of inflammation and its role in understanding stress and the immune system. It has long been understood that the inflammatory response has its place in aiding the body in overcoming unwanted pathogens. Inflammation helped organisms fight infection and is necessary in helping heal wounds. However, given the ever-changing environment, what was once adaptive for survival has now become a problem provided the chronic stress that many people experience (Miller & Raison, 2018).

One issue that has evolved from the intersection of the inflammation process and modern living has been the association between inflammation and mood disorder symptoms. Increased inflammation in response to stress has been associated with depressive symptoms and has been found in increased levels of inflammatory bio-markers within blood. Contrastingly, immune cells, such as effector T cells, which are activated and increased during stressful events have been linked to decreased depressive and anxiety behaviors in mice models (Dantzer et al., 2008; Miller & Raison, 2018). Similar findings have been shown in human research regarding inflammation and depression (Bekhbat et al., 2020). Some research has also indicated that women may be more prone to experience depressogenic and anxiogenic effects of inflammation than men, but further research is needed to better understand sex differences of neuroinflammation during stress (Bekhbat & Neigh, 2018).

The interconnection between stress, inflammation, and illness has been noted in populations outside of mood disorders as well, such as cancer populations. For example, one meta-analysis found a significant association between stress-prone personality and poor quality of life on increased incidence of cancer and decreased cancer survival rates (Chida et al., 2008). While there are studies examining inflammation as a risk factor for cancer development (Ambatipudi et al., 2018; De Visser & Coussens, 2005; Liu et al., 2020), recent research has also emphasized the role of stress as a risk factor for cancer. Research indicates that individuals who have experienced chronic stress and stressful life events and those who lack social support are at risk for developing cancer as well as having a poorer prognosis (Feller et al., 2019).

An additional frontier in research regarding health and inflammation is examining the role of the microbiome. The microbiome has been defined as the collection of microorganisms in the intestines. The microbiome changes when the body is exposed to chronic stress and does not function as it typically should (Gilbert et al., 2018). Recent research has indicated that the microbiome plays a role in regulating neurotransmitters and hormones within the digestive system that influence inflammatory processes within the body. Research on the microbiome and inflammation has expanded into various populations such as cognitive decline (Komanduri et al., 2019), depression (Pereira et al., 2019) and recovery from spinal cord injury (Noller et al., 2017). It is clear that the immune system plays a vital role in stress and health given the breadth of processes encompassed within the system: from inflammation to the microbiome. The bidirectional nature of stress and immune system processes has been associated with poor mental and physical health, solidifying its place as an important area of future stress research.

Conclusion

The history of stress research has evolved from animal models to human subjects and has spanned broad and narrow understanding of the various systems involved in helping an organism adapt to acute and chronic stress. Early models outlined the evolutionary necessity of the system activation to aid an animal in surviving a stressor through Walter Cannon's fight-or-flight theory. Hans Selye's GAS theory was a cornerstone in the field's understanding that an organism has finite resources available to handle stress, and detrimental effects can occur if the animal experiences chronic stress. The shift from animal to human subjects largely began with Lazarus's work on the transactional model of stress and cognitive appraisal and then Holmes and Rahe's Social Readjustment Rating Scale, which pioneered research examining the role of stress on human health and illness. Early research on stress and human health paved the way for research into allostasis and the two main theories built around the process, allostatic load model and allostatic calibration model, both of which indicate short-term benefits but long-term damages to the organism when chronic stress is experienced. From that shift, research examining the independent role of the cardiovascular, neuroendocrine, and immune systems in stress provided more information on the mechanism of change in stress and health, particularly the role of stress and cardiovascular disease, changes in stress hormones, as well as stress's role in creating inflammation, all of which may increase mortality and decrease quality of life of the individual. However, it should be evidently clear that while studying the independent role of these systems is important, the systems are interconnected. Future studies aim to examine the dynamic and longitudinal effects of stress on health. Advances in methodologies and data analyses may provide researchers additional tools to better understand the mechanisms of stress and how to slow and prevent stress's effect on health through clinical practices, such as mindfulness-based interventions.

References

Ader, R., & Cohen, N. (1975). Behaviorally-conditioned immunosuppression. *Psychosomatic Medicine, 37*(4), 333–340.

Ambatipudi, S., Langdon, R., Richmond, R. C., Suderman, M., Koestler, D. C., Kelsey, K. T., … Ring, S. M. (2018). NA methylation derived systemic inflammation indices are associated with head and neck cancer development and survival. *Oral Oncology, 85*, 87–94.

Bekhbat, M., & Neigh, G. N. (2018). Sex differences in the neuro-immune consequences of stress: Focus on depression and anxiety. *Brain, Behavior, and Immunity, 67*, 1–12.

Bekhbat, M., Treadway, M., Goldsmith, D. R., Woolwine, B. J., Haroon, E., Miller, A. H., & Felger, J. C. (2020). Gene signatures in peripheral blood immune cells related to insulin resistance and low tyrosine metabolism define a sub-type of depression with high CRP and anhedonia. *Brain, Behavior, and Immunity*. In Press.

Besedovsky, H., Del Rey, A., Sorkin, E., & Dinarello, C. A. (1986). Immunoregulatory feedback between interleukin-1 and glucocorticoid hormones. *Science, 233*(4764), 652–654.

Blasco-Fontecilla, H., Delgado-Gomez, D., Legido-Gil, T., De Leon, J., Perez-Rodriguez, M. M., & Baca-Garcia, E. (2012). Can the Holmes-Rahe Social Readjustment Rating Scale (SRRS) be used as a suicide risk scale? An exploratory study. *Archives of Suicide Research, 16*(1), 13–28.

Bruner, C. A., Acuña, L., & Gallardo, L. M. (1994). The Social Readjustment Rating Scale (SRRS) of Holmes and Rahe in Mexico. *Revista Latinoamericana de Psicología, 26*(2), 253–269.

Cannon, W. (1929). Cannon, W. B. (1916). *Bodily changes in pain, hunger, fear, and rage: An account of recent researches into the function of emotional excitement.* Appleton-Century-Crofts.

Carroll, D., Smith, G. D., Sheffield, D., Shipley, M. J., & Marmot, M. G. (1995). Pressor reactions to psychological stress and prediction of future blood pressure: Data from the Whitehall II study. *BMJ, 310*(6982), 771–775.

Carter, C. S. (1998). Neuroendocrine perspectives on social attachment and love. *Psychoneuroendocrinology, 23*(8), 779–818.

Chida, Y., Hamer, M., Wardle, J., & Steptoe, A. (2008). Do stress-related psychosocial factors contribute to cancer incidence and survival? *Nature Clinical Practice Oncology, 5*(8), 466–475.

Cosley, B. J., McCoy, S. K., Saslow, L. R., & Epel, E. S. (2010). S compassion for others stress buffering? Consequences of compassion and social support for physiological reactivity to stress. *Journal of Experimental Social Psychology, 46*(5), 816–823.

Dantzer, R. (2018). Neuroimmune interactions: From the brain to the immune system and vice versa. *Physiological Reviews, 98*(1), 477–504.

Dantzer, R., O'Connor, J. C., Freund, G. G., Johnson, R. W., & Kelley, K. W. (2008). From inflammation to sickness and depression: When the immune system subjugates the brain. *Nature Reviews Neuroscience, 9*(1), 46–56.

De Visser, K. E., & Coussens, L. M. (2005). The interplay between innate and adaptive immunity regulates cancer development. *Cancer Immunology Immunotherapy, 54*(11), 1143–1152.

De Vogli, R., Chandola, T., & Marmot, M. G. (2007). Negative aspects of close relationships and heart disease. *Archives of Internal Medicine, 167*(18), 1951–1957.

Dekker, D. J., & Webb, J. T. (1974). Relationships of the social readjustment rating scale to psychiatric patient status, anxiety and social desirability. *Journal of Psychosomatic Research, 18*, 125–130.

Del Giudice, M., Ellis, B. J., & Shirtcliff, E. A. (2011). The adaptive calibration model of stress responsivity. *Neuroscience & Biobehavioral Reviews, 35*(7), 1562–1592.

Dong, M., Giles, W., Felitti, V., Dube, S., Williams, J., Chapman, D., & Anda, R. (2004). Insights into causal pathways for ischemic heart disease: Adverse childhood experiences study. *Circulation, 110*(3), 1761–1766.

Elenkov, I. J., Webster, E. L., Torpy, D. J., & Chrousos, G. P. (1999). Stress, corticotropin-releasing hormone, glucocorticoids, and the immune/inflammatory response: Acute and chronic effects. *Annals of the New York Academy of Sciences, 876*(1), 1–13.

Ellis, B., & Del Giudice, M. (2014). Beyond allostatic load: Rethinking the role of stress in regulating human development. *Development and Psychopathology, 26*(1), 1–20.

Feldman, R., Weller, A., Zagoory-Sharon, O., & Levine, A. (2007). Evidence for a neuroendocrinological foundation of human affiliation: Plasma oxytocin levels across pregnancy and the postpartum period predict mother-infant bonding. *Psychological Science, 18*(11), 965–970.

Feller, L., Khammissa, R. A., Ballyram, R., Chandran, R., & Lemmer, J. (2019). Chronic psychosocial stress in relation to cancer. *Middle East Journal of Cancer, 10*(1), 1–8.

Gerin, W., Peiper, C., Levyu, R., & Picering, T. G. (1992). Social support in social interaction: A moderator of cardiovascular reactivity. *Psychosomatic Medicine, 54*, 324–336.

Gilbert, J. A., Blaser, M. J., Caporaso, J. G., Jansson, J. K., Lynch, S. V., & Knight, R. (2018). Current understanding of the human microbiome. *Nature Medicine, 24*(4), 392.

Ginty, A. T., & Conklin, S. M. (2011). High perceived stress in relation to life events is associated with blunted cardiac reactivity. *Biological Psychology, 86*(3), 383–385.

Gunnar, M. R. (2001). Effects of early deprivation: Findings from orphanage-reared infants and children. In *Handbook of developmental cognitive neuroscience* (Vol. 114, issue 8).

Habib, K. E., Gold, P. W., & Chrousos, G. P. (2001). Neuroendocrinology of stress. *Endocrinology and Metabolism Clinics, 30*(3), 695–728.

Hellhammer, D. H., Wüst, S., & Kudielka, B. M. (2009). Salivary cortisol as a biomarker in stress research. *Psychoneuroendocrinology, 34*(2), 163–171.

Hennessy, M. B., Tai, F., Carter, K. A., Watanasriyakul, W. T., Gallimore, D. M., Molina, L. A., & Schiml, P. A. (2019). Central oxytocin alters cortisol and behavioral responses of Guinea pig pups during isolation in a novel environment. *Physiology & Behavior, 212,* 112710.

Hobson, C., & Delunas, L. (2001). National norms and life-event frequencies for the revised social readjustment rating scale. *International Journal of Stress Management, 8*(4), 299–314.

Holmes, T., & Rahe, R. (1967). The social readjustment rating scale. *Journal of Psychosomatic Research.*

Hostinar, C. E., & Gunnar, M. R. (2013). The developmental effects of early life stress: An overview of current theoretical frameworks. *Current Directions in Psychological Science, 22*(5), 400–406.

Isherwood, J., & Adam, K. S. (1976). The social readjustment rating scale: A cross-cultural study of new Zealanders and Americans. *Journal of Psychosomatic Research, 20*(3), 211–214.

Johnson, A. E., Bruce, J., Tarullo, A. R., & Gunnar, M. R. (2011). Growth delay as an index of allostatic load in young children: Predictions to disinhibited social approach and diurnal cortisol activity. *Development and Psychopathology, 23*(3), 859–871.

Juster, R. P., McEwen, B. S., & Lupien, S. J. (2010a). Allostatic load biomarkers of chronic stress and impact on health and cognition. *Neuroscience & Biobehavioral Reviews, 35*(1), 2–16.

Juster, R., McEwen, B., & Lupien, S. (2010b). Allostatic load biomarkers of chronic stress and impact on health and cognition. *Neuroscience & Biobehavioral Reviews, 35*(1), 2–16.

Kaltsas, G., & Chrousos, G. (2007). The neuroendocrinology of stress. In J. T. Cacioppo, L. G. Tassinary, & G. G. Berntson (Eds.), *Handbook of psychophysiology* (3rd ed., pp. 303–318). Cambridge University Press.

Karelina, K., Stuller, K. A., Jarrett, B., Zhang, N., Wells, J., Norman, G. J., & DeVries, A. C. (2011). Oxytocin mediates social neuroprotection after cerebral ischemia. *Stroke, 42*(12), 3606–3611.

Kiecolt-Glaser, J., & Glaser, R. (1995). Measurement of immune response. In S. Cohen, R. Kessler, & L. Gordon (Eds.), *Measuring stress.* Oxford University Press.

Kirschbaum, C., Pirke, K. M., & Hellhammer, D. H. (1993). The 'Trier social stress test'–a tool for investigating psychobiological stress responses in a laboratory setting. *Neuropsychobiology, 28*(1–2), 76–81.

Komanduri, M., Gondalia, S., Scholey, A., & Stough, C. (2019). He microbiome and cognitive aging: A review of mechanisms. *Psychopharmacology,* 1–13.

Kudielka, B. M., Gierens, A., Hellhammer, D. H., Wüst, S., & Schlotz, W. (2012). Salivary cortisol in ambulatory assessment – Some dos, some don'ts, and some open questions. *Psychosomatic Medicine, 74*(4), 418–431.

Lazarus, R. (1966). *Psychological stress and the coping process.* McGraw-Hill.

Lazarus, R., & Folkman, S. (1984). *Stress, appraisal, and coping* (pp. 150–153). Springer.

Lazarus, R., & Monat, A. (1974). Cognitive coping and processes in emotion. In *Fifty years of the research and theory of RS Lazarus; An analysis of historical and perrennial issues* (pp. 70–84).

Liu, Y., Li, L., Li, Y., & Zhao, X. (2020). Research progress on tumor-associated macrophages and inflammation in cervical cancer. *BioMed Research International,* 1–6.

Mason, J. (1971). A re-evaluation of the concept of "non-specificity" in stress theory. *Journal of Psychiatric Research, 8,* 323–333.

McEwen, B. S., & Seeman, T. (1999). Protective and damaging effects of mediators of stress: Elaborating and testing the concepts of allostasis and allostatic load. *Annals of the New York Academy of Sciences, 896*(1), 30–47.

McEwen, B. S., Bowles, N. P., Gray, J. D., Hill, M. N., Hunter, R. G., Karatsoreos, I. N., & Nasca, C. (2015). Mechanisms of stress in the brain. *Nature Neuroscience, 18*(10), 1353–1363.

Miller, A. H., & Raison, C. L. (2018). The role of inflammation in depression: From evolutionary imperative to modern treatment target. *Nature Reviews Immunology, 16*(1), 22.

Murray, C. J., & Lopez, A. D. (1996). Evidence-based health policy – Lessons from the global burden of disease study. *Science, 274*(5288), 740–743.

Noller, C. M., Groah, S. L., & Nash, M. S. (2017). Inflammatory stress effects on health and function after spinal cord injury. *Topics in Spinal Cord Injury Rehabilitation, 23*(3), 207–217.

Panaite, V., Salomon, K., Jin, A., & Rottenberg, J. (2015). Cardiovascular recovery from psychological and physiological challenge and risk for adverse cardiovascular outcomes and all-cause mortality. *Psychosomatic Medicine, 77*(3), 215–226.

Pereira, J. D., Rea, K., Nolan, Y. M., O'Leary, O. F., Dinan, T. G., & Cryan, J. F. (2019). Depression's unholy trinity: Dysregulated stress, immunity, and the microbiome. *Annual Review of Psychology, 71*, 49–78.

Phillips, A. C., Ginty, A. T., & Hughes, B. M. (2013). The other side of the coin: Blunted cardiovascular and cortisol reactivity are associated with negative health outcomes. *International Journal of Psychophysiology, 90*(1), 1–7.

Potier, F., Degryse, J. M., & de Saint-Hubert, M. (2018). Impact of caregiving for older people and pro-inflammatory biomarkers among caregivers: A systematic review. *Aging Clinical and Experimental Research, 30*(2), 119–132.

Rahe, R., Meyer, M., Smith, M., Kjaer, G., & Holmes, T. (1964). Social stress and illness onset. *Journal of Psychosomatic Research, 8*, 35–44.

Raymond, C., Marin, M. F., Majeur, D., & Lupien, S. (2018). Early child adversity and psychopathology in adulthood: HPA axis and cognitive dysregulations as potential mechanisms. *Progress in Neuro-Psychopharmacology and Biological Psychiatry, 85*, 152–160.

Romero, L. (2004). Physiological stress in ecology: Lessons from biomedical research. *Trends in Ecology & Evolution, 19*(5), 249–255.

Saban, K. L., Mathews, H. L., Bryant, F. B., Tell, D., Joyce, C., DeVon, H. A., & Janusek, L. W. (2018). Perceived discrimination is associated with the inflammatory response to acute laboratory stress in women at risk for cardiovascular disease. *Brain, Behavior, and Immunity, 73*, 625–632.

Selye, H. (1946). The general adaptation syndrome and the diseases of adaptation. *The Journal of Clinical Endocrinology, 6*(2), 117–230.

Selye, H. (1951). The general-adaption-syndrome. *Annual Review of Medicine, 2*(1), 327–342.

Slavich, G. M., & Shields, G. S. (2018). Assessing lifetime stress exposure using the stress and adversity inventory for adults (adult STRAIN): An overview and initial validation. *Psychosomatic Medicine, 80*(1), 17.

Taylor, S. (2010). Mechanisms linking early life stress to adult health outcomes. *Proceedings of the National Academy of Sciences, 107*(19), 8507–8512.

Thoits, P. A. (2010). Stress and health: Major findings and policy implications. *Journal of Health and Social Behavior, 51*(1), S41–S53.

Treiber, F. A., Kamarck, T., Schneiderman, N., Sheffield, D., Kapuku, G., & Taylor, T. (2003). Cardiovascular reactivity and development of preclinical and clinical disease states. *Psychosomatic Medicine, 65*(1), 46–62.

Yahiro, K., Inoue, M., & Nozawa, Y. (1993). An examination on the social readjustment rating scale (Holmes et al) by Japanese subjects. *Japanese Journal of Health Psychology, 6*(1), 18–32.

Chapter 2
The Neurobiology of Stress

Olena Kleshchova and Mariann R. Weierich

Introduction

Physical and psychological health is critically dependent on active maintenance of **allostasis**, or a stable internal milieu, which is mediated by multiple systems in the body and regulated by the brain. **Stressors**, or stimuli or events that present an actual or perceived threat to allostasis (e.g., Radley, 2012), elicit multiple behavioral, physiological, and psychological changes in an organism. These changes collectively comprise the **stress response**, which is aimed at correction or prevention of an actual or anticipated homeostatic imbalance and promotion of short-term survival (e.g., Smith & Vale, 2006). The brain is both a regulator and a target organ of the stress response (Fig. 2.1), as it determines the "stressfulness" of sensory stimuli, coordinates compensatory responses to acute stressors, and promotes adaptation to chronic stressors (McEwen & Gianaros, 2011). In addition, the brain is itself influenced by the products of the stress response, or **stress mediators**, which include signaling molecules that are released by the nervous, endocrine, and immune systems. These products mediate physiological, cognitive, and behavioral responses to stress and can lead to structural and functional changes in the brain following prolonged or repeated exposure.

In this chapter, we discuss the neurobiology of stress, including the role of the brain in the generation and regulation of the stress response, and the effects of acute and chronic stress exposure on brain structure and function. We cover two broad themes: first, the neurobiological underpinnings of the stress response, and second, the neurobiological effects of acute and chronic stress exposure. We dedicate the first half of the chapter to the discussion of the peripheral and central components of the stress system, including the **peripheral stress effector systems** that bring about physiological changes throughout the body as part of the stress response and

O. Kleshchova · M. R. Weierich (✉)
University of Nevada, Reno, NV, USA
e-mail: mweierich@unr.edu

© Springer Nature Switzerland AG 2021
H. Hazlett-Stevens (ed.), *Biopsychosocial Factors of Stress, and Mindfulness for Stress Reduction*, https://doi.org/10.1007/978-3-030-81245-4_2

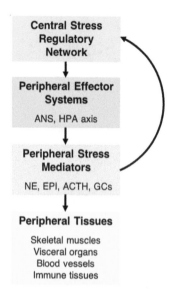

Fig. 2.1 The brain is both a regulator and a target organ of the stress response. The central stress-regulatory network regulates the activity of peripheral neural and endocrine effector systems which mediate physiological responses to stress through direct neural projections to target tissues or via release of signaling molecules (i.e., peripheral stress mediators) into the bloodstream. Peripheral stress mediators alter the functioning of various organ systems in the body, as well as the functioning of the brain itself. ANS = autonomic nervous system; HPA axis = hypothalamic-pituitary-adrenal axis; NE = norepinephrine; EPI = epinephrine; ACTH = adrenocorticotropic hormone; GCs = glucocorticoids

the **central stress-regulatory network** that is involved in stressor detection and stress response generation, regulation, and termination. We dedicate the second half of the chapter to the discussion of the effects of stress exposure on brain function and structure, and consequently on cognition and behavior. We also discuss the neurobiological effects of short-term vs. long-term exposure to elevated levels of stress mediators during acute and chronic stress, respectively.

The primary purpose of this chapter is to present a general overview of the neurobiology of stress based on a selective review of the available literature. Given such a broad scope, we assert several caveats. Most of what is currently known about the anatomical organization of the central stress-regulatory network and the biological mechanisms by which stress mediators impact brain function comes from research on rodents. Although the stress-relevant neural circuits are believed to be highly conserved across mammalian species (e.g., Hariri & Holmes, 2015), we must exercise caution when extrapolating to humans. Human research on the neural mechanisms of stress has largely focused on noninvasive neuroimaging techniques, which are relatively limited in temporal and spatial resolution. In addition, ethical considerations preclude experimental induction of certain types of stressors in humans (e.g., physical, chronic, and traumatic stressors), which are commonly implemented in rodents. Interpretation of human neuroimaging results therefore relies heavily upon comparison with the mechanistic biological models developed through animal research. Much of the information in this chapter draws on evidence from nonhuman animals. For comprehensive reviews of human neuroimaging studies involving acute stress induction, we refer interested readers to other suggested reading (see list at the end of chapter).

In addition to between-species differences, individual factors, such as age, sex, genotype, and early life experiences, can contribute to differences in the organization of the stress-relevant circuits and stress response patterns. For example, most rodent research on the mechanisms of stress is conducted on young male animals. However, accumulating evidence suggests that some results have limited generalizability to females and older animals (e.g., Luine et al., 2007; Lupien et al., 2009; McEwen et al., 2015; Ter Horst et al., 2009). In addition, the organization and function of specific stress-relevant circuits is influenced by genetic makeup and life history (e.g., Gunnar & Quevedo, 2007). With these caveats, we present a general overview of the neurobiological mechanisms of the stress response and the neurobiological effects of acute and chronic stress exposure.

Neurobiological Underpinnings of the Stress Response

"Stress response" is an umbrella term that refers to the entire range of processes initiated by an organism in response to an actual or perceived threat to homeostasis. A typical stress response includes behavioral, physiological, and psychological processes that allow an individual to cope with the stressor (Fig. 2.2). **Behavioral responses** to stressors include voluntary and reflexive motor actions (e.g., attack,

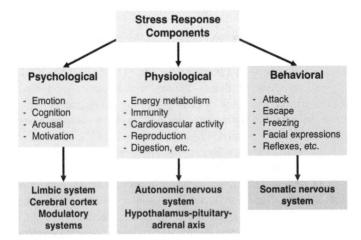

Fig. 2.2 Components of the stress response. A typical stress response entails behavioral, physiological, and psychological changes, which are respectively mediated by peripheral neural and endocrine effector systems and central cognitive, affective, and neuromodulatory networks

escape, defensive posturing) that are mediated by the somatic nervous system through direct control of skeletal muscles. **Physiological responses** to stressors entail widespread changes in the functioning of virtually every organ system in the body and include enhanced cardiac and respiratory activity, energy mobilization, and inhibition of vegetative functions like digestion, reproduction, and tissue repair (e.g., Sapolsky, 2000). Physiological stress responses are mediated by peripheral neural and endocrine systems that regulate the activity of target organs either via direct neural projections or via the release of hormones into the bloodstream, respectively (McEwen, 1998). Finally, **psychological responses** to stressors include elevated alertness, enhanced sensory processing and memory encoding, and emotional responses, such as fear and anxiety, which are mediated by specific brain networks that are involved in cognitive, affective, and neuromodulatory functions (Arnsten, 2009).

The specific pattern of behavioral, physiological, and psychological changes produced in response to a stressor depends on the nature of the stressor. Stressors are commonly classified into one of two categories: physical or psychological (e.g., Herman et al., 2016). **Physical stressors** include stimuli or events that overwhelm specific homeostatic systems and require prompt, "reactive," and largely reflexive compensatory responses to restore homeostasis (Herman et al., 2003). Examples of physical stressors include blood loss, imbalances in fluid or energy metabolism, immunologic challenge, and physical trauma. In contrast, **psychological stressors** include stimuli or events that are *perceived* to present *potential* threat to homeostasis, physical integrity, and/or social status, based on innate species-specific programs or learned contingencies (e.g., Goldstein, 2010; LeDoux, 2012). Examples of psychological stressors in rodents include predator smell, immobilization, exposure to open-field areas, and social defeat (Herman et al., 2003). In humans, psychological stressors include not only (potentially) threatening stimuli in the environment but also internal cognitive processes, such as trauma memories, aversive emotional experiences, perceived loss of control or predictability, and lack of social support (Sapolsky, 2015). In contrast to physical stressors which elicit **"reactive" stress responses**, psychological stressors trigger **"anticipatory" stress responses**, which are produced in anticipation of a homeostatic threat (Herman et al., 2003).

In the next two sections, we discuss the neural and neuroendocrine systems that produce the behavioral, physiological, and psychological responses to physical and psychological stressors. We begin with a description of the **peripheral stress effector systems** that mediate the physiological component of the stress response (section "Peripheral Stress Effector Systems"). We then discuss the organization of **central stress-regulatory network** which controls the peripheral effector systems, including the specific neural circuits involved in stressor detection (section "Neural Circuitry of Stressor Detection"), stress response generation (section "Neural Circuitry of Stress Response Generation"), regulation (section "Neural Circuitry of Stress Response Regulation"), and termination (section "Neural Circuitry of Stress Response Termination"), as well as the neural circuits that mediate the psychological components of the stress response, including changes in alertness, sensory function, affective processing, and cognition (section "Neural Circuitry of the

Psychological Components of the Stress Response"). Because physical and psychological stressors are detected by distinct neural systems and engage different circuits within the central stress-regulatory network (Herman et al., 2003), throughout this chapter we discuss the neural circuits involved in stressor detection and stress response generation separately for physical and psychological stressors.

Peripheral Stress Effector Systems

Physiological changes in the body that are triggered in response to a stressor are mediated by peripheral neural and neuroendocrine effector systems and regulated by the central stress-regulatory network. The two main peripheral stress effector systems include the **autonomic nervous system**, which regulates the activity of target tissues via peripheral nerves (section "Autonomic Nervous System"), and the **hypothalamic-pituitary-adrenal axis**, which is a neuroendocrine system that mediates physiological responses to stressors via the release of hormones into the bloodstream (section "Hypothalamic-Pituitary-Adrenal Axis"). The endocrine products of peripheral stress effector systems (i.e., peripheral stress mediators) impact the function of not only various peripheral tissues but also the central stress-regulatory network itself and can influence cognition and behavior both in the short and long term (McEwen et al., 2015), which is a topic that we discuss in the second half of this chapter.

Autonomic Nervous System

The autonomic nervous system (ANS) is a subdivision of the peripheral nervous system that controls the heart, visceral organs, blood vessels, and various glands and is not under voluntary control. The ANS consists of **preganglionic neurons**, which are located in the spinal cord and brainstem and project to **postganglionic neurons**, which are located in autonomic ganglia and project to visceral organs, blood vessels, and glands. The ANS is organized into the **sympathetic** and **parasympathetic** branches which produce mutually opposing effects on target tissues. For example, sympathetic innervation speeds up the heart, whereas parasympathetic innervation slows it down. Sympathetic innervation of the **adrenal medulla** represents an exception to the preganglionic-postganglionic organization of the ANS, because the adrenal medulla receives direct innervation from preganglionic neurons in the spinal cord rather than from postganglionic neurons in the ganglia. Thus, the sympathetic branch of the ANS can be further subdivided into the **sympathetic nervous system (SNS)**, which includes the sympathetic ganglia, and the **sympathetic adrenomedullary (SAM) system**, which includes the adrenal medulla (Goldstein & Kopin, 2008; Ulrich-Lai & Herman, 2009).

The ANS mediates a rapid response to acute stressors, which typically involves activation of the sympathetic branch and inhibition of the parasympathetic branch and results in a physiological state known as the **fight-or-flight response** (Sapolsky, 2000). Within seconds of stressor onset, sympathetic nerves that are part of the SNS release **norepinephrine (NE)** onto their target organs. Some of the released NE also enters the bloodstream and contributes ~70% of circulating plasma NE (Kvetnansky et al., 2009). In addition, the adrenal medulla, which is part of the SAM system, releases **catecholamines** (norepinephrine, epinephrine, and dopamine) into systemic bloodstream. The adrenal medulla contributes ~95% of circulating plasma **epinephrine (EPI)** and ~ 30% of circulating plasma NE (Kvetnansky et al., 2009). Both NE and EPI bind to adrenergic receptors throughout the body and mediate rapid metabolic effects that are part of the "fight-or-flight" response, including energy mobilization; enhanced cardiovascular function; decreased blood flow to the skin and abdominal and pelvic organs; increased blood flow to skeletal muscles, the heart, and the brain; and enhanced arousal and vigilance (Sapolsky, 2000). Once the stressor is over, sympathetic activation is quickly terminated by reflexive parasympathetic activation, which promotes vegetative functions, such as digestion, tissue repair, energy storage, and growth (Ulrich-Lai & Herman, 2009).

Hypothalamic-Pituitary-Adrenal Axis

Unlike the ANS, which mediates rapid responses to stress through direct innervation of target organs, the hypothalamic-pituitary-adrenal (HPA) axis mediates slower and longer-lasting endocrine responses to stressors. The three effector regions that constitute the HPA axis include the **paraventricular nucleus of the hypothalamus (PVN)**, the **anterior pituitary**, and the **adrenal cortex**. Activation of the HPA axis begins with the release of the **corticotropin-releasing hormone (CRH)** and **arginine vasopressin (AVP)** by neuroendocrine cells in the PVN into the **hypophyseal portal system**, a network of blood vessels, through which CRH and AVP are transported to the anterior pituitary. In the anterior pituitary, CRH and AVP synergistically promote the release of the **adrenocorticotropic hormone (ACTH)** into systemic circulation. ACTH is then transported to the adrenal glands, where it binds to receptors in the middle layer of the adrenal cortex and stimulates the synthesis and release of steroid hormones termed **glucocorticoids (GCs)**. **Cortisol** is the main GC in humans, whereas **corticosterone** is the main GC in rodents. The HPA axis responds to stressors slower than the ANS not only because endocrine signals travel slower than neural signals but also because GCs must be synthesized de novo upon ACTH stimulation, as they cannot be stored in lipid vesicles due to their high lipid solubility. As a result, there is a 3–5-min lag before circulating GC levels begin to increase after stressor onset (Spencer & Deak, 2017). Consequently, circulating GC levels peak with a delay of 20–40 min after stressor onset and return to baseline within 1–1.5 h of stressor offset (Dickerson & Kemeny, 2004).

In addition to producing phasic responses to stressors, the HPA axis also displays basal **circadian** and **ultradian** rhythmic activity, which includes fluctuations in HPA axis hormones with a period of about 24 h and 60 min, respectively (Spencer & Deak, 2017). Basal GC levels peak at the onset of the active phase of the circadian cycle and decline steadily until they reach a nadir during the resting phase. Circadian fluctuations in HPA activity correspond to fluctuations in metabolic demands associated with daily activity and are thought to regulate energy homeostasis and ensure energy availability (Herman et al., 2016). In addition, the HPA axis displays activity which consists of intrinsic surges of ACTH and GCs occurring around every 60 min. Although the functional significance of this HPA activity is not clear, evidence suggests that hourly GC pulses help maintain optimal HPA responsiveness to stress (Joëls et al., 2012). Thus, phasic HPA axis responses to stressors occur against the backdrop of the circadian and ultradian HPA rhythms, and the magnitude of the HPA stress response is influenced by the concurrent phases of the circadian and ultradian cycles.

GCs bind to receptors throughout the body and the brain and mediate a variety of metabolic, cardiovascular, immune, and cognitive effects in a time- and dose-dependent manner. For example, at basal concentrations, GCs facilitate the onset of the stress response and show synergistic effects with NE and EPI during the initial stages of the stress response, whereas at higher concentrations, GCs oppose the effects of NE and EPI and promote stress response termination and recovery (Sapolsky, 2000). These dose-dependent effects of GCs are mediated by different receptors, **mineralocorticoid receptors (MRs)** and **glucocorticoid receptors (GRs)**, which bind GCs with different affinity. Whereas MRs have high affinity for GCs and are 90% occupied at basal GC concentrations, GRs have about ten times lower affinity for GCs and become occupied only at elevated GC levels (de Kloët et al., 2005; Spencer & Deak, 2017). As a result, MR signaling predominates during the initial phase of the stress response, which is characterized by low, basal GC levels and promotes the onset of the stress response. In contrast, GR signaling predominates during the later phase, which is characterized by high, stress-induced GC levels and promotes stress response termination and recovery, as well as preparation for future stress exposure (de Kloët et al., 2005).

Temporally, GCs can exert both rapid effects on cell function and long-term changes in gene transcription depending on the receptors they activate (Spencer & Deak, 2017). Due to their hydrophobic nature, GCs can easily diffuse across the cell membrane and bind to intracellular receptors. Intracellular receptors regulate the transcription of a large number of genes (1–2% of total genes), including genes involved in cell metabolism, neurotransmission, membrane conductance, cellular communication, and cell structure (de Kloët et al., 2005; Spencer & Deak, 2017). Thus, whereas non-genomic effects of GCs manifest themselves within seconds to a few minutes, genomic effects of GCs take at least 60 min to develop and can last for hours to days (Spencer & Deak, 2017).

Central Stress-Regulatory Network

Regulation of behavioral, physiological, and psychological processes both in response to and in anticipation of homeostatic threat is mediated by a large-scale brain system, which encompasses multiple neural circuits (Herman et al., 2003;

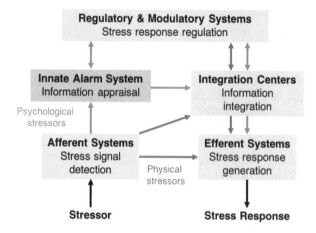

Fig. 2.3 Organization of the central stress-regulatory network. Stressor detection and regulation of the peripheral stress effector systems is mediated by the central stress-regulatory network. This network consists of complexly interconnected neural circuits involved in stress signal detection, information appraisal and integration, and stress response generation and regulation. Physical and psychological stressors are detected by distinct afferent systems and elicit peripheral stress responses via different neural pathways. Whereas physical stressors can directly activate efferent systems involved in stress response generation, psychological stressors undergo additional processing by the innate alarm system and information integration centers. Responses to both physical and psychological stressors can be regulated by higher-order cognitive and affective circuits

Kleckner et al., 2017). This central stress-regulatory network (Fig. 2.3) consists of complex and largely reciprocal connections between (1) sensory regions involved in stressor detection; (2) effector regions involved in generation of the appropriate physiological and behavioral responses via activation of peripheral effector systems; (3) limbic and cortical structures involved in information appraisal and regulation of the stress response; (4) widespread neuromodulatory projections involved in modulation of arousal, affect, motivation, cognitive function, and sensory processing; and (5) integration centers involved in integration of numerous ascending and descending inputs, including peripheral stress signals, regulatory corticolimbic projections, and other modulatory inputs (Ulrich-Lai & Herman, 2009).

We begin the discussion of the organization of the central stress-regulatory network with the description of the **afferent systems** involved in stressor detection and affective appraisal of sensory information (section "Neural Circuitry of Stressor Detection"). We then discuss the organization of **efferent circuits** that generate the physiological and behavioral stress responses via activation of peripheral effector systems (section "Neural Circuitry of Stress Response Generation"). Next, we describe the **cortical and limbic circuits** involved in stress response regulation (section "Neural Circuitry of Stress Response Regulation") and the mechanisms of stress response termination (section "Neural Circuitry of Stress Response Termination"). Finally, we conclude this section with the discussion of **neuromodulatory systems** involved in the psychological aspects of the stress response (section "Neural Circuitry of the Psychological Components of the Stress Response").

Neural Circuitry of Stressor Detection

Consistent with the notion of stressor specificity, distinct neural circuits are involved in detection of physical vs. psychological stressors. Physical stressors are detected by **interoceptive sensory systems** that monitor the internal milieu and body integrity. In contrast, psychological stressors, such as external threats, are detected by **exteroceptive sensory systems** that process sensory information from the external environment and the "**innate alarm system**" that performs a rapid and crude analysis of the biological significance of the incoming sensory information. Both interoceptive and exteroceptive sensory systems perform rapid and unconscious processing of sensory information by subcortical nuclei followed by slower processing by sensory cortices, which in turn gives rise to conscious awareness of the sensory stimuli. However, physical stressors, as well as certain psychological threats, can be detected even at the early, unconscious information processing stage, and can trigger rapid and automatic stress responses without the need for, and well ahead of, conscious awareness.

Neural Circuitry of Detection of Physical Stressors

Physical stressors constitute a genuine homeostatic challenge that is detected by interoceptive sensory systems via specialized receptors throughout the body (Radley, 2012). For example, tissue damage is detected by visceral or somatic pain receptors; deviations in blood volume and blood pressure are detected by **arterial baroreceptors**; deviations in blood oxygenation and pH are detected by **chemoreceptors** in the brainstem; and electrolyte imbalances are detected by **osmoreceptors** in the forebrain. Homeostatic stress signals can reach the brain via afferent projections from the spinal cord or cranial nerves, or via the humoral route, whereby specialized structures in the brain can detect stress-signaling molecules in the bloodstream and cerebrospinal fluid.

The first and main integration center for **viscerosensory stress signals** in the brain is the **nucleus of the solitary tract (NTS)**, a functionally heterogenous and neurochemically diverse structure in the medulla oblongata (Myers et al., 2017). The NTS receives information about the current status of various physiological parameters from arterial baroreceptors and chemoreceptors (e.g., blood pressure), cardiorespiratory receptors (e.g., heart activity) and gastrointestinal receptors (e.g., nutrient levels), as well as visceral and somatic pain receptors via cranial nerves and projections from the spinal cord (Sun, 1995; Zoccal et al., 2014). In response to signals of homeostatic threat, the NTS can generate compensatory physiological responses, such as simple visceral reflexes and more complex patterned responses (e.g., cardiovascular and respiratory reflexes) through projections to effector nuclei that control the activity of the ANS and the HPA axis (Kvetnansky et al., 2009; Saper, 2002; Sun, 1995). In addition, the NTS relays viscerosensory information to integration centers in the brainstem and forebrain that coordinate autonomic, endocrine, and behavioral responses to physical stressors (Iversen et al., 2000). For example, the **parabrachial nuclei (PBN)** receive viscerosensory inputs from the NTS and constitute another integration center for homeostatic information (Myers et al., 2017). The PBN relay viscerosensory information from the NTS to higher-order brainstem and forebrain structures for further processing and integration (Sun, 1995).

Humoral homeostatic signals, such as nutrient levels, hormones, and toxins, are detected by **circumventricular organs**, which are specialized structures in the brain that are therefore sensitive to various signaling molecules in the bloodstream (Saper, 2002). For example, humoral stress signals related to fluid and electrolyte imbalance (e.g., hemorrhage, dehydration) are detected by circumventricular organs of the **lamina terminalis system**. Humoral stress signals related to energy metabolism (e.g., low blood sugar) are detected by hypothalamic feeding centers, such as the **arcuate nucleus**, which are sensitive to circulating levels of various hunger and satiety signals, such as glucose, insulin, and ghrelin (Smith & Vale, 2006). The circumventricular organs convey information about humoral stress signals to the hypothalamus, which can initiate appropriate physiological and behavioral responses via projections to effector nuclei that control the ANS and the HPA axis (Smith & Vale, 2006).

Neural Circuitry of Detection of Psychological Stressors

In contrast to physical stressors, which constitute an *actual* threat to homeostasis, psychological stressors constitute *perceived* threat and therefore require additional processing by neural circuits involved in sensory perception and appraisal of biological and affective significance. External threat signals, such as the sight or smell of a predator, are detected and processed by exteroceptive sensory systems (e.g., visual, olfactory, and auditory). Sensory information initially undergoes rapid but crude processing by subcortical nuclei in the brainstem and thalamus, followed by slower but more complex and accurate analysis of sensory features by the sensory cortices. Whereas subcortical processing is unconscious, cortical processing is accompanied by conscious awareness and gives rise to subjective perception. Rapid assessment of the biological and emotional significance of sensory stimuli (i.e., "threat," "reward," or "neutral") occurs during the early, unconscious stage of sensory processing and is thought to be mediated by the **innate alarm system**, a neural network which monitors the environment for external threat and initiates the appropriate physiological and behavioral responses (Lanius et al., 2017; Liddell et al., 2005; Öhman, 2005).

The **amygdala** is a key structure within the innate alarm system, which is critically involved in assessing the salience and valence of external stimuli, initiation of physiological and behavioral responses to potential threats, and formation of emotional memories (e.g., Janak & Tye, 2015; Méndez-Bértolo et al., 2016; Rodrigues et al., 2009). The amygdala is a structure in the medial temporal lobe, which consists of multiple nuclei with distinct functions and connectivity patterns, some of which are primarily involved in threat detection, and others in stress response generation. For example, the **basolateral amygdala (BLA)** acts as a gateway for multimodal sensory inputs, including external threat signals, from the thalamus and the sensory cortices, as well as highly processed sensory and contextual information from the association cortices and the hippocampus (Janak & Tye, 2015). In contrast, the **central nucleus of the amygdala (CeA)** serves as the primary output region, which controls autonomic, endocrine, and behavioral responses to threat via projections to the brainstem, hypothalamus, and forebrain (Rodrigues et al., 2009). For example, CeA projections to the hypothalamus promote activation of the ANS and HPA axis; CeA projections to brainstem nuclei generate behavioral responses, such as freezing; and CeA projections to nuclei in the brainstem promote arousal and vigilance (Rodrigues et al., 2009; Ulrich-Lai & Herman, 2009).

Although the amygdala responds to both physical and psychological stressors (Herman et al., 2003), bilateral amygdala damage impairs the ability to generate fear responses to external threats, but not to homeostatic stressors (Feinstein et al., 2013). This evidence suggests that affective appraisal of sensory information by the amygdala is necessary for an appropriate stress response to psychological stressors specifically. Signals of potential threat reach the amygdala via two pathways: the rapid, **"low road" pathway**, which transmits crude sensory information from the thalamus, and the slow, **"high road" pathway**, which transmits highly processed sensory information from the sensory cortices (LeDoux, 2012). Certain sensory stimuli related to species-specific threats (e.g., fearful faces, predator smell) can rapidly activate the amygdala via the "low road" pathway without the need for cortical processing, directed attention, or conscious awareness (LeDoux, 2012; Méndez-Bértolo et al., 2016). When potential threat is detected, the amygdala initiates reflexive physiological and behavioral responses through activation of effector nuclei that control the peripheral stress effector systems and central neuromodulatory systems that promote arousal, vigilance, and enhanced sensory processing. The rapid amygdala-initiated responses to coarse sensory information that is conveyed via the "low road" pathway can promote short-term survival but are also error-prone due to the speed/accuracy trade-off (e.g., mistaking a stick for a snake). After a short delay, the amygdala receives highly processed sensory information via the "high road" pathway, as well as memory and context information from the hippocampus and association cortices, to allow for a more accurate appraisal of the significance of external stimuli (LeDoux, 2012).

Neural Circuitry of Stress Response Generation

When a stressor is detected by the afferent systems, physiological responses can be generated through activation of the peripheral effector systems (i.e., the ANS and the HPA axis), which alter the functioning of target tissues throughout the body. The activity of peripheral effector systems is controlled by a complex and hierarchically organized **central efferent system** (Fig. 2.4). At the lowest level of the hierarchy, **central effectors** directly control **peripheral effectors** (i.e., autonomic ganglia, adrenal glands, and anterior pituitary), which in turn produce physiological stress

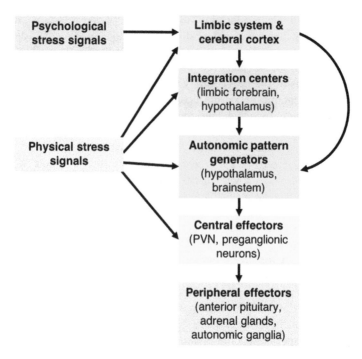

Fig. 2.4 Hierarchical organization of the central efferent pathways involved in generation of physiological responses to stress

responses via the release of peripheral stress mediators (i.e., hormones and neurotransmitters) into the bloodstream and directly onto target organs.

Central effectors that regulate the ANS include sympathetic preganglionic neurons in the spinal cord and parasympathetic preganglionic neurons in the spinal cord and brainstem, which control the activity of autonomic ganglia and the adrenal medulla (see section "Autonomic Nervous System"). Preganglionic neurons are themselves controlled by **preautonomic neurons** in the hypothalamus and brainstem. For example, preganglionic sympathetic neurons in the spinal cord receive direct projections from a number of lower brainstem regions, including the NTS; the rostral ventrolateral medulla; noradrenergic, adrenergic, and serotoninergic brainstem neurons; and hypothalamic nuclei, including the PVN, arcuate nucleus, and lateral hypothalamus (Kvetnansky et al., 2009; Saper, 2002). Groups of preautonomic neurons can function as **autonomic pattern generators**, which produce patterned and stereotypical physiological responses (e.g., cardiovascular and respiratory reflexes) via projections to distinct groups of preganglionic neurons. The hypothalamus can be thought of as the highest-level pattern generator, as it orchestrates complex autonomic, endocrine, and behavioral responses via coordinated activation of lower-level autonomic pattern generators (see section "Neural Circuitry of Generation of Complex Coordinated Responses"). The hypothalamus is in turn regulated by the limbic system and the cerebral cortex in accord with ongoing behavioral, psychological, and homeostatic needs. Central effectors that regulate the HPA axis include neuroendocrine neurons in the PVN, which control the activity of the anterior pituitary (and ultimately the adrenal cortex) via the release of CRH and AVP into circulation, which in turn promote the release of ACTH by the anterior pituitary (see section "Hypothalamic-Pituitary-Adrenal Axis"). The neuroendocrine neurons in the PVN are regulated by direct and indirect excitatory and inhibitory projections from other hypothalamic nuclei, the circumventricular organs, the NTS, monoaminergic brainstem nuclei, the limbic system, and the cerebral cortex (see section "Neural Circuitry of Stress Response Regulation").

Both physical and psychological stressors elicit physiological responses by activating the ANS and HPA axis; however, the neural circuits through which physical and psychological stressors activate peripheral effector systems differ (e.g., Radley, 2012). Physical stress signals can trigger reactive physiological responses via *direct* projections from interoceptive afferent systems onto central effectors, as well as more complex and patterned responses via afferent projections to autonomic pattern generators and higher-order limbic and cortical areas. Psychological stressors, on the other hand, elicit anticipatory responses by activating central effectors *indirectly*, via intervening relay and integration centers in the limbic forebrain, hypothalamus, and brainstem (Kvetnansky et al., 2009; Ulrich-Lai & Herman, 2009). In addition, psychological stressors recruit numerous other neural circuits, including cognitive and affective networks that regulate higher-order processing, neuromodulatory systems that regulate neurotransmission throughout the brain and promote arousal and vigilance, and the motivation system that regulates goal-directed instrumental and habitual behavior (LeDoux, 2012). Thus, anticipatory responses to psychological stressors can be viewed as being built upon reactive responses to physical stressors and involving more complex regulation (Herman et al., 2003).

Neural Circuitry of Generation of Simple Reflexive Responses

Complex physiological responses to stressors that involve changes in multiple organ systems (e.g., cardiovascular, skeletomuscular) can be viewed as a combination of simple autonomic and neuroendocrine responses that are produced by specific groups of central effectors and coordinated by higher-order autonomic pattern generators. The simplest and most stereotypical response to a homeostatic challenge is a **visceral reflex.** Visceral reflexes are mediated by neuronal circuits in the brainstem and spinal cord and can be directly initiated by viscerosensory nuclei that detect physical stress signals, such as the NTS (see section "Neural Circuitry of Detection of Physical Stressors"; Iversen et al., 2000). As a major integration center for homeostatic information, the NTS is involved in both detection of physical stressors and generation of reactive stress responses. For example, the NTS can elicit rapid activation of the ANS via projections to preautonomic and preganglionic neurons (Herman et al., 2003; Ulrich-Lai & Herman, 2009). In addition, the NTS is the main source of direct excitatory inputs to the PVN, which are both necessary and sufficient for HPA axis activation in response to physical stressors (Myers et al., 2017; Radley, 2012).

An example of a simple, reflexive autonomic response to a physical stressor initiated by the NTS is the **baroreflex**. When blood pressure drops (e.g., due to a hemorrhage or dehydration), the resulting relaxation of blood vessels is detected by arterial baroreceptors and conveyed to the NTS via cranial nerves. The NTS in turn initiates compensatory responses by increasing sympathetic outflow and decreasing parasympathetic outflow to the heart and blood vessels. Specifically, the NTS disinhibits the **rostral ventrolateral medulla (RVLM)**, a brainstem nucleus involved in the regulation of cardiovascular activity and maintenance of adequate blood pressure and organ perfusion. The RVLM in turn activates preganglionic sympathetic neurons in the spinal cord, which then stimulate the heart and promote vasoconstriction. In addition to disinhibiting the cardio-excitatory neurons in the RVLM, the NTS also inhibits the cardio-inhibitory parasympathetic neurons in the medulla oblongata (Sun, 1995; Zoccal et al., 2014). The resulting increase in sympathetic outflow raises the blood pressure via vasoconstriction and elevated heart rate, which effectively dampens or removes the original homeostatic stress signal and shuts down the compensatory response.

Neural Circuitry of Generation of Complex Coordinated Responses

Complex physiological responses to stressors are produced by autonomic pattern generators located in the hypothalamus and throughout the brainstem, such as the ventrolateral medulla (e.g., cardiovascular), the rostral medullary raphe (e.g., thermogenic), and the periaqueductal gray matter (e.g., defense; Saper, 2002). Higher-level autonomic pattern generators (e.g., in the midbrain and hypothalamus) can regulate the activity of multiple lower-level autonomic pattern generators (e.g., in the pons and medulla). For example, the **periaqueductal gray matter (PAG)** is an important autonomic pattern generator involved in "defense" reactions to various stressors, including predator exposure and restraint (Myers et al., 2017). The PAG is a functionally heterogenous nucleus in the midbrain that contains distinct cell populations involved in different aspects of the stress response, including autonomic (e.g., cardiovascular) and behavioral (e.g., freezing) defense reactions. For example, experimental stimulation of the dorsolateral PAG produces a flight reaction, elevates blood pressure via direct projections to the RVLM, and increases circulating GC levels via projections to the PVN. In contrast, stimulation of the ventrolateral PAG produces freezing and decreases heart rate and blood pressure via indirect projections to the RVLM (Saper, 2002; Sun, 1995).

The hypothalamus is the principal stress signal integration and response coordination center (Saper, 2002). It generates the appropriate patterns of autonomic and endocrine responses to specific stressors by coordinating the activities of multiple autonomic pattern generators and central effectors. The PVN is a particularly important hypothalamic nucleus involved in stress response generation. The PVN is not only a central effector of the HPA axis but also a critical autonomic pattern generator that regulates ANS activity. Whereas neuroendocrine cells within the medial division of the PVN activate the HPA axis via the release of CRH and AVP into the

bloodstream (see section "Hypothalamic-Pituitary-Adrenal Axis"), a separate population of PVN neurons controls ANS activity via direct projections to preganglionic
neurons in the brainstem and spinal cord (Myers et al., 2017; Ulrich-Lai & Herman,
2009). In addition, the PVN projects to brainstem nuclei that integrate viscerosensory information (e.g., NTS, PBN), controls autonomic responses to homeostatic
challenges and regulates arousal, and is therefore well-positioned to regulate cardiovascular, respiratory, gastrointestinal, and metabolic responses to physical stressors
(Geerling et al., 2010; Saper, 2002; Sun, 1995).

Distinct hypothalamic nuclei are involved in specific regulatory circuits and are
sensitive to specific types of stressors. For example, magnocellular paraventricular
and supraoptic nuclei are sensitive to disruptions of water and electrolyte homeostasis, the preoptic area controls body temperature and is sensitive to thermal stressors,
and the dorsolateral hypothalamus is involved in energy homeostasis (Kvetnansky
et al., 2009). Each hypothalamic control center and the lower-level autonomic pattern generators it controls form distinct "**survival circuits**" which regulate processes essential for survival, including maintenance of energy and fluid balance,
thermoregulation, and defense reactions (LeDoux, 2012). Each survival circuit is
activated by specific types of stimuli – either genetically programmed or learned
through experience – and mediates behavioral and physiological responses to biologically significant events, including stressors. For example, the "**defense circuit**"
is activated by psychological stressors, such as external threats. The defense circuit
consists of projections from sensory systems to the amygdala that detects external
threat, and from the amygdala to specific hypothalamic nuclei that coordinate a
defense response by activating specific cell populations in the PAG and other autonomic pattern generators. The PAG, in turn, projects to lower-order autonomic pattern generators, such as the RVLM, which then generate physiological and
behavioral defense responses via projections to specific central effectors
(LeDoux, 2012).

Neural Circuitry of Stress Response Regulation

Autonomic pattern generators and central effectors that constitute the central efferent system can be regulated in a **bottom-up** manner by ascending stress signals or
in a **top-down** manner by descending inputs from cortical and limbic structures,
which are involved in the processing of context, memory, affect, and motivation.
Ascending homeostatic information and descending corticolimbic inputs are integrated at multiple sites throughout the brainstem, hypothalamus, and limbic forebrain, allowing for reciprocal modulatory effects between pathways that transmit
bottom-up and top-down signals. For example, reactive responses to physical stressors can be modulated by contextual information, which is conveyed via descending
cortical projections, whereas anticipatory responses to psychological stressors can
be modulated by the current homeostatic state (Gunnar & Quevedo, 2007).

Regulation of the Stress Response by the Limbic System

The **limbic system** plays an important role in the regulation of ANS and HPA responses (Fig. 2.5), particularly to psychological stressors (e.g., Herman et al., 2005). The limbic system encompasses a number of cortical and subcortical structures within the forebrain, including the medial prefrontal cortex, cingulate cortex, insula, lateral septum, medial temporal cortex, extended amygdala, and hippocampal formation. Limbic projections to autonomic pattern generators and central effectors are a route by which psychological processes, such as affective states, cognitions, and memories, can elicit and modulate physiological stress responses. However, unlike nuclei that detect physical stressors, limbic structures make few direct connections with the central autonomic and neuroendocrine effectors (Beissner et al., 2013; Dum et al., 2016; Westerhaus & Loewy, 2001). Rather, the limbic system

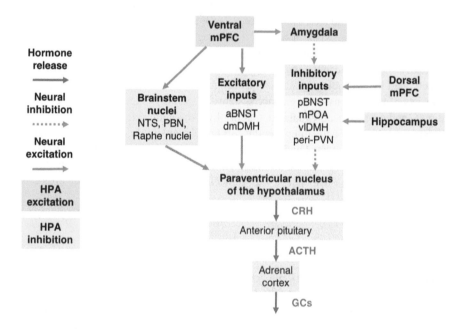

Fig. 2.5 Corticolimbic regulation of HPA axis responses to stressors. Higher-order cortical and limbic structures indirectly regulate HPA activity via relay and integration centers that target the PVN. For example, ventral mPFC and the amygdala promote HPA activation by stimulating direct excitatory inputs (aBNST, dmDMH, NTS, PBN) or suppressing inhibitory inputs (pBNST, mPOA, vlDMH, peri-PVN) to the PVN, respectively. In contrast, dorsal mPFC and the ventral hippocampus inhibit HPA activity by stimulating inhibitory inputs to the PVN. PVN = paraventricular nucleus of the hypothalamus; CRH = corticotropin-releasing hormone; ACTH = adrenocorticotropic hormone; GCs = glucocorticoids; NTS = nucleus of the solitary tract; PBN = parabrachial nuclei; a/pBNST = anterior/posterior bed nucleus of the stria terminalis; dm/vlDMH = dorsomedial/ventrolateral division of the dorsomedial nucleus of the hypothalamus; mPOA = medial preoptic area; mPFC = medial prefrontal cortex

influences the stress response indirectly via intervening **integration centers**, which in turn target central effectors (e.g., Radley, 2012; Ulrich-Lai & Herman, 2009).

Effector nuclei that control the activity of the ANS and HPA axis receive a large number of regulatory inputs that promote or inhibit the activity of the peripheral stress effector systems. For example, the PVN receives direct excitatory projections from brainstem viscerosensory nuclei, circumventricular organs, and other hypothalamic nuclei through which homeostatic stress signals can trigger reactive ANS and HPA axis responses. The PVN also receives indirect cortical and limbic projections, which regulate PVN activity via integration centers, including the **bed nucleus of the stria terminalis (BNST)**, the **medial preoptic area of the hypothalamus (mPOA)**, the **dorsomedial nucleus of the hypothalamus (DMH)**, and the **peri-PVN region** (e.g., Radley, 2012; Ulrich-Lai & Herman, 2009). Cortical and limbic structures target specific cell populations within these integration centers, which in turn either promote or suppress PVN activity via direct excitatory or inhibitory projections (e.g., Ulrich-Lai & Herman, 2009). Ultimately, the integration of the incoming excitatory and inhibitory signals by the PVN determines the net drive on the peripheral stress effector systems that the PVN controls.

The **amygdala** is an important source of stimulatory limbic projections to the PVN. Rather than stimulating the PVN directly, the amygdala lifts the "brake" on the HPA axis by suppressing inhibitory relay neurons in the BNST, mPOA, DMH, and the peri-PVN region that target the PVN (Smith & Vale, 2006). Accordingly, experimental stimulation of the amygdala results in increased HPA activity, whereas amygdala lesions reduce HPA responses to stress (Herman et al., 2005). Although the amygdala responds to both physical and psychological stressors, distinct amygdala nuclei are activated by different stressors. For example, the BLA and the **medial nucleus (MeA)** receive exteroceptive information about external threats through their dense connections with various sensory regions and show preferential activation to psychological stressors, such as social threat and restraint (Herman et al., 2003). In contrast, the CeA receives viscerosensory and nociceptive information from the brainstem and viscerosensory cortex and shows preferential activation to physical stressors, such as hypoxia, or immunologic challenge (Herman et al., 2003). Thus, the BLA and MeA promote primarily anticipatory responses to psychological stressors, whereas the CeA promotes primarily reactive responses to physical stressors (Herman et al., 2005).

Unlike the amygdala, which stimulates PVN activity, the **hippocampus** inhibits the PVN. The hippocampus is a limbic structure within the medial temporal lobe that plays an important role in both tonic inhibition of HPA activity and suppression of HPA responses to stressors (Radley, 2012). Accordingly, experimental stimulation of the hippocampus inhibits the activity of the neurosecretory PVN neurons and attenuates GC production, whereas hippocampal lesions result in elevated basal GC levels and delayed HPA response termination (Herman et al., 2005). Similar to the amygdala, the hippocampus suppresses PVN activity indirectly, by stimulating inhibitory relays in the hypothalamus and limbic forebrain (Herman et al., 2003). Hippocampal regulation of the HPA response is stressor-specific; hippocampal lesions enhance the HPA response to psychological but not physical stressors (Smith

& Vale, 2006). The hippocampus also sends excitatory projections to the **lateral septum**, a limbic structure adjacent to the septum pellucidum, which in turn suppresses HPA and ANS responses to acute stressors via inhibitory relays that target the PVN (Ulrich-Lai & Herman, 2009; Kvetnansky et al., 2009). Lesions of the lateral septum produce a constellation of symptoms of extreme irritability, defensiveness, and aggression, known as the "septal rage" (Albert & Chew, 1980).

Another major source of inputs to the PVN is the **medial prefrontal cortex (mPFC)**, which is critically involved in regulating autonomic and endocrine responses, particularly to psychological stressors (McKlveen et al., 2015). Specific mPFC regions that regulate ANS and HPA activity have been collectively termed the **visceromotor cortex** (Vertes, 2004; Price, 1999). The visceromotor cortex receives afferent inputs from various limbic and viscerosensory structures and coordinates the appropriate physiological responses via outputs to effector nuclei in the brainstem and hypothalamus, including the PVN (Kvetnansky et al., 2009). The mPFC regulates the ANS and HPA activity in a region-specific manner (Herman et al., 2003). For example, **dorsal mPFC** inhibits the HPA axis and the sympathetic outflow, promotes the parasympathetic outflow, and contributes to the termination of the HPA stress response. In contrast, **ventral mPFC** promotes sympathetic and HPA axis activation (Ulrich-Lai & Herman, 2009; Radley et al., 2006). Consistent with this functional specificity, dorsal and ventral mPFC regions project to different relays that target the PVN (McKlveen et al., 2015). For example, dorsal mPFC suppresses the PVN by activating inhibitory neurons in the BNST, DMH, mPOA, and peri-PVN, whereas ventral mPFC stimulates the PVN by activating excitatory neurons in the BNST, DMH, amygdala, and NTS (Herman et al., 2005).

Regulation of the Stress Response by the Endocannabinoid System

The endocannabinoid system plays a crucial role in HPA axis regulation both under basal conditions and in response to stressors (Hill et al., 2011). **Endocannabinoids (eCB)** constitute a family of lipid-based neurotransmitters that act on the same receptors as psychoactive compounds in cannabis. eCB receptors are abundantly expressed in key nodes of the stress-regulatory circuitry, including the PVN, amygdala, hippocampus, and mPFC. Unlike many "classical" neurotransmitters, eCB act in a retrograde manner, whereby they diffuse from the postsynaptic neuron and bind to eCB receptors on the presynaptic axon terminals. Activation of presynaptic eCB receptors inhibits neurotransmitter release from the presynaptic neuron and thus gates the inputs that the postsynaptic neuron receives. The two major eCB neurotransmitters, **anandamide (AEA)** and **2-arachidonoyl glycerol (2-AG)**, are differentially engaged in HPA axis regulation and play distinct roles in tonic and acute suppression of HPA axis activity. Whereas AEA maintains tonic inhibition of the HPA axis under unstressed conditions, 2-AG promotes termination of the HPA axis response to stressors (see section "Neural Circuitry of Stress Response Termination").

Tonic inhibition of the HPA axis in the absence of stressors is mediated by sustained AEA signaling in the amygdala (Hill & Tasker, 2012). Tonic release of AEA in the BLA gates the incoming excitatory inputs and suppresses amygdala activity under unstressed conditions. When faced with a stressor, AEA levels in the BLA rapidly drop, which results in increased BLA sensitivity to excitatory inputs and promotes downstream activation of the PVN. Thus, rapid loss of AEA signaling in the amygdala facilitates the HPA stress response (Hill & Tasker, 2012). When GCs reach elevated concentrations, they promote AEA signaling in the BLA and thus help reinstate tonic inhibition of the HPA axis after the stressor is over. Conversely, experimental disruption of eCB signaling produces a physiological and behavioral phenotype associated with the stress response, including HPA activation, anxiety, hypervigilance, arousal, suppressed feeding, and impaired cognitive flexibility (Morena et al., 2016).

Neural Circuitry of Stress Response Termination

Mounting a stress response is energetically costly, and inappropriate (i.e., excessive, prolonged, or repeated) activation of peripheral stress effectors can result in allostatic load, which inflicts wear and tear on the body and can lead to the development of pathological states and stress-related disorders (McEwen, 1998). Prompt termination of the stress response once the stressor is gone is essential, as is suppression of stress responses to stimuli that have been learned to be innocuous. The sympathetic response is typically short-lived and terminated by reflexive activation of the parasympathetic branch; however, HPA response termination involves two types of negative feedback, rapid and delayed (see section "Hypothalamic-Pituitary-Adrenal Axis").

Rapid negative feedback occurs at the level of the PVN and the anterior pituitary, as circulating GCs quickly shut down CRH and ACTH production, and hence the HPA axis, by activating membrane GRs (Tasker & Herman, 2011). **Delayed negative feedback**, on the other hand, involves indirect inhibition of the PVN by cortical and limbic structures, such as the mPFC and the hippocampus (see section "Neural Circuitry of Stress Response Regulation"; Herman et al., 2016; Myers et al., 2017; Ulrich-Lai & Herman, 2009). Both the mPFC and the hippocampus are rich in GRs and are therefore sensitive to stress-induced elevations in circulating GCs (Herman et al., 2003; Ulrich-Lai & Herman, 2009). Consistent with their role in HPA axis inhibition, lesions of the dorsal mPFC and ventral hippocampus impair negative feedback, prolong ACTH and GC release, delay HPA response termination, and promote anxiety in response to psychological stressors (Herman et al., 2016; Jones et al., 2011; Radley, 2012; Radley et al., 2006; Sullivan & Gratton, 2002). In contrast to rapid negative feedback, which is mediated by rapid, nongenomic effects of membrane GRs that develop within a few minutes, delayed negative feedback is mediated by slow, genomic effects of intracellular GRs that take at least an hour to develop (see section "Hypothalamic-Pituitary-Adrenal Axis"). Activation of intracellular GRs in the mPFC and hippocampus enhances the inhibitory drive to the PVN and terminates the HPA response, particularly to psychological stressors.

Both types of negative feedback critically depend on the release of eCB in the hypothalamus and the corticolimbic circuitry. Elevated GC levels stimulate 2-AG release in the PVN, mPFC, and hippocampus, which in turn promotes termination of the HPA response via rapid or delayed negative feedback (Morena et al., 2016). In the PVN, activation of membrane GRs causes rapid synthesis and release of eCB, which in turn inhibit the incoming excitatory inputs to the PVN (Herman et al., 2016). As a result, the excitatory drive on the PVN is reduced and the HPA axis response shuts down via rapid negative feedback (Herman et al., 2016). Similarly, activation of intracellular GRs in the mPFC and hippocampus promotes the release of eCB, which in turn block the incoming inhibitory inputs and effectively "lift the brake" on mPFC and hippocampus activity. Once disinhibited, the mPFC and hippocampus indirectly suppress PVN activity by stimulating inhibitory relays in the BNST and the peri-PVN region, which ultimately leads to HPA response termination via delayed negative feedback (Hill et al., 2011).

Neural Circuitry of the Psychological Components of the Stress Response

In the preceding sections, we described the circuits within the central stress-regulatory network that regulate physiological responses to stressors by controlling the peripheral stress effector systems, such as the ANS and HPA axis. In addition to physiological changes in the body, the stress response also includes important psychological components, including changes in arousal, sensory processing, cognitive function, and affect. In this section, we discuss the neuromodulatory systems that impact aspects of cognition, and consequently behavior, with the aim of promoting survival and short-term adaptation to acute stressors.

The Psychological Components of the Stress Response

Acute stressors elicit a number of changes in sensory, cognitive, and affective functioning that are aimed at promoting adaptive behavioral coping responses. For example, acute stress increases alertness and vigilance, enhances sensory processing and memory encoding, and promotes dissipated attention, preferential orientation to salient stimuli, and facilitated disengagement from ongoing tasks (Joëls et al., 2018). These cognitive and behavioral changes are adaptive in high-threat environments, because they promote short-term survival by increasing an organism's responsiveness to external stimuli (e.g., danger signals or coping opportunities) and contribute to long-term survival by promoting the formation of stressor-related memories, which could prove beneficial in the future. At the same time, acute stress biases behavior toward simple, reflexive, and stereotypical response strategies, such as escape or attack, at the expense of higher-order cognitive functions, such as focused attention, cognitive flexibility, working memory, declarative memory retrieval, contextual processing, and delayed gratification (Joëls et al., 2018; Plessow et al., 2012; Shields et al., 2016).

Although higher-order cognitive functions are useful strategies in safe environments when sufficient cognitive resources can be devoted to the task at hand, in stressful and dangerous environments, focused attention on a single task or target at the expense of vigilance for potential threat can have deleterious consequences. In addition, because higher-order cognitive functions, such as reasoning and decision-making, are slower, more energetically costly, and more computationally intense than the less flexible habitual and conditioned behaviors, higher-order cognition is less effective in highly stressful situations which require quick and efficient responses (Arnsten, 2015). Thus, acute stress promotes a switch from "reflective" functions mediated by the hippocampus (e.g., declarative memory retrieval, contextual processing) and the prefrontal cortex (e.g., focused attention, cognitive flexibility) to "reactive" functions mediated by the amygdala (e.g., threat detection, emotional memory) and the striatum (e.g., habitual behaviors, conditioned responses; Arnsten, 2015; Joëls et al., 2018).

Central Norepinephrine and Psychological Components of the Stress Response

Changes in arousal, sensory function, and cognition in response to stress are mediated by widespread neuromodulatory projections that originate from small clusters of neurons in the brainstem and modulate neurotransmission in their target circuits, including sensory, cognitive, and affective networks. These projections can influence global brain function (and hence cognition and behavior) via widespread release of neurotransmitters, including norepinephrine (NE), dopamine (DA), and serotonin (Kvetnansky et al., 2009). These neuromodulatory projections are part of the **ascending arousal system**, which promotes arousal, wakefulness, and alertness by stimulating the cerebral cortex directly, or via the thalamus, hypothalamus, and basal forebrain (Edlow et al., 2012). Central catecholamine projections, particularly the noradrenergic system, are key to the modulation of arousal, cognition, sensory function, and behavior in response to stress (de Kloët et al., 2005; Gunnar & Quevedo, 2007).

Within seconds to a few minutes of stressor onset, rapid activation of the central NE system promotes reallocation of resources from higher-order cognitive networks, such as the **executive control network** to neural circuits involved in threat detection and habitual responding, such as the **salience network**, which consists of the amygdala, anterior cingulate cortex, anterior insula, as well as a number of striatal, thalamic, hypothalamic, and brainstem nuclei (Hermans et al., 2014; Seeley et al., 2007). In this way, central NE signaling is thought to promote a rapid switch from the "reflective" functions, which are mediated by higher-order cognitive networks centered on the **prefrontal cortex (PFC)**, to the "reflexive" functions, which are mediated by the salience network centered on the amygdala (Schwabe, 2017).

About 70% of the brain's total NE is produced by the **locus coeruleus (LC)**, a small nucleus in the dorsolateral pons. The LC receives sensory afferents that convey information about physical and psychological stressors from the brainstem, hypothalamus, and amygdala, and sends widespread neuromodulatory projections to all major forebrain areas and the cerebellum (Kvetnansky et al., 2009). The LC is the sole source of NE projections to the cerebral cortex and densely innervates areas involved in sensory processing (e.g., thalamus, sensory cortex), attention regulation (e.g., posterior parietal cortex, pulvinar, superior colliculus), memory (e.g., amygdala, hippocampus), executive function (e.g., PFC), and motor responses (e.g., primary motor cortex, striatum; Wood & Valentino, 2017). In addition, the LC sends direct and indirect projections to autonomic preganglionic neurons, including those that innervate the adrenal medulla (Kvetnansky et al., 2009; Aston-Jones & Cohen, 2005; Morilak et al., 2005). Due to its unique connectivity, the LC is well-positioned to regulate both physiological and psychological aspects of the stress response via projections to effector neurons and to the cerebral cortex and the limbic system, respectively (Wood & Valentino, 2017).

LC neurons exhibit three modes of activity, which are associated with different levels of arousal and cognitive function: (1) low tonic activity (low NE release), (2) moderate tonic activity with phasic bursts (moderate NE release), and (3) high tonic activity in the absence of phasic activity (high NE release). Tonic LC activity correlates with physiological, neural, and behavioral measures of arousal, including elevated sympathetic tone, and anxiety-like behaviors (McCall et al., 2015; Myers et al., 2017). In addition to arousal, the LC also modulates higher-order cognitive functions, including attention allocation, cognitive flexibility, learning, and working memory, due to its widespread projections to sensory and cognitive networks. The relation between modes of LC activity and cognitive function follows an inverted U-shape (Aston-Jones & Cohen, 2005). For example, both low and high tonic LC activity are associated with poor performance on tasks that engage the PFC and require focused attention, due to low arousal and drowsiness, or high arousal and distractibility, respectively. In contrast, moderate tonic LC activity coupled with phasic bursting in response to task-relevant stimuli is associated with moderate arousal and optimal cognitive performance (Aston-Jones & Cohen, 2005).

These dose-dependent effects of NE signaling are mediated by distinct adrenergic receptors which display different affinities for NE. Thus, phasic LC activity, which is associated with moderate NE release, promotes working memory, cognitive flexibility, and sustained attention via activation of higher-affinity α2-adrenoreceptors on target neurons, particularly in the PFC (Aston-Jones & Cohen, 2005). In contrast, high tonic LC activity, which is associated with elevated NE release, impairs PFC function by activating lower-affinity α1-adrenoceptors (Winklewski et al., 2017). The effects of NE on PFC function are further reinforced by concurrent DA release from midbrain projections to the PFC (Arnsten, 2015). At moderate levels, NE and DA enhance the signal-to-noise ratio within PFC circuits via complementary effects on PFC function. Whereas NE signaling improves "signal" by promoting the sensitivity of PFC neurons to task-relevant inputs, DA reduces the "noise" by decreasing PFC sensitivity to task-irrelevant inputs. In contrast, excessive PFC NE and DA release results in lower signal-to-noise ratio and enhanced sensitivity to both task-relevant stimuli and irrelevant distractors (Arnsten, 2015).

Stressors promote LC activity and stimulate central NE release (Valentino & Van Bockstaele, 2008). In response to acute stress, LC activity shifts toward high tonic firing, which inhibits focused attention and promotes disengagement from the ongoing task; increases arousal, alertness, and vigilance; and enhances sensitivity to environmental stimuli, all of which are adaptive behavioral strategies in high-threat environments (Wood & Valentino, 2017). High tonic LC activity also promotes habitual and conditioned behaviors mediated by the amygdala and the striatum at the expense of higher-order cognitive and executive functions mediated by the PFC and the hippocampus (Arnsten, 2015). In contrast to higher-order cognitive functions, which are impaired by elevated NE signaling, emotional memory formation is enhanced by increased NE release in response stress (e.g., Rodrigues et al., 2009). For example, NE released by the LC increases neuronal excitability and facilitates synaptic plasticity in the amygdala, hippocampus, and limbic cortex, and thus promotes encoding and consolidation of emotional memories, particularly for stressor-related information (Krugers et al., 2012; Joëls & Baram, 2009). Thus, high tonic LC activity is thought to mediate the cognitive aspect of the stress response by promoting the shift from PFC-mediated "reflective" behaviors to amygdala-mediated "reactive" strategies (Arnsten, 2015; Wood & Valentino, 2017).

Regulation of Central Norepinephrine Signaling

Switching between different modes of LC activity is regulated by cortical, limbic, and brainstem projections (Valentino & Van Bockstaele, 2008). For example, excitatory projections from the amygdala bias LC activity toward the high tonic firing mode through the release of CRH onto LC dendrites (e.g., McCall et al., 2015; Valentino & Van Bockstaele, 2008). These CRH projections originate from the CeA and are thought to constitute the main route through which physical and psychological stressors detected by the amygdala activate the LC and give rise to the psychological component of the stress response (Van Bockstaele et al., 1998; Snyder et al., 2012; Morilak et al., 2005). In contrast to the CRH projections from the amygdala, which promote high tonic LC activity, opioid projections from the **nucleus paragigantocellularis (PGi)** in the medulla oblongata have the opposite effect on LC firing (Wood & Valentino, 2017). **Opioid** neurotransmitters released by the PGi decrease the tonic firing rate of LC neurons and shift their activity toward the phasic firing mode. In this way, opioid signaling from the PGi promotes focused and selective attention to relevant stimuli, maintains ongoing behavior, and resets LC activity to baseline upon stressor termination (Valentino & Van Bockstaele, 2008).

The ability to shift between high tonic and phasic activity of LC neurons has been associated with differences in behavioral coping with social stress (Reyes et al., 2015). For example, following repeated exposure to social stress, some animals develop an adaptive coping strategy that is characterized by increased latency to subordination, whereas other animals maintain their initial maladaptive strategy, which is characterized by short latency to subordination. In animals with the adaptive coping strategy, the initial increase in CRH signaling from the amygdala gradually decreases, whereas opioid signaling from the PGi is maintained, leading to a gradual normalization of LC activity. In contrast, animals that fail to develop the adaptive coping strategy maintain amygdala CRH signaling and lose opioid signaling from the PGi, leading to sustained high tonic LC activity (Reyes et al., 2015). Thus, appropriate switching between amygdala-initiated high tonic LC activity and PGi-initiated moderate tonic LC activity is important for efficient behavioral responses to both acute and repeated stress.

Neurobiological Effects of Acute and Chronic Stress Exposure

The central stress-regulatory network described in "Central stress-regulatory network" regulates the physiological, cognitive, and behavioral components of the stress response through a carefully timed release of central and peripheral stress mediators (e.g., neurotransmitters and hormones) directly onto target cells or into the bloodstream. Stress mediators bind to their receptors throughout the body and the brain and bring about a coordinated physiological and behavioral response, which is aimed at promoting both short-term and long-term adaptation. The effects of stress mediators on their target cells are dose- and location-specific and depend on the timing and duration of exposure. For example, the rapid effects of the stress mediators that are released within seconds to minutes of stressor onset promote the physiological and behavioral changes necessary for immediate survival and coping with the stressor. In contrast, the slow effects of stress mediators help terminate the stress response and promote recovery. The effects of stress mediators on target tissues also depend on the duration of exposure to these stress mediators. Acute stressors that last minutes to a few hours at most elicit a transient increase in stress mediators, which produce short-lived and adaptive physiological and behavioral changes and a prompt return to baseline after the stressor is over. In contrast, chronic stressors that span weeks to months or longer typically result in prolonged or repeated exposure to elevated levels of stress mediators, which can have deleterious cumulative effects on various organ systems and are associated with increased allostatic load and morbidity (e.g., Lupien et al., 2018).

The brain is an important target of various stress mediators, which can influence cognition and behavior, as well as regulate their own production via feedback mechanisms that target the central stress-regulatory network. Key nodes of the central stress-regulatory network co-express receptors for multiple peripheral and central stress mediators and are therefore highly sensitive to fluctuations in their levels (Joëls & Baram, 2009). Such sensitivity is necessary for the ability of the central stress-regulatory network to regulate the peripheral stress effector systems and calibrate the neuroendocrine stress response in order to promote adaptation to both acute and chronic stressors. In addition, stress mediators can produce profound changes in sensory function, cognition, and affect by acting on specific circuits that mediate the psychological component of the stress response. However, enhanced sensitivity of the central stress-regulatory network to stress mediators also renders it vulnerable to chronic stress, which is characterized by prolonged or repeated elevations in stress mediators. As a result, chronic stressors can produce sustained structural and functional changes across the central stress-regulatory network that are not typically seen after acute stressors and that are associated with a number of cognitive and behavioral deficits (e.g., Lupien et al., 2018).

In this second half of the chapter, we focus on the neurobiological effects of acute (section "Neurobiological Effects of Acute Stress") and chronic (section "Neurobiological Effects of Chronic Stress") stress, which are respectively mediated by short-term and long-term exposure to elevated levels of stress mediators. In section "Neurobiological Effects of Acute Stress", we describe the temporal dynamics of the neuroendocrine response to an acute stressor and the mechanisms through which stress mediators promote physiological and behavioral adaptation via their time-, dose-, and location-dependent effects in the body and the brain. In section "Neurobiological Effects of Chronic Stress", we describe the neurobiological effects of chronic or repeated stress. Specifically, we discuss the structural and functional changes in the peripheral effector systems and the central stress-regulatory network due to prolonged or cumulative exposure to stress mediators, and the behavioral and cognitive consequences of these stress-induced neuroplastic changes.

Neurobiological Effects of Acute Stress

Temporally, a typical response to an acute stressor can be subdivided into two phases, each dominated by the effects of different stress mediators and characterized by specific physiological, psychological, and behavioral changes (Joëls et al., 2018). The **initial phase** begins within seconds of stressor onset and constitutes a rapid response aimed at promoting immediate survival. The **later phase** of the stress response begins 1–2 h post stressor onset and is thought to promote recovery, restore homeostasis, and prepare for future stress exposure (Joëls & Baram, 2009). The physiological and behavioral changes observed during the later phase are largely the opposite of the changes observed during the initial phase of the stress response. During the initial phase, peripheral stress effector systems, such as the SNS, the SAM system, and the HPA axis, become activated and release various neuroendocrine stress mediators (e.g., NE, EPI, CRH, AVP, ACTH, GCs), which exert their initial effects on target tissues via rapid, non-genomic effects that rely on second messenger signaling (Sapolsky et al., 2000; Joëls & Baram, 2009). In addition to these fast effects, stress mediators can also produce genomic effects that take many minutes to hours to develop and therefore predominate during the later phase of the stress response (Sapolsky et al., 2000).

We begin this section by describing the neuroendocrine events that occur during the initial "alarm" phase of the stress response, including the rapid release of central and peripheral catecholamines and the delayed release of GCs. We discuss the rapid, non-genomic effects of catecholamines and GCs and their interactive effects on peripheral tissues and the brain. We then outline the later "recovery" phase of the stress response, and in particular the physiological, cognitive, and behavioral changes mediated by the slow, genomic effects of GCs.

Initial Phase of the Stress Response

Within seconds of exposure to a typical mammalian psychological stressor, such as a predator, the stress-regulatory network initiates a cascade of neuroendocrine events with the aim of inhibiting ongoing processes that are not essential for immediate survival (e.g., reproduction, digestion) and diverting energy to systems that promote coping, such as escape or attack (e.g., the cardiovascular, respiratory, and skeletomuscular systems). The **first wave** of the neuroendocrine cascade includes the release of (1) catecholamines (NE and EPI) by the sympathetic nerves and the adrenal medulla, (2) CRH and AVP by the PVN, (3) ACTH by the anterior pituitary, and (4) other hormones (e.g., growth hormone, opioids, glucagon) by the pituitary and peripheral glands (Sapolsky et al., 2000). Several minutes later, the **second wave** of the neuroendocrine cascade begins, with slow elevations of circulating GCs released by the adrenal cortex and inhibition of steroid hormone synthesis in the gonads (Sapolsky et al., 2000).

Because the ANS responds to stress immediately, whereas GC levels rise with a lag of several minutes and peak 20–40 min later, the initial phase of the stress response is primarily dominated by the non-genomic effects of catecholamines which are released by the sympathetic nerves and the adrenal medulla (i.e., NE and EPI). Catecholamines act on various tissues throughout the body and trigger physiological changes necessary to cope with the stressor, including energy mobilization; enhanced cardiovascular and respiratory activity; immune system activation; reduced blood flow to the gut, skin, and reproductive organs; enhanced blood flow to the exercising muscles and the brain; and increased cerebral glucose utilization (Sapolsky et al., 2000). These rapid effects of catecholamines are facilitated or reinforced by GCs, which are initially present at basal concentrations due to a delay in synthesis and delivery (Myers et al., 2017). For example, basal GCs augment cardiovascular activation, prime the immune system, stimulate energy mobilization, and inhibit energy storage, all of which are the primary functions of catecholamines and other mediators of the first wave of the neuroendocrine cascade (Sapolsky et al., 2000). Thus, basal GCs prime stress response mechanisms and help mediate the backbone of the stress response to most generalized stressors via permissive effects on catecholamine signaling (Sapolsky et al., 2000).

Rapid Neurobiological Effects of Catecholamines

In addition to the physiological changes across multiple organ systems that help the body cope with a stressor, stress mediators that are released during the initial phase of the stress response also impact brain function to promote adaptive behavioral responses. For example, the initial phase of the stress response is characterized by enhanced arousal, vigilance, sensory processing, and memory encoding, as well as impaired cognitive flexibility, memory retrieval, and higher-order cognitive functions (de Kloët et al., 2005; Gunnar & Quevedo, 2007; see section "The Psychological

Components of the Stress Response"). These cognitive and behavioral changes are largely mediated by the central catecholamine systems, and in particular widespread NE projections from the LC (de Kloët et al., 2005; Gunnar & Quevedo, 2007; see section "Central Norepinephrine and Psychological Components of the Stress Response"). Peripheral catecholamines released into the bloodstream by the sympathetic nerves and the adrenal medulla also can contribute to the behavioral and cognitive response to stress (McIntyre et al., 2012). Although peripheral catecholamines cannot cross the blood-brain barrier, they impact cognition and behavior by indirectly stimulating the central NE system (McGaugh & Roozendaal, 2002). For example, peripheral EPI can bind to receptors on the vagus nerve, which projects to the NTS, a major viscerosensory integration center (see section "Neural Circuitry of Detection of Physical Stressors"). Upon stimulation by the vagal afferents, the NTS activates its target regions in the forebrain and brainstem. One important target of the NTS is the LC, which sends widespread NE projections throughout the limbic system, cerebral cortex, brainstem, and cerebellum (Roosevelt et al., 2006). Thus, peripheral catecholamines stimulate NE release throughout the brain and might contribute to the cognitive component of the stress response by promoting arousal, modulating attention, and enhancing memory formation (McIntyre et al., 2012). For example, peripheral EPI promotes memory encoding, particularly for stressor-related information, by triggering NE release in the BLA, which is necessary for enhancement in emotional memory (Rodrigues et al., 2009).

Rapid Neurobiological Effects of Glucocorticoids

In contrast to peripheral catecholamines, GCs can easily cross the blood-brain barrier and impact brain function directly; however, it takes 20–60 min for GCs to elevate in the brain. The effects of GCs on neuronal function depend on the number and types of GC receptors (MRs vs. GRs) a given brain region expresses (see section "Hypothalamic-Pituitary-Adrenal Axis"). GC receptors differ in their localization, affinity for GCs, and signaling mechanisms, which accounts for the target-, dose-, and time-specific effects of GCs, respectively. For example, **membrane GC receptors** mediate the rapid, non-genomic effects of GCs, which set in around the peak in circulating GCs and include regulation of cell excitability, synaptic transmission, and neuroplasticity, either directly or via eCB signaling. Thus, the effects of elevated GCs that are mediated by membrane receptors manifest during the initial phase of the stress response and can overlap with the effects of the first wave of the neuroendocrine cascade (e.g., NE and EPI). In contrast, **intracellular GC receptors** mediate the slow, genomic effects of GCs, which include regulation of gene expression and manifest only during the later phase of the stress response (i.e., after GC levels return to baseline). In addition to the difference in signaling mechanisms, membrane and intracellular GC receptors also differ in their affinity for GCs. For example, intracellular MRs have high affinity for GCs and are substantially occupied even when GC levels are low. In contrast, membrane MRs and both types of GRs have lower affinity for GCs and therefore serve as sensors of GC elevations.

In addition to the differences in affinity, GC receptors also differ in their distribution. GRs are expressed virtually in every cell in the body but are particularly abundant in key nodes of the central stress-regulatory network, suggesting enhanced sensitivity of these regions to GC elevations. In contrast, MRs are expressed in a limited number of structures, with the highest concentrations in the hippocampus and lateral septum, suggesting a role of GCs in regulating the basal functioning of these regions (Joëls & Baram, 2009).

Due to differences in localization, affinity, and signaling mechanisms of different GC receptors, GCs can exert distinct effects on the functioning of GC-sensitive brain circuits that span both the initial and the later phases of the stress response. For example, at basal GC concentrations, the effects of GCs on neuronal function are mediated by the high-affinity **intracellular MRs (iMR)**, which regulate gene transcription in a slow and largely continuous fashion and are primarily expressed in the hippocampus (e.g., Joëls et al., 2007). Although not directly involved in the stress response, basal iMR signaling sets the threshold for the HPA stress response and is necessary for the viability and excitability of hippocampal neurons. When GC levels rise in response to a stressor, the low-affinity membrane GC receptors and **intracellular GRs (iGR)** become activated and mediate the rapid, non-genomic and slow, genomic effects of GCs, respectively. The slow, genomic effects of iGR signaling include termination of the stress response, normalization of neuronal function, and memory consolidation (see section "Later Phase of the Stress Response"). In contrast, the rapid, non-genomic effects of GCs in the brain are implicated in behavioral and cognitive changes that are characteristic of the initial phase of the stress response, including a shift from deliberate, reflective cognition toward automatic and intuitive information processing (Margittai et al., 2016), enhanced emotional interference with selective attention (Henckens et al., 2012), impaired contextualization (Van Ast et al., 2013) and retrieval of declarative memory (Pruessner et al., 2010; De Quervain et al., 2003), and enhanced appraisal of novel situations, selection of response strategies (Joëls et al., 2012), and emotional and habitual forms of learning (Schwabe et al., 2010; Joëls et al., 2012).

These rapid effects of elevated GCs on cognitive function are thought to be mediated by distinct types of membrane GC receptors that are differentially distributed across the central stress-regulatory network. In general, activation of **membrane GRs (mGR)** has largely inhibitory effects on neuronal function, whereas activation of **membrane MRs (mMR)** has largely excitatory effects on neuronal function. For example, activation of mGR inhibits excitatory neurotransmission in the PVN and promotes termination of the HPA stress response via rapid negative feedback (see section "Neural Circuitry of Stress Response Termination"). Similarly, activation of mGR reduces neuronal excitability and impairs neuroplasticity in the PFC and **dorsal hippocampus** (Maggio & Segal, 2012; Joëls, 2018). Given the roles of the PFC and dorsal hippocampus in higher-order cognitive functions, such as context processing, declarative memory, and executive function, impaired plasticity in these structures due to elevated GCs might account for the cognitive deficits during the initial phase of the stress response (Maggio & Segal, 2012; Joëls & Baram, 2009; Rodrigues et al., 2009). Conversely, activation of mMR enhances neuronal excitability, stimulates neuroplasticity, and promotes long-term potentiation in the amygdala and **ventral hippocampus** (Popoli et al., 2012). In contrast to the dorsal hippocampus, which is primarily involved in spatial and declarative memory, the ventral hippocampus is primarily involved in affective processing and stress response regulation (Fanselow & Dong, 2010). Thus, enhanced neuroplasticity in the ventral hippocampus and amygdala could underlie enhanced consolidation of emotional memories and stressor-related information following stress exposure (Maggio & Segal, 2012).

Interactions Between Catecholamines and Glucocorticoids in the Brain

Similar to the permissive effects of GCs on catecholamine signaling in peripheral tissues, GCs also interact with catecholamines (particularly NE) in the brain, and these interactive effects underlie many cognitive and behavioral effects of acute stress (Roozendaal et al., 2009). Within seconds to a few minutes of stressor onset, the widespread release of NE by the central noradrenergic system is thought to promote activation of the salience network and suppress the executive control network (see section "Neural Circuitry of the Psychological Components of the Stress Response"; Schwabe, 2017). A few minutes later, rising GC levels begin to exert rapid, non-genomic effects via membrane receptors, which synergize with NE to further promote the activity of neural circuits involved in threat detection, habitual responses, and gist-like information processing, which rely on the amygdala and dorsal striatum, at the expense of higher-order cognitive functions that rely on the dorsal hippocampus and PFC. For example, GCs interact with NE to promote amygdala reactivity to threat-relevant stimuli, such as fearful faces (Krugers et al., 2012). Similarly, concurrent catecholamine and GC signaling in the PFC impairs working memory (Barsegyan et al., 2010; Henckens et al., 2011), disrupts flexible goal-oriented behavior, and renders behavior habitual (Schwabe et al., 2012).

Interactions between NE and GCs have also been implicated in the effects of acute stress on various memory processes (e.g., Rodrigues et al., 2009). For example, interactions between NE and GCs in the BLA are crucial for enhanced **memory formation**, particularly for stressor-related information (Rodrigues et al., 2009; Krugers et al., 2012). Although the BLA is not directly involved in memory storage, it can modulate synaptic plasticity in different memory systems, including the hippocampal, striatal, and cortical circuits, which support different types of learning and memory (Roozendaal et al., 2009). NE signaling and the rapid, non-genomic effects of GCs in the BLA are thought to shift specific memory systems into a **"memory formation mode,"** which is characterized by enhanced encoding of stressor-related information and suppression of competing processes, such as memory retrieval and encoding of stressor-unrelated information (Schwabe, 2017). Specifically, NE increases the excitability and synaptic plasticity of BLA neurons, whereas GCs enhance these effects through non-genomic actions either directly in the BLA or by promoting NE release from the NTS and LC onto the BLA (Joëls & Baram, 2009; McGaugh & Roozendaal, 2002). Importantly, instead of stimulating all memory systems equally, the BLA might function as a switch between different memory systems (Roozendaal et al., 2009). For example, mMR signaling in the BLA in response to acute stress mediates a shift from declarative and contextual memory that is dependent on the hippocampus toward habitual and procedural stimulus-response memory that is dependent on the dorsal striatum (Schwabe et al., 2010; Schwabe et al., 2013; Vogel et al., 2017).

Later Phase of the Stress Response

In contrast to the initial phase of the stress response, which promotes immediate physiological and psychological coping with an acute stressor, the later phase promotes stress response termination and recovery (Joëls et al., 2018). The later phase of the stress response begins 1–2 h post stressor onset and is dominated by the slow, genomic effects of the stress mediators that were released during the initial phase, primarily GCs. GC levels peak 20–40 min post stressor onset and return to baseline within 1–1.5 h of stressor offset (Spencer & Deak, 2017). Stress-induced GC elevations can trigger changes in gene expression via activation of iGRs, which take at least an hour to develop (typically around 3 h) and last hours to days, long after HPA activity returns to baseline (Dickerson & Kemeny, 2004). In contrast to the basal GCs, which permissively promote the onset of the stress response and act synergistically with the first wave of stress mediators, elevated GCs produce delayed genomic effects that suppress the initial stress response, promote recovery, and help prepare for future stressors (de Kloët et al., 2005; Joëls et al., 2018). For example, basal levels of GCs prime the immune system response to stress; however, stress-induced GC elevations have immunosuppressive effects that prevent the immune response from overshooting (Sapolsky et al., 2000). Similarly, basal levels of GCs promote local cerebral glucose utilization in the brain, but stress-induced GC elevations reduce regional cerebral blood flow and inhibit glucose transport in brain cells (Sapolsky et al., 2000). The slow effects of GCs also aid preparation for future stressors or promote adaptation to chronic stressors. For example, GCs help replenish glycogen stores in the liver, stimulate appetite, and inhibit reproduction (Sapolsky et al., 2000).

In addition to the physiological changes, the later phase of the stress response also includes specific changes in cognition and behavior that oppose the cognitive and behavioral effects of the initial phase (see section "Initial Phase of the Stress Response"), and are largely mediated by the slow, genomic effects of GCs (Joëls et al., 2018). During the initial phase of the stress response, the rapid effects of GCs and NE promote activation of the salience network and suppress activation of the executive control network (Hermans et al., 2014). In contrast, during the later phase, when the genomic effects of GCs set in, the salience network is suppressed, and the executive control network activity normalizes (Hermans et al., 2014). Similarly, during the initial phase, **working memory** is impaired through synergistic effects of elevated NE and mGR signaling in the PFC, whereas during the later phase, working memory becomes enhanced due to the delayed genomic effects of elevated GCs, which promote excitatory neurotransmission in the PFC via iGR signaling (Popoli et al., 2012; Henckens et al., 2011; Yuen et al., 2011; Yuen et al., 2009). The rapid effects of GCs augment the effects of NE signaling within limbic structures and promote the "**memory formation mode**," characterized by enhanced information encoding and impaired memory retrieval. In contrast, the delayed genomic effects of GCs oppose the facilitatory effects of NE on neuroplasticity and promote the transition to the "**memory storage mode**," characterized by enhanced memory consolidation and suppressed encoding of new information to protect consolidation of stressor-related information (Krugers et al., 2012).

Thus, in contrast to the initial phase of the stress response, which is characterized by enhanced memory encoding and impaired higher-order cognitive functions, such as focused attention and memory retrieval (see section "Initial Phase of the Stress Response"), the later phase of the stress response is characterized by improved executive function and working memory, diminished distractibility, and enhanced memory consolidation and contextualization (Joëls et al., 2018; van Ast et al., 2013). Cognitive improvements during the later phase constitute not only merely a reversal of cognitive deficits that occur during the initial phase but also improvements above baseline function (e.g., Henckens et al., 2011). Whereas the cognitive changes during the initial phase promote short-term survival and coping, cognitive changes during the later phase promote contextualization of the stress-related situation and elaboration of more complex and efficient solutions to be remembered in the long term (Joëls et al., 2018).

Neurobiological Effects of Chronic Stress

In section "Neurobiological Effects of Acute Stress", we described the dynamics of a typical response to an acute stressor: an initial "alarm" phase that includes rapid physiological and behavioral changes necessary for coping with the stressor, followed by a later "recovery" phase that restores homeostasis and promotes long-term adaptation in preparation for future stressors. The acute stressor-initiated physiological and behavioral changes are adaptive in the short term; they promote immediate survival and coping and are quickly reversed when the stressor is over. However, repeated or prolonged stress response activation due to chronic exposure to stressors over weeks or longer constitutes sustained challenge to homeostasis and can have deleterious effects on physical and mental health (McEwen, 1998). The detrimental impact of chronic stress might be due to the cumulative effects of stress mediators, particularly GCs (McEwen & Gianaros, 2011). Prolonged exposure to GCs produces allostatic load across multiple organ systems, including immune, cardiovascular, reproductive, and nervous systems, increases the risk of neurological and psychiatric disorders, and adversely affects cognitive function (Lupien et al., 2009).

Unlike acute stressors, which by definition are transient and sporadic, chronic stressors elicit repeated or sustained activation of the stress system and produce a distinct set of physiological adaptations, such as recalibration of the neuroendocrine stress response (e.g., Herman et al., 2016). When a chronic stressor is predictable, mild, and controllable, it might be beneficial to reduce the magnitude of the neuro-endocrine response to the repeated stressor, to minimize the detrimental cumulative effects of prolonged exposure to stress mediators. In contrast, chronic exposure to variable, severe, and uncontrollable stressors might signal a dangerous and unpre-dictable environment in which it is adaptive to be constantly on alert and ready to initiate a full-blown stress response. Adaptation to chronic variable stressors might require a primed and reactive stress system with increased biosynthesis of stress mediators, decreased sensitivity to negative feedback, and enhanced sensitivity to potential threat.

In this section, we focus on the neurobiological effects of chronic stressors on the functioning of the peripheral stress effector systems and the central stress-regulatory network. We discuss the functional and structural changes in stress-relevant neural circuitry that result from chronic stress exposure and are not typically seen after an acute stressor (e.g., Lupien et al., 2018). We conclude this section with a discussion of the functional significance of the neuroplastic changes in the central stress-regulatory network that are produced by chronic stress, including their relation to the recalibration of the neuroendocrine stress response and the cognitive and behav-ioral consequences of chronic stress.

Effects of Chronic Stress on Peripheral Effector Systems

The effects of chronic stress on the functioning of the peripheral stress effector sys-tems depend on specific stressor characteristics, such as type, duration, severity, controllability, and predictability. For example, repeated exposure to the same (i.e., **homotypic**) stressor leads to a distinct set of physiological adaptations compared to exposure to chronic variable (i.e., **heterotypic**) stressors. Repeated exposure to a homotypic stressor over a period of weeks or longer results in the same pattern of changes in the peripheral effector systems (the SNS, SAM, and HPA axis), which includes (1) response **habituation** to the same stressor, (2) response **sensitization** to novel stressors, and (3) elevated baseline activity. For example, chronic stress leads to reduced catecholamine release upon subsequent exposure to the same stressor and also to elevated basal levels of plasma catecholamines, suggesting SNS and SAM system habituation along with elevated baseline sympathetic tone (Kvetnansky et al., 2009). Similarly, chronic stress can lead to changes in both reac-tive and basal HPA axis functioning, including enhanced HPA reactivity to novel stressors, flattened diurnal HPA rhythm, increased frequency of ultradian HPA pulses, and elevated basal GC levels due to impaired negative feedback (Miller et al., 2007; Lightman, 2008; Uschold-Schmidt et al., 2012). In contrast to chronic homotypic stressors, chronic heterotypic stress typically leads to stress system hyperactivity and sensitization (Kopp et al., 2013; Herman et al., 2016). Importantly, both homotypic and heterotypic chronic stressors produce facilitation of the HPA response to novel stressors, which includes faster HPA response onset and higher peak (Herman et al., 2016).

Stressor duration, type, and severity also influence the effects of chronic stress on peripheral stress effector systems. Repeated cold exposure over several days enhances SAM system activity, whereas prolonged cold exposure over the span of a month leads to SAM system habituation (Kvetnansky et al., 2009). In contrast to cold stress, however, enhanced SAM system activity persists during exposure to chronic immobilization stress, suggesting that habituation of the SAM system depends not only on the **duration** but also the **type** of chronic stressor (Kvetnansky et al., 2009). Similarly, chronic exposure to **severe, unpredictable**, and/or **uncontrollable** homotypic stressors tends to result in sustained SNS and HPA axis hyperactivity rather than habituation (Herman et al., 2016; Ulrich-Lai & Herman, 2009). In contrast, chronic exposure to **mild** and **predictable** stressors tends to result in habituation of the stress response and a more efficient energy allocation through selective and transient activation of specific peripheral stress effector systems (Koolhaas et al., 2011). When naïve rats are subjected to stressful forced-swim training, they initially show large SAM system and HPA axis responses. However, with repeated training, the neuroendocrine response changes, such that trained rats begin to show greater SNS activity, lower SAM activity, and faster HPA axis recovery when subjected to forced swimming compared to naïve rats. These results suggest that with training, the neuroendocrine response becomes more efficient and limited to what is necessary to meet the metabolic demands of a swim (Koolhaas et al., 2011). Thus, although homotypic chronic stressors are typically habituating, some homotypic stressors, such as those that are more severe, unpredictable, uncontrollable, or ethologically significant, do not produce stress response habituation and can in fact produce HPA response sensitization (e.g., Figueiredo et al., 2003). Although heterotypic chronic stressors are typically non-habituating, exposure to chronic variable stress produces some degree of adaptation (Herman et al., 2016).

The mechanisms through which chronic stress impacts the functioning of peripheral effector systems include changes at the level of peripheral effectors (e.g., autonomic ganglia and adrenal glands), as well as changes at the level of the central stress-regulatory network. For example, chronic stress promotes the expression of enzymes necessary for catecholamine biosynthesis in sympathetic ganglia and the adrenal medulla, which results in enhanced catecholamine synthesis capacity and might account for elevated basal catecholamine levels and sensitization of the SNS and SAM system response to novel stressors (Kvetnansky et al., 2009). Sensitization of the HPA axis response as a result of chronic stress, on the other hand, can be due to enhanced adrenal sensitivity to ACTH (Uschold-Schmidt et al., 2012) or adrenal hypertrophy (Ulrich-Lai et al., 2006; Kopp et al., 2013), both of which result in elevated basal GCs and increased GC output upon ACTH stimulation. Chronic stress also can increase the bioavailability of circulating GCs. Because 95% of GCs in the bloodstream are bound to transport proteins, such as the **corticosteroid-binding protein**, which limit the access of circulating GCs to target tissues, only 3–5% of circulating GCs are unbound and thus biologically active. Chronic stress downregulates the expression of the corticosteroid-binding protein and thus increases the bioavailability of GCs (Herman et al., 2016).

Effects of Chronic Stress on the Central Stress-Regulatory Network

Regulation of the peripheral stress effector systems during chronic stress engages a distinct set of neural structures compared to acute stress (section "Neural Circuitry of Responses to Acute vs. Chronic Stress"). In addition to changes at the level of peripheral effectors, chronic stress exposure can also produce structural and functional changes within the central stress-regulatory network, which can have downstream effects on peripheral stress system functioning and the neuroendocrine stress response. For example, chronic stress can lead to hyperactivity and sensitization of stress-excitatory circuits (section "Effects of Chronic Stress on Stress-Excitatory Circuits"), along with hypoactivity and decreased sensitivity of stress-inhibitory circuits (section "Effects of Chronic Stress on Stress-Inhibitory Circuits"). These functional changes are at least partly due to the neuroplastic changes within key areas of the central stress-regulatory network following chronic stress exposure. These structural and functional effects of chronic stress on the central stress-regulatory network have been linked to sustained elevations in stress mediators, particularly GCs, and are at least partially reversible following a period of reduced stress.

Neural Circuitry of Responses to Acute vs. Chronic Stress

Evidence suggests that several structures within the central stress-regulatory network differentially respond to acute vs. chronic stress, indicating potentially distinct mechanisms for stress response regulation. For example, the **paraventricular thalamic nucleus (PVT)** has been implicated in HPA response calibration during chronic but not acute stress (Herman et al., 2016). PVT lesions prevent both HPA response habituation and sensitization, suggesting that the PVT is involved in representing stressor chronicity. Conversely, hippocampal lesions enhance the HPA response to acute but not chronic stressors, suggesting that the hippocampus is involved in the suppression of the HPA response to acute stressors specifically (Herman et al., 2016). Similarly, ventral mPFC has been implicated in HPA response termination following acute stress only, whereas dorsal mPFC responds to acute and chronic stressors alike (McKlveen et al., 2015). Anterior BNST also has a distinct role in HPA axis regulation in response to acute vs. chronic stressors. Anterior BNST lesions reduce HPA axis responses to acute stress, which is consistent with its involvement in the excitatory HPA drive (see section "Regulation of the Stress Response by the Limbic System"). At the same time, anterior BNST lesions promote sensitization of the HPA response to novel stressors, suggesting an inhibitory role of anterior BNST during chronic stress (Herman et al., 2016). There are also differences in neural responses to chronic homotypic vs. heterotypic stressors (Flak et al., 2012). Although both homotypic and heterotypic stressors lead to sustained activation of the DMH, amygdala, and hippocampus, only chronic heterotypic stress selectively increases activity in the NTS, posterior hypothalamic nucleus, and mPFC, suggesting that these structures are involved in chronic HPA drive and stress system sensitization due to chronic variable stress (Flak et al., 2012).

Effects of Chronic Stress on Stress-Excitatory Circuits

Changes in the function of peripheral stress effector systems due to chronic stress, such as response sensitization to novel stressors and elevated baseline activity, have been linked to structural and functional changes at the level of the PVN, which is a critical regulator of ANS and HPA axis activity (see section "Neural Circuitry of Generation of Complex Coordinated Responses"). Chronic stress increases the number of excitatory projections that target CRH-producing neurons, upregulates the expression of glutamate and CRH receptors, and downregulates the expression of GABA and eCB receptors in the PVN (Bains et al., 2015). These structural changes result in enhanced excitatory and reduced inhibitory drive on the PVN and increased PVN sensitivity to excitatory inputs, which might account for the increased CRH and AVP production and contribute to sensitization of the HPA axis response to novel stressors (Bains et al., 2015). Chronic stress also impairs PVN sensitivity to rapid negative feedback by reducing GR expression and eCB signaling (Myers et al., 2017). Diminished PVN sensitivity to negative feedback might contribute to HPA axis sensitization, and to elevated basal HPA activity following chronic stress (Ulrich-Lai & Herman, 2009).

The amygdala is another key region of the central stress-regulatory network that undergoes structural and functional changes as a result of chronic stress exposure. Specifically, chronic stress increases dendritic branching and synaptic density in the BLA, the main input nucleus of the amygdala, which results in greater excitability and responsiveness of BLA neurons and might account for enhanced amygdala sensitivity to threat signals (Henckens et al., 2015; Rodrigues et al., 2009). In addition, chronic stress reduces the number of tonic inhibitory inputs to the BLA, which leads to sustained amygdala disinhibition and downstream activation of the amygdala targets, such as the PVN (Liu et al., 2014; see section "Regulation of the Stress Response by the Limbic System"). In contrast to the BLA, chronic stress does not seem to induce neuronal hypertrophy in the CeA, the main output nucleus of the amygdala (Ulrich-Lai & Herman, 2009). Rather, chronic stress promotes CeA hyperactivity and CRH production, which in turn result in downstream disinhibition of the PVN (see section "Neural Circuitry of Stress Response Regulation") and can thus contribute to HPA and possibly ANS response sensitization to novel stressors (Ulrich-Lai & Herman, 2009).

In addition to the PVN, the CeA also targets the LC, another important node of the stress-regulatory network that regulates the physiological and psychological components of the stress response via its projections to preganglionic autonomic neurons and the corticolimbic circuitry (see section "Central Norepinephrine and Psychological Components of the Stress Response"). Chronic stress is associated with increased expression of enzymes necessary for NE biosynthesis in the LC and enhanced NE release onto LC targets, such as the PFC and hippocampus (Ulrich-Lai & Herman, 2009). Chronic stress also leads to increased stimulation of the LC by the CeA and enhanced LC sensitivity to CRH, which is a neurotransmitter released by the CeA onto LC neurons (see section "Regulation of Central Norepinephrine Signaling"; Morilak et al., 2005; Ulrich-Lai & Herman, 2009). Enhanced stimulation of the LC by the CeA promotes high activity of LC neurons and sensitization of LC firing rate to novel stressors, which might contribute to downstream potentiation of ANS reactivity and sustained hyperarousal following chronic stress (Herman et al., 2005; Ulrich-Lai & Herman, 2009). In addition, the LC sends excitatory feedback projections to the CeA, which further promote CeA hyperactivity (Ulrich-Lai & Herman, 2009). The CeA and LC form a positive feed-back loop, stimulating each other via mutually excitatory projections, and this reciprocal stimulation is enhanced by chronic stress (de Kloët et al., 2005).

Effects of Chronic Stress on Stress-Inhibitory Circuits

In contrast to the stress-excitatory structures, such as the PVN, amygdala, and LC, which become sensitized and hyperactive as a result of chronic stress exposure, stress-inhibitory structures, such as the PFC and hippocampus, show the opposite pattern of structural and functional changes. For example, in principal PFC neurons, chronic stress leads to reduced dendritic length and complexity, as well as a net loss of excitatory synapses (de Kloët et al., 2005; Radley, 2012; Henckens et al., 2015). Chronic stress also promotes the hypertrophy of inhibitory interneurons, which regulate information flow in the PFC by gating the activity of the principal PFC neurons (McKlveen et al., 2019). In addition, chronic stress downregulates GR expression in the PFC, and this decrease in GR signaling shifts the balance of exci-tation and inhibition within the PFC toward increased inhibition of principal neu-rons (McKlveen et al., 2019). Thus, the retraction of PFC dendrites and loss of excitatory inputs lead to diminished excitatory drive on the PFC, whereas decreased GR signaling and hypertrophy of inhibitory interneurons lead to increased PFC inhibition, both of which result in reduced regulatory outflow from the PFC to stress-excitatory structures, such as the PVN and the amygdala (McKlveen et al., 2019; see section "Neural Circuitry of Stress Response Regulation"). For example, reduced PFC activity following chronic unpredictable stress leads to BLA hyperex-citability due to diminished PFC stimulation of the inhibitory interneurons in the BLA (Wei et al., 2018). In addition, given the role of the mPFC in delayed negative feedback of the HPA axis (see section "Neural Circuitry of Stress Response Termination"), downregulation of GRs in the mPFC might contribute to diminished sensitivity to elevated GCs and consequently impaired negative feedback and ele-vated baseline HPA activity (McKlveen et al., 2019).

Chronic stress also leads to neuroplastic changes in the hippocampus. For exam-ple, chronic stress leads to reduced dendritic length and branching and decreased spine density in hippocampal **CA3 pyramidal neurons** (e.g., Henckens et al., 2015). Given that the hippocampus is involved in tonic suppression of the HPA axis, as well as in termination of the HPA stress response via indirect inhibition of the PVN (see section "Regulation of the Stress Response by the Limbic System"), dendritic retraction in hippocampal neurons could lead to reduced regulatory drive from the hippocampus to the PVN. Reduced outflow from the hippocampus, together with reduced outflow from the mPFC and enhanced outflow from the amygdala, might contribute to PVN overstimulation and elevated HPA axis activity following chronic stress (Radley, 2012). In addition to dendritic atrophy, chronic stress also promotes the downregulation of GRs in the hippocampus, which is thought to contribute to decreased sensitivity to negative feedback and elevated basal GC levels (McEwen & Gianaros, 2011; Ulrich-Lai & Herman, 2009; Spencer & Deak, 2017).

In addition to the morphological changes in hippocampal neurons, chronic stress influences the relative rates of **neurogenesis** (i.e., cell proliferation) and **apoptosis** (i.e., programmed cell death) in the **dentate gyrus**, which is a structure within the hippocampal formation that supports lifelong neurogenesis (de Kloët et al., 2005; Mirescu & Gould, 2006). The effects of chronic stress on neurogenesis critically depend on the severity and predictability of the stressor. For example, repeated exposure to mild and predictable stressors promotes hippocampal neurogenesis (Parihar et al., 2011), whereas chronic exposure to severe and unpredictable stressors suppresses hippocampal neurogenesis and stimulates apoptosis of progenitor cells in the dentate gyrus (Roman et al., 2005). Because mild and predictable stressors typically produce HPA response habituation, whereas severe and unpredictable stressors typically produce HPA response sensitization and elevated baseline GC levels, stressor-specific effects on neurogenesis might be mediated by differences in the HPA response. Low-to-moderate GC concentrations are necessary for normal synaptic transmission and maintenance of neuronal viability in the hippocampus, whereas chronically elevated GCs suppress neurogenesis and promote neurodegeneration in the hippocampus by increasing the vulnerability of hippocampal neurons to various neurological insults (Lupien et al., 2018).

Endocannabinoid signaling plays a crucial role in the regulation of the HPA axis, including tonic inhibition of the HPA axis under unstressed conditions (see section "Regulation of the Stress Response by the Endocannabinoid System") and termination of the HPA axis response to acute stressors via rapid negative feedback in the PVN and delayed negative feedback in the mPFC (see section "Neural Circuitry of Stress Response Termination"). Whereas acute GC elevations stimulate eCB signaling as part of the HPA response termination mechanism, chronic GC elevations reduce eCB signaling, and this reduction in eCB levels has been linked to impaired negative feedback (Hill et al., 2010). In addition, mounting evidence suggests that eCB signaling is also involved in elevated baseline HPA axis activity and HPA response habituation to repeated stress, with the two main eCB neurotransmitters, AEA and 2-AG, having distinct roles in HPA axis regulation (Hill et al., 2010). Elevated baseline HPA activity following exposure to chronic stress can result from diminished AEA levels within key structures of the central stress-regulatory network, including the amygdala, PFC, hippocampus, and hypothalamus (Hill et al., 2010). In contrast, habituation of the HPA response to a repeated stressor is mediated by transient increases in 2-AG release in the amygdala (Hill et al., 2010). 2-AG might function as an inhibitory retrograde neurotransmitter that suppresses excitatory inputs to the amygdala, reducing the activity of the amygdala and downstream targets such as the PVN (Hill et al., 2010). The resulting loss of excitatory drive from the amygdala would lead to diminished PVN response and lower HPA activation to a repeated stressor. In this way, two distinct eCB neurotransmitters contribute to elevated basal HPA axis activity and habituation of the HPA axis response to chronic stress. Chronic stress also downregulates eCB receptors in the hippocampus and the PVN, which might further diminish sensitivity to negative feedback and elevated excitatory drive on the HPA axis (Bains et al., 2015; Morena et al., 2016).

Mechanisms of Stress-Induced Neuroplasticity

Increased dendritic branching in the amygdala and dendritic retraction in the hippocampus and PFC are well-documented neuroplastic changes associated with chronic stress exposure (e.g., McEwen & Gianaros, 2011). However, recent research suggests that stress-induced neuroplasticity is not limited to these three structures. Chronic unpredictable stress in rats can lead to widespread structural atrophy spanning multiple cortical and subcortical nodes of the stress-regulatory network (Magalhães et al., 2018). The mechanism through which chronic stress leads to these neuroplastic changes might involve elevated GC signaling; enhanced release of excitatory neurotransmitters, such as glutamate; and diminished eCB neurotransmission. Many of the structural and functional effects of chronic stress, including dendritic atrophy, synaptic loss, and downregulation of GRs in the hippocampus (de Kloët et al., 2005; McEwen & Gianaros, 2011), increased activity and CRH production in the CeA (Ulrich-Lai & Herman, 2009), downregulation of eCB receptors in the PVN and hippocampus (Morena et al., 2016), and decreased excitatory neurotransmission in the PFC (Yuen et al., 2012), have been linked to the genomic effects of chronically elevated GCs.

Elevated GC levels promote excitatory neurotransmission and extracellular glutamate levels, particularly in the hippocampus and mPFC, by stimulating glutamate release and reducing synaptic glutamate clearance (Popoli et al., 2012). Blocking either glutamatergic or GC signaling prevents dendritic atrophy in the hippocampus and mPFC following chronic stress, suggesting that both GCs and glutamate are necessary for the stress-induced neuroplastic changes (McEwen et al., 2015). Prolonged elevations in extracellular glutamate increase neuronal vulnerability to neurological insults and can result in neuronal damage or even cell loss due to **excitotoxicity**. Therefore, dendritic retraction in response to chronic stress has been proposed to represent an adaptation that protects hippocampal and mPFC neurons from GC-mediated overstimulation and excitotoxic damage (McEwen et al., 2015). In contrast to glutamate neurotransmission, which is enhanced by chronic stress, eCB neurotransmission is impaired by chronic stress (Hill et al., 2010). Together with elevated glutamate levels, diminished eCB signaling might also contribute to neuroplastic changes in corticolimbic neurons (McEwen et al., 2015). For example, blocking of eCB signaling leads to the hypertrophy of amygdala dendrites and hypotrophy of PFC dendrites, suggesting that eCB signaling might buffer against glutamate-induced neuroplasticity due to chronic stress (McEwen et al., 2015).

Some of the structural and functional changes induced by chronic stress are reversible. For example, stress-induced suppression of neurogenesis and dendritic shrinkage in hippocampal and PFC neurons can be reversed after a period of reduced stress (McEwen & Gianaros, 2011). However, the neurons' ability to recover declines with age (Bloss et al., 2010). Interestingly, although recovery from chronic stress promotes regrowth of hippocampal dendritic arbors, the dendrites that grow back are more proximal than the ones that underwent atrophy, suggesting that recovery from chronic stress involves structural remodeling of neurons rather than a mere reversal of stress-induced changes (McEwen et al., 2015). In addition, despite structural recovery from chronic stress, hippocampal neurons do not return to pre-stress functioning, but display a novel pattern of gene expression in response to subsequent stressors (McEwen et al., 2016). Together, this evidence suggests that dendritic atrophy induced by chronic stress is largely reversible in that the retracted dendrites can grow back; however, recovery from chronic stress involves structural remodeling and functional changes in cortical and limbic neurons, suggesting that chronic stress leaves lasting traces in the stress-regulatory circuitry. In contrast to dendritic atrophy in the hippocampus and PFC, increased dendritic branching in the BLA is maintained after a 21-day stress-free recovery period, suggesting that structural remodeling of the BLA is more enduring (Vyas et al., 2004).

Effects of Chronic Stress on Cognition and Behavior

Structural and functional alterations in the central stress-regulatory network due to chronic stress not only impact the functioning of the peripheral stress effector systems but have also been linked to changes in cognitive and affective function. The cognitive and behavioral effects of chronic stress resemble the short-term cognitive and behavioral changes observed during the initial phase of an acute stress response (see section "Initial Phase of the Stress Response"). For example, chronic stress impairs contextual processing and higher-order cognitive functions, which are mediated by the PFC and hippocampus, in favor of habitual, procedural, and instinctual behaviors, which are mediated by the amygdala and striatum (Arnsten, 2015). However, in the case of acute stress, these cognitive and behavioral changes are short-lived and become reversed within 1–2 h of stress exposure, i.e., during the later phase of the stress response (see section "Later Phase of the Stress Response"). In contrast, cognitive impairments due to chronic stress persist throughout the duration of the stressor and can lead to sustained periods of reduced productivity and diminished quality of life.

Similar to acute stress, chronic stress impairs **working memory** and **cognitive flexibility**, and these deficits correlate with the extent of dendritic atrophy (Cerqueira et al., 2008; McEwen & Gianaros, 2011), decreased excitatory neurotransmission (McEwen et al., 2016; Holmes & Wellman, 2009; McKlveen et al., 2016; Yuen et al., 2012), and increased **microglial** activation (Hinwood et al., 2012) in the PFC. Similarly, prolonged psychosocial stress impairs **attentional control**, which is associated with disrupted functional connectivity within a frontoparietal network that mediates attention shifting (Liston et al., 2009). In addition to impairing flexible, goal-directed behaviors, chronic stress also facilitates **inflexible, habitual behaviors** (Schwabe et al., 2008), possibly via structural changes in the striatum (Taylor et al., 2014). Chronic stress also leads to impaired **episodic, contextual,** and **spatial memory**; and these memory deficits have been linked to dendritic retraction, suppressed neurogenesis, and impaired neuroplasticity in the hippocampus (Radley, 2012; McEwen & Gianaros, 2011; Roman et al., 2005; Joëls et al., 2012). Finally, chronic stress also impacts **affective function**. For example, chronic stress promotes depressive, aggressive, and anxiety-like behaviors, facilitates fear conditioning, and impairs fear extinction and voluntary affect regulation (e.g., Golkar et al., 2014; McEwen & Gianaros, 2011; Mitra & Sapolsky, 2008; Akirav & Maroun, 2007; Anderson et al., 2019). The effects of chronic stress on affective processing have been associated with structural and functional changes in the corticolimbic circuitry, including increased dendritic branching in the BLA (Vyas et al., 2002; Vyas et al., 2004), amygdala hyperactivity (Wei et al., 2018), enhanced neuroplasticity in the ventral hippocampus (Joëls et al., 2012), and reduced mPFC activity (Arnsten, 2015; Roozendaal et al., 2009).

Similar to their adaptive function in response to acute stress, behavioral and cognitive changes associated with chronic stress exposure are thought to represent adaptations to an unpredictable and dangerous environment, as signified by chronic stress, in which rapid and instinctual behavioral responses are more adaptive than time-consuming, metabolically costly, and error-prone complex cognition (Reser, 2016). Thus, chronic stress-induced neuroplasticity can be thought of as an adaption that prevents neuronal damage due to excitotoxicity and promotes adaptive behavioral strategies, such as enhanced sensitivity to threat and fear learning. These behavioral strategies might increase chances of survival in dangerous and unpredictable environments at the expense of the less effective higher-order cognitive functions, such as contextual memory, cognitive flexibility, and affect regulation (McEwen et al., 2016). However, when the stress response is repeatedly activated in safe environments (e.g., due to chronic intermittent psychological stress), neuroplastic changes in the corticolimbic circuitry can lead to persistent impairments in cognitive and affective functioning, which not only interfere with an individual's productivity and quality of life but might also create risk of physical and psychological stress-related disorders (Lupien et al., 2018). However, evidence suggests that when chronic stress is followed by a stress-free period, stress-induced changes can become at least partially reversed, and cognitive function is restored (Liston et al., 2009).

Summary

We presented a broad overview of the role of the brain as both a regulator and a target of the stress response. The brain initiates and regulates physiological, behavioral, and psychological responses to stressors via the release of stress mediators, which in turn can impact brain structure and function both in the short and long run. The physiological components of the stress response are mediated by the peripheral stress effector systems, such as the ANS and the HPA axis. The ANS mediates a rapid fight-or-flight response to acute stressors by activating the sympathetic branch, which in turn promotes adaptive changes throughout the body via the release of catecholamines. The HPA axis mediates a slower, neuroendocrine response to stressors via the release of neuropeptides and GCs. The activity of the peripheral stress effector systems is regulated by the central stress-regulatory network, which is a complex system that includes sensory regions, effector regions, limbic and cortical structures, neuromodulatory projections, and integration centers that are involved in stressor detection and stress response generation, regulation, and termination.

Distinct neural pathways detect and respond to physical stressors, which represent *actual* threat to homeostasis, and psychological stressors, which represent *perceived* threat to homeostasis. Physical stressors are detected by interoceptive sensory systems, whereas psychological stressors, such as external threats, are detected by exteroceptive sensory systems and the innate alarm system, which is involved in appraisal of biological significance of sensory information. Physical stressors trigger "reactive" stress responses via direct afferent projections to central effectors, whereas psychological stressors trigger "anticipatory" responses via indirect projections to central effectors through relays in the limbic forebrain, hypothalamus, and brainstem. In addition to physiological changes in the body, the stress response also includes a psychological component that involves adaptive changes in arousal, sensory processing, cognition, and affect, which are primarily mediated by widespread neuromodulatory monoaminergic projections. When the stressor is over, the sympathetic response is quickly terminated by reflexive activation of the parasympathetic branch, whereas the HPA axis response is terminated via rapid and delayed negative feedback by elevated GCs.

Temporally, the stress response consists of the initial "alarm" phase, which is dominated by the rapid, non-genomic effects of various stress mediators, followed by the later "recovery" phase, which is dominated by the slow, genomic effects of stress mediators. The rapid physiological changes throughout the body that promote immediate survival and coping in response to an acute stressor are mediated by the rapid effects of catecholamines and the permissive effects of basal GCs. Interactions between the rapid effects of NE and GCs in the brain underlie many cognitive and behavioral changes that occur during the initial phase of the stress response, including heightened sensitivity to salient stimuli, reliance on habitual behaviors, enhanced memory encoding, and impaired higher-order cognition. The later phase of the stress response is dominated by the genomic effects of GCs, which oppose the effects of the first wave of stress mediators, suppress the stress response, promote recovery, and help prepare for future stressors. The genomic effects of elevated GCs during the later phase of the stress response reverse the cognitive and behavioral changes that are produced during the initial phase of the stress response and promote contextualization of the stress-related situation and elaboration of more efficient solutions to be remembered in the long term.

Whereas acute stress elicits a transient coping response, which is adaptive in the short run, chronic stress constitutes a sustained challenge to homeostasis and elicits repeated activation of the stress system, which can result in increased allostatic load in multiple organ systems in the long run. Chronic stress exposure produces functional and structural changes at the level of the peripheral stress effector systems and the central stress-regulatory network, which might represent adaptations designed to optimize stress system functioning within the context of a repeated, predictable stressor or a chronically stressful environment. For example, chronic stress exposure can produce a recalibration of the neuroendocrine stress response, including stress response facilitation, habituation of the stress response to a repeated stressor, sensitization of the stress response to novel stressors, and elevated baseline levels of stress mediators. In addition, chronic stress can lead to neuroplastic changes in the central stress-regulatory network, including sensitization of stress-excitatory circuits, hypoactivity of stress-inhibitory circuits, and impaired negative feedback. Behaviorally, chronic stress is associated with impairments in higher-order cognitive functions and affects dysregulation, as well as a bias toward habitual and instinctual behaviors at the expense of cognitive flexibility. Behavioral, cognitive, and neuroplastic changes due to chronic stress might represent adaptations to a stressful environment and can become at least partially reversed after a stress-free period or following mindfulness-based interventions.

Neuroimaging of Acute Stress: Suggested Reading

Reviews and Meta-Analyses

Dedovic, K., D'Aguiar, C., & Pruessner, J. C. (2009). What stress does to your brain: A review of neuroimaging studies. *The Canadian Journal of Psychiatry, 54*(1), 6–15.

Dedovic, K., Duchesne, A., Andrews, J., Engert, V., & Pruessner, J. C. (2009). The brain and the stress axis: The neural correlates of cortisol regulation in response to stress. *NeuroImage, 47*(3), 864–871.

Hermans, E. J., Henckens, M. J., Joëls, M., & Fernández, G. (2014). Dynamic adaptation of large-scale brain networks in response to acute stressors. *Trends in Neurosciences, 37*(6), 304–314.

Kogler, L., Müller, V. I., Chang, A., Eickhoff, S. B., Fox, P. T., Gur, R. C., & Derntl, B. (2015). Psychosocial versus physiological stress — Meta-analyses on deactivations and activations of the neural correlates of stress reactions. *NeuroImage, 119*, 235–251.

Pruessner, J. C., Dedovic, K., Khalili-Mahani, N., Engert, V., Pruessner, M., Buss, C., ... Lupien, S. (2008). Deactivation of the limbic system during acute psychosocial stress: Evidence from positron emission tomography and functional magnetic resonance imaging studies. *Biological Psychiatry, 63*(2), 234–240.

Van Oort, J., Tendolkar, I., Hermans, E. J., Mulders, P. C., Beckmann, C. F., Schene, A. H., ... van Eijndhoven, P. F. (2017). How the brain connects in response to acute stress: A review at the human brain systems level. *Neuroscience & Biobehavioral Reviews, 83*, 281–297.

References

Akirav, I., & Maroun, M. (2007). The role of the medial prefrontal cortex-amygdala circuit in stress effects on the extinction of fear. *Neural Plasticity, 2007*, 30873.

Albert, D. J., & Chew, G. L. (1980). The septal forebrain and the inhibitory modulation of attack and defense in the rat. A review. *Behavioral and Neural Biology, 30*(4), 357–388.

Anderson, E. M., Gomez, D., Caccamise, A., McPhail, D., & Hearing, M. (2019). Chronic unpredictable stress promotes cell-specific plasticity in prefrontal cortex D1 and D2 pyramidal neurons. *Neurobiology of Stress, 10*, 100152.

Arnsten, A. F. (2009). Stress signalling pathways that impair prefrontal cortex structure and function. *Nature Reviews Neuroscience, 10*(6), 410–422.

Arnsten, A. F. (2015). Stress weakens prefrontal networks: Molecular insults to higher cognition. *Nature Neuroscience, 18*(10), 1376–1385.

Aston-Jones, G., & Cohen, J. D. (2005). An integrative theory of locus coeruleus-norepinephrine function: Adaptive gain and optimal performance. *Annual Review of Neuroscience, 28*, 403–450.

Bains, J. S., Cusulin, J. I. W., & Inoue, W. (2015). Stress-related synaptic plasticity in the hypothalamus. *Nature Reviews Neuroscience, 16*(7), 377–388.

Barsegyan, A., Mackenzie, S. M., Kurose, B. D., McGaugh, J. L., & Roozendaal, B. (2010). Glucocorticoids in the prefrontal cortex enhance memory consolidation and impair working memory by a common neural mechanism. *Proceedings of the National Academy of Sciences, 107*(38), 16655–16660.

Beissner, F., Meissner, K., Bär, K. J., & Napadow, V. (2013). The autonomic brain: An activation likelihood estimation meta-analysis for central processing of autonomic function. *Journal of Neuroscience, 33*(25), 10503–10511.

Bloss, E. B., Janssen, W. G., McEwen, B. S., & Morrison, J. H. (2010). Interactive effects of stress and aging on structural plasticity in the prefrontal cortex. *Journal of Neuroscience, 30*(19), 6726–6731.

Cerqueira, J. J., Almeida, O. F., & Sousa, N. (2008). The stressed prefrontal cortex. Left? Right! *Brain, Behavior, and Immunity, 22*(5), 630–638.

De Kloet, E. R., Joëls, M., & Holsboer, F. (2005). Stress and the brain: From adaptation to disease. *Nature Reviews Neuroscience, 6*(6), 463–475.

De Quervain, D. J. F., Henke, K., Aerni, A., Treyer, V., McGaugh, J. L., Berthold, T., ... Hock, C. (2003). Glucocorticoid-induced impairment of declarative memory retrieval is associated with reduced blood flow in the medial temporal lobe. *European Journal of Neuroscience, 17*(6), 1296–1302.

Dickerson, S. S., & Kemeny, M. E. (2004). Acute stressors and cortisol responses: A theoretical integration and synthesis of laboratory research. *Psychological Bulletin, 130*(3), 355–391.

Dum, R. P., Levinthal, D. J., & Strick, P. L. (2016). Motor, cognitive, and affective areas of the cerebral cortex influence the adrenal medulla. *Proceedings of the National Academy of Sciences, 113*(35), 9922–9927.

Edlow, B. L., Takahashi, E., Wu, O., Benner, T., Dai, G., Bu, L., ... Folkerth, R. D. (2012). Neuroanatomic connectivity of the human ascending arousal system critical to consciousness and its disorders. *Journal of Neuropathology & Experimental Neurology, 71*(6), 531–546.

Fanselow, M. S., & Dong, H. W. (2010). Are the dorsal and ventral hippocampus functionally distinct structures? *Neuron, 65*(1), 7–19.

Feinstein, J. S., Buzza, C., Hurlemann, R., Follmer, R. L., Dahdaleh, N. S., Coryell, W. H., ... Wemmie, J. A. (2013). Fear and panic in humans with bilateral amygdala damage. *Nature Neuroscience, 16*(3), 270–272.

Figueiredo, H. F., Bodie, B. L., Tauchi, M., Dolgas, C. M., & Herman, J. P. (2003). Stress integration after acute and chronic predator stress: Differential activation of central stress circuitry and sensitization of the hypothalamo-pituitary-adrenocortical axis. *Endocrinology, 144*(12), 5249–5258.

Flak, J. N., Solomon, M. B., Jankord, R., Krause, E. G., & Herman, J. P. (2012). Identification of chronic stress-activated regions reveals a potential recruited circuit in rat brain. *European Journal of Neuroscience, 36*(4), 2547–2555.

Geerling, J. C., Shin, J. W., Chimenti, P. C., & Loewy, A. D. (2010). Paraventricular hypothalamic nucleus: Axonal projections to the brainstem. *Journal of Comparative Neurology, 518*(9), 1460–1499.

Goldstein, D. S. (2010). Adrenal responses to stress. *Cellular and Molecular Neurobiology, 30*(8), 1433–1440.

Goldstein, D. S., & Kopin, I. J. (2008). Adrenomedullary, adrenocortical, and sympathoneural responses to stressors: A meta-analysis. *Endocrine Regulations, 42*(4), 111–119.

Golkar, A., Johansson, E., Kasahara, M., Osika, W., Perski, A., & Savic, I. (2014). The influence of work-related chronic stress on the regulation of emotion and on functional connectivity in the brain. *PLoS One, 9*(9), e104550.

Gunnar, M., & Quevedo, K. (2007). The neurobiology of stress and development. *Annual Review of Psychology, 58*, 145–173.

Hariri, A. R., & Holmes, A. (2015). Finding translation in stress research. *Nature Neuroscience, 18*(10), 1347–1352.

Henckens, M. J., van der Marel, K., van der Toorn, A., Pillai, A. G., Fernández, G., Dijkhuizen, R. M., & Joëls, M. (2015). Stress-induced alterations in large-scale functional networks of the rodent brain. *NeuroImage, 105*, 312–322.

Henckens, M. J., van Wingen, G. A., Joëls, M., & Fernández, G. (2012). Time-dependent effects of cortisol on selective attention and emotional interference: A functional MRI study. *Frontiers in Integrative Neuroscience, 6*, 66.

Henckens, M. J., van Wingen, G. A., Joëls, M., & Fernández, G. (2011). Time-dependent corticosteroid modulation of prefrontal working memory processing. *Proceedings of the National Academy of Sciences, 108*(14), 5801–5806.

Herman, J. P., Figueiredo, H., Mueller, N. K., Ulrich-Lai, Y., Ostrander, M. M., Choi, D. C., & Cullinan, W. E. (2003). Central mechanisms of stress integration: Hierarchical circuitry controlling hypothalamo–pituitary–adrenocortical responsiveness. *Frontiers in Neuroendocrinology, 24*(3), 151–180.

Herman, J. P., McKlveen, J. M., Ghosal, S., Kopp, B., Wulsin, A., Makinson, R., ... Myers, B. (2016). Regulation of the hypothalamic-pituitary-adrenocortical stress response. *Comprehensive Physiology, 6*(2), 603–621.

Herman, J. P., Ostrander, M. M., Mueller, N. K., & Figueiredo, H. (2005). Limbic system mechanisms of stress regulation: Hypothalamo-pituitary-adrenocortical axis. *Progress in Neuro-Psychopharmacology and Biological Psychiatry, 29*(8), 1201–1213.

Hermans, E. J., Henckens, M. J., Joëls, M., & Fernández, G. (2014). Dynamic adaptation of large-scale brain networks in response to acute stressors. *Trends in Neurosciences, 37*(6), 304–314.

Hill, M. N., & Tasker, J. G. (2012). Endocannabinoid signaling, glucocorticoid-mediated negative feedback, and regulation of the hypothalamic-pituitary-adrenal axis. *Neuroscience, 204*, 5–16.

Hill, M. N., McLaughlin, R. J., Bingham, B., Shrestha, L., Lee, T. T., Gray, J. M., ... Viau, V. (2010). Endogenous cannabinoid signaling is essential for stress adaptation. *Proceedings of the National Academy of Sciences, 107*(20), 9406–9411.

Hill, M. N., McLaughlin, R. J., Pan, B., Fitzgerald, M. L., Roberts, C. J., Lee, T. T. Y., ... McEwen, B. S. (2011). Recruitment of prefrontal cortical endocannabinoid signaling by glucocorticoids contributes to termination of the stress response. *Journal of Neuroscience, 31*(29), 10506–10515.

Hinwood, M., Morandini, J., Day, T. A., & Walker, F. R. (2012). Evidence that microglia mediate the neurobiological effects of chronic psychological stress on the medial prefrontal cortex. *Cerebral Cortex, 22*(6), 1442–1454.

Holmes, A., & Wellman, C. L. (2009). Stress-induced prefrontal reorganization and executive dysfunction in rodents. *Neuroscience & Biobehavioral Reviews, 33*(6), 773–783.

Iversen, S., Iversen, L., & Saper, C. B. (2000). The autonomic nervous system and the hypothalamus. *Principles of Neural Science, 4*, 960–680.

Janak, P. H., & Tye, K. M. (2015). From circuits to behaviour in the amygdala. *Nature, 517*(7534), 284–292.

Joëls, M. (2018). Corticosteroids and the brain. *Journal of Endocrinology, 238*(3), R121–R130.

Joëls, M., & Baram, T. Z. (2009). The neuro-symphony of stress. *Nature Reviews Neuroscience, 10*(6), 459–466.

Joëls, M., Karst, H., & Sarabdjitsingh, R. A. (2018). The stressed brain of humans and rodents. *Acta Physiologica, 223*(2), e13066.

Joëls, M., Karst, H., DeRijk, R., & de Kloët, E. R. (2007). The coming out of the brain mineralocorticoid receptor. *Trends in Neurosciences, 31*(1), 1–7.

Joëls, M., Sarabdjitsingh, R. A., & Karst, H. (2012). Unraveling the time domains of corticosteroid hormone influences on brain activity: Rapid, slow, and chronic modes. *Pharmacological Reviews, 64*(4), 901–938.

Jones, K. R., Myers, B., & Herman, J. P. (2011). Stimulation of the prelimbic cortex differentially modulates neuroendocrine responses to psychogenic and systemic stressors. *Physiology & Behavior, 104*(2), 266–271.

Kleckner, I. R., Zhang, J., Touroutoglou, A., Chanes, L., Xia, C., Simmons, W. K., … Barrett, L. F. (2017). Evidence for a large-scale brain system supporting allostasis and interoception in humans. *Nature Human Behaviour, 1*(5), 1–14.

Koolhaas, J. M., Bartolomucci, A., Buwalda, B., de Boer, S. F., Flügge, G., Korte, S. M., … Richter-Levin, G. (2011). Stress revisited: A critical evaluation of the stress concept. *Neuroscience & Biobehavioral Reviews, 35*(5), 1291–1301.

Kopp, B. L., Wick, D., & Herman, J. P. (2013). Differential effects of homotypic vs. heterotypic chronic stress regimens on microglial activation in the prefrontal cortex. *Physiology & Behavior, 122*, 246–252.

Krugers, H. J., Karst, H., & Joels, M. (2012). Interactions between noradrenaline and corticosteroids in the brain: From electrical activity to cognitive performance. *Frontiers in Cellular Neuroscience, 6*, 15.

Kvetnansky, R., Sabban, E. L., & Palkovits, M. (2009). Catecholaminergic systems in stress: Structural and molecular genetic approaches. *Physiological Reviews, 89*(2), 535–606.

Lanius, R. A., Rabellino, D., Boyd, J. E., Harricharan, S., Frewen, P. A., & McKinnon, M. C. (2017). The innate alarm system in PTSD: Conscious and subconscious processing of threat. *Current Opinion in Psychology, 14*, 109–115.

LeDoux, J. (2012). Rethinking the emotional brain. *Neuron, 73*(4), 653–676.

Liddell, B. J., Brown, K. J., Kemp, A. H., Barton, M. J., Das, P., Peduto, A., … Williams, L. M. (2005). A direct brainstem–amygdala–cortical 'alarm' system for subliminal signals of fear. *NeuroImage, 24*(1), 235–243.

Lightman, S. L. (2008). The neuroendocrinology of stress: A never ending story. *Journal of Neuroendocrinology, 20*(6), 880–884.

Liston, C., McEwen, B. S., & Casey, B. J. (2009). Psychosocial stress reversibly disrupts prefrontal processing and attentional control. *Proceedings of the National Academy of Sciences, 106*(3), 912–917.

Liu, Z. P., Song, C., Wang, M., He, Y., Xu, X. B., Pan, H. Q., … Pan, B. X. (2014). Chronic stress impairs GABAergic control of amygdala through suppressing the tonic GABAA receptor currents. *Molecular Brain, 7*(1), 1–14.

Luine, V. N., Beck, K. D., Bowman, R. E., Frankfurt, M., & Maclusky, N. J. (2007). Chronic stress and neural function: Accounting for sex and age. *Journal of Neuroendocrinology, 19*(10), 743–751.

Lupien, S. J., Juster, R. P., Raymond, C., & Marin, M. F. (2018). The effects of chronic stress on the human brain: From neurotoxicity, to vulnerability, to opportunity. *Frontiers in Neuroendocrinology, 49*, 91–105.

Lupien, S. J., McEwen, B. S., Gunnar, M. R., & Heim, C. (2009). Effects of stress throughout the lifespan on the brain, behaviour and cognition. *Nature Reviews Neuroscience, 10*(6), 434–445.

Magalhães, R., Barrière, D. A., Novais, A., Marques, F., Marques, P., Cerqueira, J., … Jay, T. M. (2018). The dynamics of stress: A longitudinal MRI study of rat brain structure and connectome. *Molecular Psychiatry, 23*(10), 1998–2006.

Maggio, N., & Segal, M. (2012). Steroid modulation of hippocampal plasticity: Switching between cognitive and emotional memories. *Frontiers in Cellular Neuroscience, 6*, 12.

Margittai, Z., Nave, G., Strombach, T., van Wingerden, M., Schwabe, L., & Kalenscher, T. (2016). Exogenous cortisol causes a shift from deliberative to intuitive thinking. *Psychoneuroendocrinology, 64*, 131–135.

McCall, J. G., Al-Hasani, R., Siuda, E. R., Hong, D. Y., Norris, A. J., Ford, C. P., & Bruchas, M. R. (2015). CRH engagement of the locus coeruleus noradrenergic system mediates stress-induced anxiety. *Neuron, 87*(3), 605–620.

McEwen, B. S. (1998). Stress, adaptation, and disease: Allostasis and allostatic load. *Annals of the New York Academy of Sciences, 840*(1), 33–44.

McEwen, B. S., & Gianaros, P. J. (2011). Stress-and allostasis-induced brain plasticity. *Annual Review of Medicine, 62*, 431–445.

McEwen, B. S., Gray, J. D., & Nasca, C. (2015). 60 years of neuroendocrinology: Redefining neuroendocrinology: Stress, sex and cognitive and emotional regulation. *Journal of Endocrinology, 226*(2), T67–T83.

McEwen, B. S., Nasca, C., & Gray, J. D. (2016). Stress effects on neuronal structure: Hippocampus, amygdala, and prefrontal cortex. *Neuropsychopharmacology, 41*(1), 3–23.

McGaugh, J. L., & Roozendaal, B. (2002). Role of adrenal stress hormones in forming lasting memories in the brain. *Current Opinion in Neurobiology, 12*(2), 205–210.

McIntyre, C. K., McGaugh, J. L., & Williams, C. L. (2012). Interacting brain systems modulate memory consolidation. *Neuroscience & Biobehavioral Reviews, 36*(7), 1750–1762.

McKlveen, J. M., Moloney, R. D., Scheimann, J. R., Myers, B., & Herman, J. P. (2019). "Braking" the prefrontal cortex: The role of glucocorticoids and interneurons in stress adaptation and pathology. *Biological Psychiatry, 86*(9), 669–681.

McKlveen, J. M., Morano, R. L., Fitzgerald, M., Zoubovsky, S., Cassella, S. N., Scheimann, J. R., … Baccei, M. L. (2016). Chronic stress increases prefrontal inhibition: A mechanism for stress-induced prefrontal dysfunction. *Biological Psychiatry, 80*(10), 754–764.

McKlveen, J. M., Myers, B., & Herman, J. P. (2015). The medial prefrontal cortex: Coordinator of autonomic, neuroendocrine and behavioural responses to stress. *Journal of Neuroendocrinology, 27*(6), 446–456.

Méndez-Bértolo, C., Moratti, S., Toledano, R., Lopez-Sosa, F., Martínez-Alvarez, R., Mah, Y. H., … Strange, B. A. (2016). A fast pathway for fear in human amygdala. *Nature Neuroscience, 19*(8), 1041–1049.

Miller, G. E., Chen, E., & Zhou, E. S. (2007). If it goes up, must it come down? Chronic stress and the hypothalamic-pituitary-adrenocortical axis in humans. *Psychological Bulletin, 133*(1), 25.

Mirescu, C., & Gould, E. (2006). Stress and adult neurogenesis. *Hippocampus, 16*(3), 233–238.

Mitra, R., & Sapolsky, R. M. (2008). Acute corticosterone treatment is sufficient to induce anxiety and amygdaloid dendritic hypertrophy. *Proceedings of the National Academy of Sciences, 105*(14), 5573–5578.

Morena, M., Patel, S., Bains, J. S., & Hill, M. N. (2016). Neurobiological interactions between stress and the endocannabinoid system. *Neuropsychopharmacology, 41*(1), 80–102.

Morilak, D. A., Barrera, G., Echevarria, D. J., Garcia, A. S., Hernandez, A., Ma, S., & Petre, C. O. (2005). Role of brain norepinephrine in the behavioral response to stress. *Progress in Neuro-Psychopharmacology and Biological Psychiatry, 29*(8), 1214–1224.

Myers, B., Scheimann, J. R., Franco-Villanueva, A., & Herman, J. P. (2017). Ascending mechanisms of stress integration: Implications for brainstem regulation of neuroendocrine and behavioral stress responses. *Neuroscience & Biobehavioral Reviews, 74*, 366–375.

Öhman, A. (2005). The role of the amygdala in human fear: Automatic detection of threat. *Psychoneuroendocrinology, 30*(10), 953–958.

Parihar, V. K., Hattiangady, B., Kuruba, R., Shuai, B., & Shetty, A. K. (2011). Predictable chronic mild stress improves mood, hippocampal neurogenesis and memory. *Molecular Psychiatry, 16*(2), 171–183.

Plessow, F., Kiesel, A., & Kirschbaum, C. (2012). The stressed prefrontal cortex and goal-directed behaviour: Acute psychosocial stress impairs the flexible implementation of task goals. *Experimental Brain Research, 216*(3), 397–408.

Popoli, M., Yan, Z., McEwen, B. S., & Sanacora, G. (2012). The stressed synapse: The impact of stress and glucocorticoids on glutamate transmission. *Nature Reviews Neuroscience, 13*(1), 22–37.

Price, J. L. (1999). Prefrontal cortical networks related to visceral function and mood. *Annals of the New York Academy of Sciences, 877*(1), 383–396.

Pruessner, J. C., Dedovic, K., Pruessner, M., Lord, C., Buss, C., Collins, L., … Lupien, S. J. (2010). Stress regulation in the central nervous system: Evidence from structural and functional neuroimaging studies in human populations-2008 Curt Richter award winner. *Psychoneuroendocrinology, 35*(1), 179–191.

Radley, J. J. (2012). Toward a limbic cortical inhibitory network: Implications for hypothalamic-pituitary-adrenal responses following chronic stress. *Frontiers in Behavioral Neuroscience, 6*, 7.

Radley, J. J., Arias, C. M., & Sawchenko, P. E. (2006). Regional differentiation of the medial prefrontal cortex in regulating adaptive responses to acute emotional stress. *Journal of Neuroscience, 26*(50), 12967–12976.

Reser, J. E. (2016). Chronic stress, cortical plasticity and neuroecology. *Behavioural Processes, 129*, 105–115.

Reyes, B. A., Zitnik, G., Foster, C., Van Bockstaele, E. J., & Valentino, R. J. (2015). Social stress engages neurochemically-distinct afferents to the rat locus coeruleus depending on coping strategy. *Eneuro, 2*(6), e0042-15.2015, 1–12.

Rodrigues, S. M., LeDoux, J. E., & Sapolsky, R. M. (2009). The influence of stress hormones on fear circuitry. *Annual Review of Neuroscience, 32*, 289–313.

Roman, V., Van der Borght, K., Leemburg, S. A., Van der Zee, E. A., & Meerlo, P. (2005). Sleep restriction by forced activity reduces hippocampal cell proliferation. *Brain Research, 1065*(1–2), 53–59.

Roosevelt, R. W., Smith, D. C., Clough, R. W., Jensen, R. A., & Browning, R. A. (2006). Increased extracellular concentrations of norepinephrine in cortex and hippocampus following vagus nerve stimulation in the rat. *Brain Research, 1119*(1), 124–132.

Roozendaal, B., McEwen, B. S., & Chattarji, S. (2009). Stress, memory and the amygdala. *Nature Reviews Neuroscience, 10*(6), 423–433.

Saper, C. B. (2002). The central autonomic nervous system: Conscious visceral perception and autonomic pattern generation. *Annual Review of Neuroscience, 25*(1), 433–469.

Sapolsky, R. M. (2000). Stress hormones: Good and bad. *Neurobiology of Disease, 5*(7), 540–542.

Sapolsky, R. M. (2015). Stress and the brain: Individual variability and the inverted-U. *Nature Neuroscience, 18*(10), 1344–1346.

Sapolsky, R. M., Romero, L. M., & Munck, A. U. (2000). How do glucocorticoids influence stress responses? Integrating permissive, suppressive, stimulatory, and preparative actions. *Endocrine Reviews, 21*(1), 55–89.

Schwabe, L. (2017). Memory under stress: From single systems to network changes. *European Journal of Neuroscience, 45*(4), 478–489.

Schwabe, L., Dalm, S., Schächinger, H., & Oitzl, M. S. (2008). Chronic stress modulates the use of spatial and stimulus-response learning strategies in mice and man. *Neurobiology of Learning and Memory, 90*(3), 495–503.

Schwabe, L., Tegenthoff, M., Höffken, O., & Wolf, O. T. (2012). Simultaneous glucocorticoid and noradrenergic activity disrupts the neural basis of goal-directed action in the human brain. *Journal of Neuroscience, 32*(30), 10146–10155.

Schwabe, L., Tegenthoff, M., Höffken, O., & Wolf, O. T. (2013). Mineralocorticoid receptor blockade prevents stress-induced modulation of multiple memory systems in the human brain. *Biological Psychiatry, 74*(11), 801–808.

Schwabe, L., Wolf, O. T., & Oitzl, M. S. (2010). Memory formation under stress: Quantity and quality. *Neuroscience & Biobehavioral Reviews, 34*(4), 584–591.

Seeley, W. W., Menon, V., Schatzberg, A. F., Keller, J., Glover, G. H., Kenna, H., ... Greicius, M. D. (2007). Dissociable intrinsic connectivity networks for salience processing and executive control. *Journal of Neuroscience, 27*(9), 2349–2356.

Shields, G. S., Sazma, M. A., & Yonelinas, A. P. (2016). The effects of acute stress on core executive functions: A meta-analysis and comparison with cortisol. *Neuroscience & Biobehavioral Reviews, 68*, 651–668.

Smith, S. M., & Vale, W. W. (2006). The role of the hypothalamic-pituitary-adrenal axis in neuroendocrine responses to stress. *Dialogues in Clinical Neuroscience, 8*(4), 383–395.

Snyder, K., Wang, W. W., Han, R., McFadden, K., & Valentino, R. J. (2012). Corticotropin-releasing factor in the norepinephrine nucleus, locus coeruleus, facilitates behavioral flexibility. *Neuropsychopharmacology, 37*(2), 520–530.

Spencer, R. L., & Deak, T. (2017). A users guide to HPA axis research. *Physiology & Behavior, 178*, 43–65.

Sullivan, R. M., & Gratton, A. (2002). Prefrontal cortical regulation of hypothalamic–pituitary–adrenal function in the rat and implications for psychopathology: Side matters. *Psychoneuroendocrinology, 27*(1–2), 99–114.

Sun, M. K. (1995). Central neural organization and control of sympathetic nervous system in mammals. *Progress in Neurobiology, 47*(3), 157–233.

Tasker, J. G., & Herman, J. P. (2011). Mechanisms of rapid glucocorticoid feedback inhibition of the hypothalamic–pituitary–adrenal axis. *Stress, 14*(4), 398–406.

Taylor, S. B., Anglin, J. M., Paode, P. R., Riggert, A. G., Olive, M. F., & Conrad, C. D. (2014). Chronic stress may facilitate the recruitment of habit-and addiction-related neurocircuitries through neuronal restructuring of the striatum. *Neuroscience, 280*, 231–242.

Ter Horst, G. J., Wichmann, R., Gerrits, M., Westenbroek, C., & Lin, Y. (2009). Sex differences in stress responses: Focus on ovarian hormones. *Physiology & Behavior, 97*(2), 239–249.

Ulrich-Lai, Y. M., & Herman, J. P. (2009). Neural regulation of endocrine and autonomic stress responses. *Nature Reviews Neuroscience, 10*(6), 397–409.

Ulrich-Lai, Y. M., Figueiredo, H. F., Ostrander, M. M., Choi, D. C., Engeland, W. C., & Herman, J. P. (2006). Chronic stress induces adrenal hyperplasia and hypertrophy in a subregion-specific manner. *American Journal of Physiology-Endocrinology and Metabolism, 291*(5), E965–E973.

Uschold-Schmidt, N., Nyuyki, K. D., Füchsl, A. M., Neumann, I. D., & Reber, S. O. (2012). Chronic psychosocial stress results in sensitization of the HPA axis to acute heterotypic stressors despite a reduction of adrenal in vitro ACTH responsiveness. *Psychoneuroendocrinology, 37*(10), 1676–1687.

Valentino, R. J., & Van Bockstaele, E. (2008). Convergent regulation of locus coeruleus activity as an adaptive response to stress. *European Journal of Pharmacology, 583*(2–3), 194–203.

van Ast, V. A., Cornelisse, S., Meeter, M., Joëls, M., & Kindt, M. (2013). Time-dependent effects of cortisol on the contextualization of emotional memories. *Biological Psychiatry, 74*(11), 809–816.

Van Bockstaele, E. J., Colago, E. E. O., & Valentino, R. J. (1998). Amygdaloid corticotropin-releasing factor targets locus coeruleus dendrites: Substrate for the co-ordination of emotional and cognitive limbs of the stress response. *Journal of Neuroendocrinology, 10*(10), 743–758.

Vertes, R. P. (2004). Differential projections of the infralimbic and prelimbic cortex in the rat. *Synapse, 51*(1), 32–58.

Vogel, S., Klumpers, F., Schröder, T. N., Oplaat, K. T., Krugers, H. J., Oitzl, M. S., … Fernández, G. (2017). Stress induces a shift towards striatum-dependent stimulus-response learning via the mineralocorticoid receptor. *Neuropsychopharmacology, 42*(6), 1262–1271.

Vyas, A., Mitra, R., Rao, B. S., & Chattarji, S. (2002). Chronic stress induces contrasting patterns of dendritic remodeling in hippocampal and amygdaloid neurons. *Journal of Neuroscience, 22*(15), 6810–6818.

Vyas, A., Pillai, A. G., & Chattarji, S. (2004). Recovery after chronic stress fails to reverse amygdaloid neuronal hypertrophy and enhanced anxiety-like behavior. *Neuroscience, 128*(4), 667–673.

Wei, J., Zhong, P., Qin, L., Tan, T., & Yan, Z. (2018). Chemicogenetic restoration of the prefrontal cortex to amygdala pathway ameliorates stress-induced deficits. *Cerebral Cortex, 28*(6), 1980–1990.

Westerhaus, M. J., & Loewy, A. D. (2001). Central representation of the sympathetic nervous system in the cerebral cortex. *Brain Research, 903*(1–2), 117–127.

Winklewski, P. J., Radkowski, M., Wszedybyl-Winklewska, M., & Demkow, U. (2017). Stress response, brain noradrenergic system and cognition. In *Respiratory system diseases* (pp. 67–74). Cham: Springer.

Wood, S. K., & Valentino, R. J. (2017). The brain norepinephrine system, stress and cardiovascular vulnerability. *Neuroscience & Biobehavioral Reviews, 74*, 393–400.

Yuen, E. Y., Liu, W., Karatsoreos, I. N., Feng, J., McEwen, B. S., & Yan, Z. (2009). Acute stress enhances glutamatergic transmission in prefrontal cortex and facilitates working memory. *Proceedings of the National Academy of Sciences, 106*(33), 14075–14079.

Yuen, E. Y., Liu, W., Karatsoreos, I. N., Ren, Y., Feng, J., McEwen, B. S., & Yan, Z. (2011). Mechanisms for acute stress-induced enhancement of glutamatergic transmission and working memory. *Molecular Psychiatry, 16*(2), 156–170.

Yuen, E. Y., Wei, J., Liu, W., Zhong, P., Li, X., & Yan, Z. (2012). Repeated stress causes cognitive impairment by suppressing glutamate receptor expression and function in prefrontal cortex. *Neuron, 73*(5), 962–977.

Zoccal, D. B., Furuya, W. I., Bassi, M., Colombari, D. S., & Colombari, E. (2014). The nucleus of the solitary tract and the coordination of respiratory and sympathetic activities. *Frontiers in Physiology, 5*, 238.

Chapter 3
The Tend and Befriend Theory of Stress: Understanding the Biological, Evolutionary, and Psychosocial Aspects of the Female Stress Response

Laura Cohen and Amy Hughes Lansing

Introduction

Tend and befriend theory was developed by Taylor and collogues as an approach to studying female stress responding from an evolutionary and biosocial perspective (Taylor et al., 2000). Historically, stress has largely been understood in terms of the fight-or-flight model that was originally developed by Walter Cannon in 1932 wherein a person either mounts an antagonistic response or flees away from a threatening or stressful circumstance (Cannon, 1932). Moreover, prior to 1995, the majority of this fight-or-flight stress research was conducted only on male subjects (Taylor et al., 2000). The exclusion of female subjects was argued as necessity due to the cyclical hormonal variations of the female reproductive system which complicate the measurement of hormonal responses to stress in females (Ganz, 2012). As contemporary stress research became more female inclusive, it was elucidated that males and females differ in the ways they respond to stress both behaviorally and biologically. For example, males and females show different physiological responses to stress in terms of levels of cortisol and testosterone as well as different behavior responding patterns (Byrd-Craven et al., 2016; Kivlighan et al., 2005; Levy et al., 2019; Smeets et al. 2009; Taylor et al., 2010). Thus, the generation of an additional explanation of stress responding, such as the tend and befriend theory, that considers female human physiology and the evolutionary history of the female sex was crucial for a comprehensive understanding of human stress.

Tend and befriend theory characterizes a social response to threat that aligns with female stress responding in both physiology and behavior (Taylor, 2002; Taylor et al., 2000). During a stressor, females are more likely to provide care and seek affiliative support from others than to engage in fight-or-flight behavior patterns (Byrd-Craven et al., 2016; Taylor, 2006). Tend and befriend theory outlines this

L. Cohen · A. H. Lansing (✉)
University of Vermont, Burlington, VT, USA
e-mail: Amy.Hughes.Lansing@uvm.edu

© Springer Nature Switzerland AG 2021   67
H. Hazlett-Stevens (ed.), *Biopsychosocial Factors of Stress, and Mindfulness for Stress Reduction*, https://doi.org/10.1007/978-3-030-81245-4_3

unique biobehavioral pattern of stress response. It is not intended to contradict the established theory of natural selection or the fight-or-flight theory of stress responding (Eisler & Levine, 2002), but to establish that stress responding can take alternative forms while still maintaining the same survival and reproductive functions. In this chapter, we will first review the major components of Taylor's tend and befriend theory. Second, we will discuss the unique characteristics of the tend and befriend stress response and the psychological and behavioral mechanisms undergirding that response as well as how tending and befriending response patterns differ between the sexes and the survival functions of these patterns. Last, we will describe the benefits of and physical and mental health outcomes associated with tending and befriending.

Tend and Befriend: A Biobehavioral Theory to Stress Responding

A coordinated biobehavioral stress response is necessary to react effectively to threats and increase survival; however, there are both biological and behavioral differences across males and females in how individuals respond to threats (Taylor et al., 2000). According to evolutionary theory, yielding to a life-threatening stimulus does not increase the chances that an individual will survive and reproduce. Thus, across species, the fight-or-flight stress response is characterized by aggressive and anxious behavior where individuals respond to a threat by mounting an attack on threatening stimuli or fleeing from it (Cannon, 1932; Steinbeis et al., 2015; Taylor, 2012). This is the primary stress response that has been examined in humans since the early 1930s. Yet, the responses described in fight-or-flight theory do not fully characterize the stress responses of human females. Early in human history, daily roles and responsibilities were often segregated across the sexes. Males would be responsible for protection and hunting and females were often responsible for childcare and foraging (Taylor, 2012). Because of this, it is difficult to imagine that a female would choose to fight or flee in response to a threat. Rather, females are more likely to respond to stress by tending to their young and engaging social resources (Byrd-Craven et al., 2016). For example, females endorse more flight and tend and befriend behaviors, while males endorse more fight responses to stress (Levy et al., 2019). These patterns emerge early in life, with female infants more likely to approach and befriend, while male infants are more likely to show behaviors closely associated with fight or flight when managing proximity toward their mothers (David & Lyons-Ruth, 2005). A robust stress response is vital to ensure the survival of the species, and beyond fight or flight, these tend and befriend behaviors are also critical to species survival and reproduction. Drawing on this broadened evolutionary perspective, the tend and befriend theory was developed to explain and account for differences in male and female stress responses across biological and behavioral systems.

A departure from SNS and HPA activation stress responding: Oxytocin and opioid systems guiding the tend and befriend response. Although both males and females demonstrate activation of the sympathetic nervous system (SNS) and hypothalamic-pituitary-adrenal (HPA) axis during a threat, there remain robust differences in the biological underpinnings of stress responding across sexes. Androgens, such as testosterone, appear to regulate a typical fight-or-flight pattern (Geary & Flinn, 2002; Taylor et al., 2000). In males, testosterone is released during stress responding and supports SNS and HPA axis activation (Cumming et al., 1986; Girdler et al., 1997; Mathur et al., 1986; Wheeler et al., 1994). Although androgens are also released during stress, but in lower concentrations in females, a stress responding pattern more consistent with tend and befriend is regulated by oxytocin and endogenous opioids (Taylor et al., 2000). In contrast, testosterone (the primary sex hormones in males) does not appear to be largely involved in the tend or befriend pattern. These neuroendocrine differences between the sexes likely contribute to differences in stress responding patterns: Fight-or-flight behaviors in males are likely modulated by androgynous hormones such as testosterone, while tend and befriend behaviors in females are likely modulated by endogenous hormones such as oxytocin.

The oxytocin system is critical to understanding the tend and befriend stress response in females. Oxytocin is produced in the magnocellular neurosecretory cells in the supraoptic and paraventricular nuclei of the hypothalamus (Cardoso et al., 2013) and can act as a neuromodulator in the brain in response to emotional and physical challenges (Engelmann et al., 2004; Neumann, 2009). Oxytocin is modulated by estrogen, the primary sex hormone in females (McCarthy, 1995). It is commonly known to promote maternal behavior, such as uterine contraction and milk ejection, and is potent during pregnancy (Eisler & Levine, 2002). In stressful situations, oxytocin is released into the blood through the posterior pituitary gland (Lang et al., 1983). In contrast to androgens, oxytocin is associated with parasympathetic functioning, or the downregulation of the SNS and HPA responses (Dreifuss et al., 1992; Sawchenko & Swanson, 1982; Taylor et al., 2000). Research has found that exogenous oxytocin can have stress-dampening effects after administration on laboratory stress challenges (Cardoso et al., 2013), including decreasing cardiovascular activity, promoting muscle relaxation, and increasing digestion (Ganz, 2012; Uvnas-Moberg et al., 2005). Although oxytocin is involved in both male and female stress responding (Taylor et al., 2000), there has been found to be greater concentrations of oxytocin in females compared to males (Jezova et al., 1996). The tend and befriend model draws on these sex differences in oxytocin and endogenous opioid activation in response to stress to help explain why females engage in fewer aggressive and hostile behaviors in response to threat (Taylor, 2012).

Response Patterns of Tending and Befriending

Tending Response Pattern Tending is described as caring for an offspring and protecting them from harm (Taylor et al., 2000). The biobehavioral mechanisms underlying the tending response overlap with those that support infant-caregiver attachment (Taylor et al., 2000). Although attachment theory does not address maternal stress responses specifically, it does provide an evolutionary and biobehavioral account of patterns of maternal bonding and child socialization that also underlie a maternal stress response of tending (Bowlby, 1988). Consistent with evolutionary and attachment theory, tending is critical to species survival and reproduction, as strong infant-caregiver bonds increase survival of young. When an offspring is threatened or in distress, they will likely experience separation distress and engage in behaviors to increase maternal tending, such as vocalizations that promote the return of their caregiver (Taylor, 2006). Maternal tending in response to infant distress is undergirded by the maternal oxytocin system, which motivates and reinforces tending behaviors to reduce maternal distress (Byrd-Craven et al., 2016; Fleming et al., 1997; Stallings et al., 1997). In addition to infant distress during separations, mothers can also experience distress at separation from offspring (Taylor et al., 2000). Tending not only downregulates infant distress, but also downregulates the stress system of the mother during the mother-infant tending exchange (Taylor, 2012). Thus, although tending behaviors have obvious benefits for offspring survival, engaging in tending behaviors may also be advantageous for stress reduction in the mother.

Tending is a biologically driven behavior pattern that decreases maternal stress and increases survival of young and that is consistent with evolutionary theory despite the incompatibility of tending with a fight-or-flight stress response pattern. For example, if a female chose to fight when a threat presents itself, the female may be putting both themselves and their offspring in danger (Levy et al., 2019; Taylor et al., 2000). Further, if a female is pregnant or nursing, they may not be able to successfully flee from a threat with their offspring due to having to carry the offspring with them (Taylor et al., 2000). Compared to species of animals that produce infants that are capable of fleeing within hours of birth (such as horses or deer), human infants rely on protection from a mature caregiver when faced with a threat (Taylor et al., 2000). In humans and other species that birth offspring that are immobile for larger periods of time, fighting or fleeing may not be the best solution to alleviate stress or manage a threat. Instead, tending behaviors may provide both mother and offspring the best chances of survival.

Sex Differences in Tending Response Patterns Tending is considered a form of parental investment, and the engagement of males and females in this behavior varies by species (Geary & Flinn, 2002). Factors argued to contribute to these differences are hormones, contextual factors (e.g., parenting roles within a family), and sexual-selection factors (e.g., species mating patterns) (Andersson, 1994; Darwin, 1871; Geary & Flinn, 2002; Trivers, 1972; Williams, 1966). For example, in species where females make a larger initial investment to the offspring through pregnancy and nursing, the female often continues to provide the majority of care to the offspring until maturity (Taylor et al., 2000). These species typically follow a breeding pattern characterized by male competition over access to females and male provision of resources that help the female raise the offspring (Geary & Flinn, 2002). There are species where the reverse pattern exists (e.g., some species of insects and birds) that result in males providing the majority of caregiving and the females competing over access to males (Geary & Flinn, 2002). The human species typically follows the high maternal parental investment pattern, and accordingly females are more likely than males to engage in tending as a stress response pattern (Taylor et al., 2000). For example, human men may show decreased motivation to care for an infant after a stressful situation, while women may show increased motivation to care for the infant (Probst et al., 2017). In addition, after a stressful workday, fathers may be more likely to withdraw from family members, while mothers may be more likely to show nurturant and care behaviors toward family and children (Repetti, 1989).

As with the broader tending response, hormones drive and contribute to differences in parental investment (Geary & Flinn, 2002). In humans, females have evolved biochemical patterns associated with increased parental investment (Eisler & Levine, 2002). Oxytocin is present at high levels in females following birth and has been shown to promote bonding and caregiving behaviors (David & Lyons-Ruth, 2005; Taylor, 2002; Taylor, 2006). In addition, low levels of testosterone may also explain a stronger tendency for females to care for offspring. Research suggests that in males, high testosterone is related to competitive behavior, while low testosterone leads to higher parental investment and pair-bonding behavior (Gray et al., 2006). Studies on sympathetic pregnancy in males found that human men with more of these symptoms were more affected by infant distress and experienced higher levels of the hormone prolactin and lower levels of testosterone (Storey et al., 2000). Given the overlap in behavioral and biological mechanisms, sex differences in parental investment behaviors can then also be accounted for by the tending pattern of stress response.

Befriending Response Pattern In addition to responding to stress by tending to an offspring, females also respond to stress by "befriending" (Taylor, 2012; Taylor et al., 2000). Befriending is defined as the creation and maintenance of social networks that provide resources and protection for both the female and her offspring during times of stress (Taylor et al., 2000). The central drive of the befriend response is the desire to affiliate with other people when under stress (Baumeister & Leary, 1995; Taylor, 2006). Affiliation in times of stress has been shown to have numerous advantages for both a female and her offspring (Taylor, 2012). For example, affiliating with others can increase the chances that both mother and offspring will be protected by other group members from threats (Taylor et al., 2000). Humans compared to other animals do not have physical advantages, such as camouflage abilities or sharp teeth, to protect themselves from a threat (Taylor, 2012). Because of this, they are more likely to seek support from one another than attack one another when an immediate threat is present (Taylor, 2012).

Perceived social support and affiliation behaviors are linked with diminished physiological stress responding. Perceived social support is defined as the perception that one is loved and cared for by others and is part of a mutual social network (Wills, 1991). The availability of social contacts and social support has been found to downregulate the HPA axis and SNS stress response (Taylor, 2006; Taylor et al., 2000). Taylor (2006, 2012) proposed that the befriending behavior pattern of seeking social contacts occurs similar to the way the body processes other appetitive needs. If an individual lacks affiliation during stress responding, the body will signal a drive to meet this "social support" need similar to the regulation of the hunger and sexual drives. The effects of social support on distress also occur in a negative direction. For example, a person's biological stress response can also become heightened if their social contacts are not supportive or are hostile (Taylor, 2006). Thus, maintaining social support needs to reduce distress likely fosters the befriending response pattern.

Physiologically, oxytocin also plays a large part in regulating the befriending response pattern. Oxytocin is involved in many different forms of social attachment and not limited to that of mother-infant attachment as described above (Carter, 1998). This includes pair-bonding between adults and within-gender friendships (Drago et al., 1986; Fahrbach et al., 1985; Panksepp, 1998, Taylor et al., 2000). Oxytocin is present during pleasurable social contacts in animals and humans alike (Eisler & Levine, 2002). In animal studies, administering oxytocin has been found to promote social approach behaviors, such as grooming, following social stress (Lukas et al., 2011). Among college students, administering exogenous oxytocin promoted feelings of trust, openness to new experiences, and extraversion (Cardoso et al., 2013). Under threat, oxytocin also reinforces females befriending by attenuating heightened biological stress responding from the SNS and HPA axis during social affiliation (Cardoso et al., 2013; Taylor, 2006; Taylor, 2012; Taylor et al., 2000). Like that of the tending response pattern, befriending during a stressful situation can improve the chances of survival of group members and deregulate heightened arousal of the biological stress responding system of females.

Sex Differences in Befriending Response Pattern A desire to affiliate with a group is at the heart of the befriending response pattern to stress. It has been observed that human females are much more likely to affiliate with others and engage in social networks under conditions of stress compared to males (Ganz, 2012; Tamres et al., 2002; Taylor, 2012; Taylor et al., 2000). In studies on rodents, crowding was found to increase male stress but have a calming effect in female rodents (Brown & Grunberg, 1995). Kivlighan et al. (2005) found that female athletes were more likely to interact cooperatively with teammates compared to males. It is suggested that women rely more on their female friends than their male spouses for social support, report mores satisfaction with contact with female family and friends, and provide more frequent social support to others compared to men (Belle, 1987; McDonald & Korabik, 1991; Taylor et al., 2000). Consistent with the befriend model, for females, affiliation behaviors are more common and help reduce distress compared to affiliation behaviors in males.

Persistent sex differences in affiliation for stress reduction in humans may be further explained by both sex differences in aggression and group formation. High levels of testosterone have been found to predict aggression in males (Girdler et al., 1997) and are frequently associated with the fight response in males (Taylor et al., 2000). Aggression in males is often used to gain power against others, defend territories, and attack threats in response to stress (Taylor et al., 2000). For example, males are more likely to act selfishly to gain an advantage after stress, while females are more likely to act cooperatively (Nickels et al. 2017). In addition to a difference in function, males have also been found to engage in more frequent physical aggression compared to females (Taylor et al., 2000). In animals, the presence of another female does not usually evoke attack behaviors in females, although the presence of another male often evokes this in males (Taylor et al., 2000). Females also are not often observed engaging in rough and tumble play (Maccoby & Jacklin, 1974). This behavior may be selected against in females due to possible reproductive costs associated with physical violence (Campbell, 1999; Geary & Flinn, 2002).

Although emitted slightly differently, forms of female aggression are still well documented. Female aggression differs from male aggression in that it is not mediated by sympathetic arousal and testosterone like it is in males (Taylor et al., 2000). Females have been observed acting in aggressive manners when defending offspring against attack (Adams, 1992; Brain et al., 1992) or competing against other females for paternal investment (Geary & Flinn, 2002). Most often, female aggression among humans is more indirect compared to male aggression (Holmstrom, 1992). For example, engaging in forms of relational aggression, such as gossiping and exclusion, is more commonly observed in human females compared to physical attack behaviors (Crick et al., 1997). Sex differences in engagement and benefits of aggression for stress reduction are also consistent with the befriend model of stress response.

Finally, there are also sex differences in group formation as a tool for stress reduction. As previously discussed, females form groups with other females in order to reduce stress and increase social support. Males, in contrast, have been observed forming gender-segregated groups for different functions. Research suggests that males often form groups for combat purposes, such as defending against or fighting off an enemy (Byrd-Craven et al., 2016; Levy et al., 2019). In species where males compete for access to females, males organize themselves into a hierarchical system of dominance (Geary & Flinn, 2002). It is also less likely that males form groups for purposes of maintaining friendships. One study found that men focused on friendship in the workplace only when it benefited their careers, while women used friendships in the workplace for social support systems during stress (Morrison, 2009). Friendship in male primates is very rare and not often observed (Nishida & Hiraiwa-Hasegawa, 1987). This may because maintaining friendship relationships with other males may be too time-consuming to outweigh other evolutionary advantageous behaviors, such as fighting off threats (Geary & Flinn, 2002).

When human males do form groups, it is most often with their own kin and does not require the same resources as non-kin relationships (Chagnon, 1988; Hamilton, 1964). Contrasted with the common roles and size of female groups, small and focused on maintaining socioemotional bonds (Cross & Madson, 1997), male groups are often much larger, organized, and task-focused (Taylor et al., 2000). Males and females affiliate differently, form groups for different reasons, and interact with their social environment in different ways. For this reason, befriending may be a more advantageous form of stress responding for females compared to other strategies such as fight or flight.

Survival Functions of Tending and Befriending The tend and befriend theory was developed with strong consideration of the evolutionary advantages of this pattern of stress responding that benefits further discussion. Taylor et al. (2000) contended that successful stress responding has been passed down to newer generations through principles of natural selection. It has been argued that tending and befriending has both survival and reproductive benefits, thus making the stress response pattern evolutionarily advantageous. First, it is vital for offspring to survive to ensure species survival (Taylor, 2006). In humans, offspring remain immature for extended periods of time and rely on protection from a caregiver. This longer development period may be related to the appearance of the tend and befriend stress responding pattern (Geary & Flinn, 2002). Abandoning an infant offspring during a stressful encounter drastically decreases the chances that the infant offspring will survive on their own (Levy et al., 2019). Due to this, the tending pattern may have evolved to protect both the self and an immature offspring (Taylor, 2012). Instead of putting an offspring in danger by fighting or fleeing, a female may engage in tending behaviors aimed to get offspring away from harm, retrieve them from harm, or calm and quiet them to reduce attention from perceived threats (Taylor et al., 2000). These tending behaviors increase the likelihood that the offspring will mature and eventually reproduce on their own.

Second, affiliation and group living also have evolutionary advantages. Social isolation has been found to increase the risk of mortality, while social support has been found to lead to many improved health outcomes, including decreased risk of mortality (Taylor, 2006). Protecting both self and offspring can be a difficult task without help. Because of this, individuals that use social groups to facilitate this task may have been more likely to survive against threats than those who did not (Taylor, 2012; Taylor et al., 2000). Groups also allow there to be more eyes watching for predators and inflict fear in predators by indicating to them that others are around to aid if one member is attacked (Janson, 1992; Rubenstein, 1978).

Moreover, affiliation in female groups has survival benefits in addition to protection from out-group threats. Many female primates develop harem group structures that consist of one dominant male and several females and offspring. For males, the harem structure has the benefit of keeping many potential female mates in close proximity (Taylor et al., 2000). However, the harem structure may also increase the protection females have against aggressive male attacks due to large female numbers (Taylor et al., 2000; Wrangham, 1980). There are other advantages to female primate groups, such as the ability to share information about the best food sites (Silk, 2000), share caretaker responsibilities among members (Wrangham, 1980), and provide premature female opportunities to observe and gain experience caring for offspring before taking on her own (Taylor et al., 2000). Similarly in humans, affiliating with others increases opportunities for sharing knowledge and experience, as well as increases the emotional and intellectual growth of group members (Taylor, 2012). Beyond social support affiliation, group formation among females has many survival functions as well as social and emotional benefits.

Tend and Befriend: Outcomes and Benefits

In addition to direct survival function when under threat, tending and befriending stress response patterns provide additional benefits maternal as well as child physical and mental health. A pattern of stress responding more consistent with tending and befriending has been found to lead to better health outcomes compared to that of the fight-or-flight pattern. Chronic stress can increase the likelihood of chronic diseases such as heart disease, type II diabetes, and hypertension, especially when combined with other genetic or acquired risks (McEwen, 1998, Taylor, 2012). Fight-or-flight behaviors have been shown to have lasting negative effects on health due to repeated and chronic activation of the HPA axis and related neuroendocrine and immune systems, which leads to increased wear and tear on and decreased cardiovascular and immune function (e.g., allostatic load; McEwen & Stellar, 1993). Earlier mortality in males has been related to increased engagement in behaviors of the fight-or-flight response including aggression, substance use, and coronary artery disease (McEwen, 1998; Taylor, 2012).

Alternatively, tend and befriend behaviors have been shown to have positive consequences for long-term health and well-being (Taylor, 2012) primarily through the mechanism of increased social support. For example, increased social support is associated with faster recovery from surgery (King et al., 1993), fewer pregnancy and birth complications in human females (Collins et al., 1993), and decreased cognitive decline in older adults (Seeman et al., 2001). The presence of social support or social contacts also predicts an increased life span by about 2.5 years (Berkman & Syme, 1979). Finally, social support has been shown to reduce psychological distress such as anxiety or depression (Fleming et al., 1982; Sarason et al., 1997). Befriending is also likely to reduce loneliness, which when elevated is associated with impaired physical and mental health (Hawkley & Cacioppo, 2010). Interventions that promote tending and befriending behaviors, such as those that encourage women to use social support and reduce tendencies to isolate from others during stressful life circumstances, are likely to support long-term physical and emotional health for individuals struggling with stress.

Tending to an infant also contributes to the development and regulation of the infant's stress system, which has lifelong implications for child health, emotion regulation, and well-being. Early contact with a caregiver is essential for the proper development of an infant's stress-regulatory system (Repetti et al., 2002). Human infants become calm and cease crying after receiving milk, being touched, or sucking on a nipple from their mother (Blass, 1997; Field et al., 1996). In addition to behaviors that promote the early health and survival of an infant, tending behaviors from mothers also support the development of a secure attachment. Consistent with Bowlby's attachment theory discussed above (1988), a secure attachment is associated with several health benefits for offspring. For example, infants that are securely attached to their mothers are less likely to have persistently elevated cortisol response to threats when compared to infants that are less securely attached (Gunnar et al., 1996; Taylor, 2012). Warm and nurturant families foster secure attachments and better stress regulation, while less nurturant families or families with more conflict can lead to learning fewer socioemotional skills for managing stress (Taylor, 2012). Thus, interventions that foster tending behaviors may not only support maternal health and well-being but also foster health and well-being in children.

Limitations

Since its original development, researchers have identified some limits of the tend and befriend theory. In writing this theory, Taylor et al. (2000) recognized that tending and befriending as described did not bring into consideration female menstruation and the cyclical hormonal variations inherent in that process, and further work is needed to integrate hormonal variations into the model. In addition, there is also research suggesting tending and befriending may be mediated by different mechanisms in males. For example, increased testosterone may be related to increased parenting behaviors in males (Byrd-Craven et al., 2016; Fleming et al. 2002; Storey et al. 2000). This has led researchers to propose modifications to the theory to describe male tending and befriending (Geary & Flinn, 2002). Further research is needed to fully understand sex differences in tending and befriending as well as the contextual and physiological conditions that lead to the emergence of this form of stress responding in both sexes.

Summary

The tend and befriend theory was developed as an approach to studying female stress responding from an evolutionary and biosocial perspective. It attempts to explain why it may not be evolutionarily advantageous for female members of mammalian species to fight or flee from threat due to the negative consequence the behaviors may have on offspring. Instead, it is proposed that females respond to stress by tending to an offspring and affiliating with other female members of their species. At the core of this response is the presence of oxytocin. Oxytocin has been found to downregulate SNS and HPA axis arousal and promote both tending and befriending behaviors. Tend and befriend theory is also consistent with identified differences in male and female parental investment patterns, sex-segregated group formation patterns, and aggressive behavior patterns in polygamous species. Finally, tending and befriending in response to stress has positive benefits on physical and mental health by diminishing allostatic load (via downregulation of SNS activation in response to stress) and increasing motivation to seek social support when under stress. Through the study of the tend and befriend stress response, we can now also identify means of intervention that may facilitate tending and befriending responses to increase both child and maternal health and well-being. For example, mindfulness interventions may be an effective tool for increasing befriending (i.e., social connection; Hutcherson et al., 2008) and, when applied to the parenting context, increasing tending behaviors (Bögels et al., 2010). Further research is needed to explore the intersection of the tend and befriend theory and mindfulness intervention as mindfulness approaches may be particularly well suited to increasing tending and befriend stress responses in everyday life.

References

Adams, D. (1992). Biology does not make men more aggressive than women. In K. Bjorkqvist & P. Niemela (Eds.), *Of mice and women: Aspects of female aggression* (pp. 17–26). Academic Press.

Andersson, M. (1994). *Sexual selection*. Princeton University Press.

Baumeister, R. F., & Leary, M. R. (1995). The need to belong: Desire for interpersonal attachments as a fundamental human motivation. *Psychological Bulletin, 117*(3), 497–529. https://doi.org/10.1037/0033-2909.117.3.497

Belle, D. (1987). Gender differences in the social moderators of stress. In R. C. Barnett, L. Biener, & G. K. Baruch (Eds.), *Gender and stress* (pp. 257–277). Free Press.

Berkman, L. F., & Syme, S. L. (1979). Social networks, host resistance, and mortality: A nine-year follow-up study of Alameda County residents. *American Journal of Epidemiology, 109*, 186–204.

Blass, E. M. (1997). Infant formula quiets crying human newborns. *Journal of Developmental and Behavioral Pediatrics, IS*, 162–165.

Bögels, S. M., Lehtonen, A., & Restifo, K. (2010). Mindful parenting in mental health care. *Mindfulness, 1*, 107–120.

Bowlby, J. (1988). *A secure base: Parent-child attachment and healthy human development*. Basic Books.

Brain, P. F., Haug, M., & Parmigiani, S. (1992). The aggressive female rodent: Redressing a "scientific" bias. In K. Bjorkqvist & P. Niemela (Eds.), *Of mice and women: Aspects of female aggression* (pp. 27–36). Academic Press.

Brown, K. I., & Grunberg, N. E. (1995). Effects of housing on male and female rats: Crowding stresses males but calms females. *Physiology and Behavior, 58*, 1085–1089.

Byrd-Craven, J., Calvi, J. L., & Kennison, S. M. (2016). Rapid cortisol and testosterone responses to sex-linked stressors: Implications for the tend-and-befriend hypothesis. *Evolutionary Psychological Science, 2*(3), 199–206. https://doi.org/10.1007/s40806-016-0053-9

Campbell, A. (1999). Staying alive: Evolution, culture and intra-female aggression. *Behavioral and Brain Sciences, 22*, 203–252.

Cannon, W. B. (1932). *The wisdom of the body*. Norton.

Cardoso, C., Ellenbogen, M. A., Serravalle, L., & Linnen, A. M. (2013). Stress-induced negative mood moderates the relation between oxytocin administration and trust: Evidence for the tend-and-befriend response to stress? *Psychoneuroendocrinology, 38*(11), 2800–2804. https://doi.org/10.1016/j.psyneuen.2013.05.006

Carter, C. S. (1998). Neuroendocrine perspectives on social attachment and love. *Psychoneuroendocrinology, 23*, 779–818.

Chagnon, N. A. (1988, February 26). Life histories, blood revenge, and warfare in a tribal population. *Science, 239*, 985–992.

Collins, N. L., Dunkel-Schetter, C., Lobel, M., & Scrimshaw, S. C. M. (1993). Social support in pregnancy: Psychosocial correlates of birth outcomes and post-partum depression. *Journal of Personality and Social Psychology, 65*, 1243–1158.

Crick, N. R., Casas, J. F., & Mosher, M. (1997). Relational and overt aggression in preschool. *Developmental Psychology, 33*, 579–588.

Cross, S. E., & Madson, L. (1997). Models of the self: Self-construals and gender. *Psychological Bulletin, 122*, 5–37.

Cumming, D. C., Brunsting, L. A., Strich, G., Ries, A. L., & Rebar, R. W. (1986). Reproductive hormone increases in response to acute exercise in men. *Medical Science in Sports and Exercise, 18*, 369–373.

Darwin, C. (1871). *The descent of man, and selection in relation to sex*. John Murray.

David, D. H., & Lyons-Ruth, K. (2005). Differential attachment responses of male and female infants to frightening maternal behavior: Tend or befriend versus fight or flight? *Infant Mental Health Journal, 26*(1), 1–18. https://doi.org/10.1002/imhj.20033

Drago, F., Pederson, C. A., Caldwell, J. D., & Prange, A. I., Jr. (1986). Oxytocin potently enhances novelty-induced grooming behavior in the rat. *Brain Research, 368*, 287–295.

Dreifuss, J. J., Dubois-Dauphin, M., Widmer, H., & Raggenbass, M. (1992). Electrophysiology of oxytocin actions on central neurons. *Annals of the New York Academy of Science, 652*, 46–57.

Eisler, R., & Levine, D. S. (2002). Nurture, nature, and caring: We are not prisoners of our genes. *Brain and Mind, 3*(1), 9–52. https://doi.org/10.1023/A:1016553723748

Engelmann, M., Landgraf, R., & Wotjak, C. T. (2004). The hypothalamic- neurohypophysial system regulates the hypothalamic-pituitary- adrenal axis under stress: An old concept revisited. *Frontiers in Neuroendocrinology, 25*, 132–149.

Fahrbach, S. E., Morrell, J. I., & Pfaff, D. W. (1985). Possible role for endogenous oxytocin in estrogen-facilitated maternal behavior in rats. *Neuroendocrinology, 40*, 526–532.

Field, T. M., Schanberg, S., Davalos, M., & Malphurs, J. (1996). Massage with oil has more positive effects on normal infants. *Pre- & Peri-Natal Psychology Journal, 11*, 75–80.

Fleming, A. S., Corter, C., Stallings, J., & Steiner, M. (2002). Testosterone and prolactin are associated with emotional responses to infant cries in new fathers. *Hormones and Behavior, 42*(4), 399–413. https://doi.org/10.1006/hbeh.2002.1840

Fleming, A. S., Steiner, M., & Corter, C. (1997). Cortisol, hedonics, and maternal responsiveness in human mothers. *Hormones and Behavior, 32*, 85–98.

Fleming, R., Baum, A., Gisriel, M. M., & Gatchel, R. J. (1982). Mediating influences of social support on stress at Three Mile Island. *Journal of Human Stress, 8*, 14–22.

Ganz, F. D. (2012). Tend and befriend in the intensive care unit. *Critical Care Nurse, 32*(3), 25–34. https://doi.org/10.4037/ccn2012903

Geary, D. C., & Flinn, M. V. (2002). Sex differences in behavioral and hormonal response to social threat: Commentary on Taylor et al. (2000). *Psychological Review, 109*(4), 745–750. https://doi.org/10.1037/0033-295X.109.4.745

Girdler, S. S., Jamner, L. D., & Shapiro, D. (1997). Hostility, testosterone, and vascular reactivity to stress: Effects of sex. *International Journal of Behavioral Medicine, 4*, 242–263.

Gray, P. B., Jeffrey Yang, C.-F., & Pope, H. G. (2006). Fathers have lower salivary testosterone levels than unmarried men and married non- fathers in Beijing, China. *Proceedings of the Royal Society B: Biological Sciences, 273*(1584), 333–339. https://doi.org/10.1098/rspb.2005.3311

Gunnar, M. R., Brodersen, L., Krueger, K., & Rigatuso, J. (1996). Dampening of adrenocortical responses during infancy: Normative changes and individual differences. *Child Development, 67*, 877–889.

Hamilton, W. D. (1964). The genetical evolution of social behaviour II. *Journal of Theoretical Biology, 7*, 17–52.

Hawkley, L. C., & Cacioppo, J. T. (2010). Loneliness matters: A theoretical and empirical review of consequences and mechanisms. *Annals of Behavioral Medicine, 40*, 218–227.

Holmstrom, R. (1992). Female aggression among the great apes: A psychoanalytic perspective. In K. Bjorkqvist & P. Niemela (Eds.), *Of mice and women: Aspects of female aggression* (pp. 295–306). Academic Press.

Hutcherson, C. A., Seppala, E. M., & Gross, J. J. (2008). Loving-kindness meditation increases social connectedness. *Emotion, 8*, 720–724.

Janson, C. H. (1992). Evolutionary ecology of primate structure. In E. A. Smith & B. Winterhalder (Eds.), *Evolutionary ecology and human behavior* (pp. 95–130). Aldine.

Jezova, D., Jurankova, E., Mosnarova, A., Kriska, M., & Skultetyova, I. (1996). Neuroendocrine response during stress with relation to gender differences. *Acta Neurobiotogae Experimental, 56*, 779–785.

King, K. B., Reis, H. T., Porter, L. A., & Norsen, L. H. (1993). Social support and long-term recovery from coronary artery surgery: Effects on patients and spouses. *Health Psychology, 12*, 56–63.

Kivlighan, K. T., Granger, D. A., & Booth, A. (2005). Gender differences in testosterone and cortisol response to competition. *Psychoneuroendocrinology, 30*(1), 58–71.

Lang, R. E., Heil, J. W. E., Ganten, D., Hermann, K., Unger, T., & Rascher, W. (1983). Oxytocin unlike vasopressin is a stress hormone in the rat. *Neuroendocrinology, 37*, 314–316.

Levy, K. N., Hlay, J. K., Johnson, B. N., & Witmer, C. P. (2019). An attachment theoretical perspective on tend-and-befriend stress reactions. *Evolutionary Psychological Science, 5*(4), 426–439. https://doi.org/10.1007/s40806-019-00197-x

Lukas, M., Toth, I., Reber, S. O., Slattery, D. A., Veenema, A. H., & Neumann, I. D. (2011). The neuropeptide oxytocin facilitates pro-social behavior and prevents social avoidance in rats and mice. *Neuropsychopharmacology, 36*, 2159–2168.

Maccoby, E. E., & Jacklin, C. H. (1974). *The psychology of set differences*. Stanford University Press.

Mathur, D. N., Toriola, A. L., & Dada, O. A. (1986). Serum cortisol and testosterone levels in conditioned male distance runners and non-athletes after maximal exercise. *Journal of Sports Medicine and Physical Fitness, 26*, 245–250.

McCarthy, M. M. (1995). Estrogen modulation of oxytocin and its relation to behavior. *Advances in Experimental Medicine and Biology, 395*, 235–245.

McDonald, L. M., & Korabik, K. (1991). Sources of stress and ways of coping among male and female managers. *Journal of Social Behavior and Personality, 6*, 185–198.

McEwen, B. S. (1998). Protective and damaging effects of stress mediators. *New England Journal of Medicine, 338*, 171–179.

McEwen, B. S., & Stellar, E. (1993). Stress and the individual. Mechanisms leading to disease. *Archieves of Internal Medicine, 153*(18), 2093–2101.

Morrison, R. L. (2009). Are women tending and befriending in the work- place? Gender differences in the relationship between workplace friendships and organizational outcomes. *Sex Roles, 60*(1–2), 1–13. https://doi.org/10.1007/s11199-008-9513-4

Neumann, I. D. (2009). The advantages of social living: Brain neuro- peptides mediate the beneficial consequences of sex and motherhood. *Frontiers in Neuroendocrinology, 30*, 483–496.

Nickels, N., Kubicki, K., & Maestripieri, D. (2017). Sex differences in the effects of psychosocial stress on cooperative and prosocial behavior: Evidence for 'flight or fight' in males and 'tend and befriend' in females. *Adaptive Human Behavior and Physiology, 3*(2). https://doi.org/10.1007/s40750-017-0062-3

Nishida, T., & Hiraiwa-Hasegawa, M. (1987). Chimpanzees and bonobos: Cooperative relationships among males. In B. B. Smuts, D. L. Cheney, R. M. Seyfarth, R. W. Wrangham, & T. T. Struhsaker (Eds.), *Primate societies* (pp. 165–177). The University of Chicago Press.

Panksepp, J. (1998). *Affective neuroscience*. Oxford University Press.

Probst, F., Meng-Hentschel, J., Golle, J., Stucki, S., Akyildiz-Kunz, C., & Lobmaier, J. S. (2017). Do women tend while men fight or flee? Differential emotive reactions of stressed men and women while viewing newborn infants. *Psychoneuroendocrinology, 75*, 213–221. https://doi.org/10.1016/j.psyneuen.2016.11.005

Repetti, R. L. (1989). Effects of daily workload on subsequent behavior during marital interactions: The role of social withdrawal and spouse support. *Journal of Personality and Social Psychology, 57*, 651–659.

Repetti, R. L., Taylor, S. E., & Seeman, T. E. (2002). Risky families: Family social environments and the mental and physical health of offspring. *Psychological Bulletin, 128*, 330–366.

Rubenstein, D. E. (1978). On predation, competition, and the advantages of group living. In P. P. G. Bateson & P. H. Klopfer (Eds.), *Perspectives in ethnology* (Vol. 3, pp. 205–231). Plenum Press.

Sarason, B. R., Sarason, I. G., & Gurung, R. A. R. (1997). Close personal relationships and health outcomes: A key to the role of social support. In S. Duck (Ed.), *Handbook of personal relationships* (pp. 547–573). Wiley.

Sawchenko, P. E., & Swanson, L. W. (1982). Immunohistochemical identification of neurons in the paraventricular nucleus of the hypothalamus that project to the medulla or to the spinal cord in the rat. *Journal of Comparative Neurology, 205*, 260–272.

Seeman, T. E., Lusignolo, T. M., Albert, M., & Berkman, L. (2001). Social relationships, social support, and patterns of cognitive aging in healthy, high-functioning older adults: MacArthur studies of successful aging. *Health Psychology, 20*, 243–255.

Silk, J. B. (2000). Ties that bond: The role of kinship in primate societies. In L. Stone (Ed.), *New directions in anthropological kinship*. Rowman and Littlefield.

Smeets, T., Dziobek, I., & Wolf, O. T. (2009). Social cognition under stress: Differential effects of stress-induced cortisol elevations in healthy young men and women. *Hormones and Behavior, 55*(4), 507–513. https://doi.org/10.1016/j.yhbeh.2009.01.011

Stallings, J. F., Fleming, A. S., Worthman, C. M., Steiner, M., Corter, C., & Coote, M. (1997). Mother/father differences in response to infant crying. *American Journal of Physical Anthropology, 24*, 217.

Steinbeis, N., Engert, V., Linz, R., & Singer, T. (2015). The effects of stress and affiliation on social decision-making: Investigating the tend-and-befriend pattern. *Psychoneuroendocrinology, 62*(2015), 138–148.

Storey, A. E., Walsh, C. J., Quinton, R. L., & Wynne-Edwards, K. E. (2000). Hormonal correlates of paternal responsiveness in new and expectant fathers. *Evolution and Human Behavior, 21*, 79–95.

Tamres, L., Janicki, D., & Helgeson, V. S. (2002). Sex differences in coping behavior: A meta-analytic review. *Personality and Social Psychology Review, 6*, 2–30.

Taylor, S. E. (2002). *The tending instinct: How nurturing is essential to who we are and how we live*. Holt.

Taylor, S. E. (2006). Tend and Befriend. *Current Directions in Psychological Science, 15*(6), 273–277. https://doi.org/10.1111/j.1467-8721.2006.00451.x

Taylor, S. E. (2012). Tend and befriend theory. In Intergovernmental Panel on Climate Change (Ed.), *Handbook of theories of social psychology: Volume 1* (pp. 32–49). Sage. https://doi.org/10.4135/9781446249215.n3

Taylor, S. E., Klein, L. C., Lewis, B. P., Gruenewald, T. L., Gurung, R. A. R., & Updegraff, J. A. (2000). Biobehavioral responses to stress in females: Tend-and-befriend, not fight-or-flight. *Psychological Review, 107*(3), 411–429. https://doi.org/10.1037/0033-295X.107.3.411

Taylor, S. E., Saphire-Bernstein, S., & Seeman, T. E. (2010). Are plasma oxytocin in women and plasma vasopressin in men biomarkers of distressed pair-bond relationships? *Psychological Science, 21*(1), 3–7. https://doi.org/10.1177/0956797609356507

Trivers, R. L. (1972). Parental investment and sexual selection. In B. Campbell (Ed.), *Sexual selection and the descent of man 1871–1971* (pp. 136–179). Aldine Publishing.

Uvnas-Moberg, K., Arn, I., & Magnusson, D. (2005). The psychobiology of emotion: The role of oxytocinergic system. *International Journal of Behavioral Medicine, 12*, 59–65.

Wheeler, G., Gumming, D., Burnham, R., Maclean, I., Sloley, B. D., Bhambhani, Y., & Steadward, R. D. (1994). Testosterone, cortisol and catecholamine responses to exercise stress and autonomic dysreflexia in elite quadriplegic athletes. *Paraplegia, 32*, 292–299.

Wrangham, R. W. (1980). An ecological model of female-bonded primate groups. *Behaviour, 75*, 262–300.

Williams, G. C. (1966). *Adaptation and natural selection: A critique of some current evolutionary thought*. Princeton University Press.

Wills, T. A. (1991). Social support and interpersonal relationships. In M. S. Clark (Ed.), *Prosocial behavior* (pp. 265–289). Sage.

Chapter 4
Psychoneuroimmunology: How Chronic Stress Makes Us Sick

Andrew W. Manigault and Peggy M. Zoccola

Introduction to Stress and the Immune System

The fact that stress is linked to poor health is well-established. Individuals who report high levels of stress tend to also display a variety of poor health outcomes. For example, stress is associated with depression (Hammen, 2005), cardiovascular disease (Black & Garbutt, 2002), childhood asthma (Bloomberg & Chen, 2005), autoimmune diseases (Elenkov & Chrousos, 2002), HIV progression (Evans et al., 1997), and cancer (Godbout & Glaser, 2006). Moreover, particularly stressful events like the death of a spouse are associated with increased mortality (Bloom et al., 1978). However, not all stressors lead to disease and death. To the contrary, some forms of stress may enhance survival and promote positive outcomes. For example, acute stress can increase blood sugar levels, thus fueling the brain and body to deal with ongoing threats. To understand how stressors can have positive and negative effects on the body, we must acknowledge that not all stressors are created equal or rely on the same mechanisms to influence health.

The aim of this chapter is to review evidence linking long-term or chronic stressors to disease via their effects on the immune system. To understand the importance of focusing on chronic stressors and immune outcomes, a few questions must first be answered: (1) How are chronic stressors different from other forms of stress, and why is this distinction important? (2) Why is the immune system so important to health? To this end, we start by reviewing the concept of stress and pathways connecting stressors to health. Next, we provide an overview of the immune system and review prior work linking chronic stress to immune outcomes, including responses to vaccination, systemic inflammation, cellular aging, and other immune functions. We conclude with a summary of extant findings and discussion of future directions, including factors that may buffer against the effects of chronic stress on immunity.

A. W. Manigault · P. M. Zoccola (✉)
Ohio University, Athens, OH, USA
e-mail: zoccola@ohio.edu

© Springer Nature Switzerland AG 2021
H. Hazlett-Stevens (ed.), *Biopsychosocial Factors of Stress, and Mindfulness for Stress Reduction*, https://doi.org/10.1007/978-3-030-81245-4_4

The Concept of Stress

The term "stress" has been simultaneously used to refer to stressful stimuli, or stressors, as well as stress responses and stress-related appraisals, thus leading to some confusion. More formally, stress has been defined as "a process in which environmental demands tax or exceed the adaptive capacity of an organism, resulting in psychological and biological changes that may place persons at risk for disease" (Cohen et al., 1995, p. 3). According to the transactional model of stress (Lazarus & Folkman, 1984), situations are appraised as irrelevant, benign/positive, or potentially stressful. A potentially stressful encounter is further evaluated with respect to its potential costs and the resources that can be allocated toward managing this encounter. When perceived costs exceed resources, the event is perceived to be more threatening. From this perspective, the degree to which an event is perceived as stressful and threatening depends on appraisals of that situation and appraisals of resources available to manage it. Perceived stress, in turn, can elicit emotional and physiological responses, which may confer risk for illness. For example, perceived stress has been linked with increased susceptibility to the common cold (Cohen et al., 1993).

Nevertheless, it is possible to study stress and health without measuring stress appraisals. For example, some researchers have focused on identifying events hypothesized to be overwhelming for the majority of individuals (i.e., the epidemiological approach; for a review, see Cohen et al. (2016) and Cohen et al. (2007a)). This approach has led to the creation of life event scales which aim to measure stress by counting the number of stressful events (e.g., death of a close family member) which occurred during a predetermined time period (e.g., 12 months; Holmes & Rahe, 1967). Moreover, this line of research has been successful in linking groups of individuals thought to be under considerable pressure to a variety of health outcomes. For example, caring for a spouse with dementia has been linked to multiple forms of immune dysfunction that can place an individual at risk for disease (Gouin et al., 2008). In summary, both individuals' perceptions of stressors and independently defined stressful events may lead to an increased risk for disease.

Pathways Linking Stress to Illness

Typically, stressors are thought to influence health via their effects on health behaviors and physiological stress response systems, including the immune system. For example, women who report a history of sexual assault also report greater substance use, increased frequency of risky sexual behaviors, and decreased exercise (Lang et al., 2003). Thus, to the extent that stressors can influence the frequency of health-related behaviors like diet, exercise, or drug use, they can influence health (see Park & Iacocca, 2014, for a more detailed discussion of this pathway).

Moreover, stress can cause disease via its effects on biological stress responses. When threatened, the body undergoes a wide variety of functional changes. These changes are mediated by the interconnection of the nervous system, the endocrine system, and the immune systems (Glaser & Kiecolt-Glaser, 2005; Sarafino et al., 2008). For example, stressors have been shown to alter heart rate, blood pressure, cellular growth, and blood sugar levels (Sarafino et al., 2008). Among the biological systems involved in stress responding, the sympathetic-adrenal-medullary (SAM) and hypothalamic-pituitary-adrenocortical (HPA) axes are considered to be so intimately connected with the stress response that some have defined stressors as "a stimulus that activates the hypothalamic–pituitary–adrenal (HPA) axis and/or the sympathetic nervous system (SNS)" (Glaser & Kiecolt-Glaser, 2005. p. 243). This is because activation of the SAM and HPA axes results in the production of catecholamines (i.e., epinephrine and norepinephrine), adrenocorticotropic hormone, cortisol, growth hormone, and prolactin (Glaser & Kiecolt-Glaser, 2005). These hormonal stress mediators interact with an array of physiological systems, including the immune system (Black, 2003), and can promote survival when properly controlled or disease susceptibility when poorly regulated.

Prominent stress-health theories posit that repeated or excessive exposure to primary stress mediators of the SAM and HPA axis (e.g., epinephrine, norepinephrine, and cortisol) is the primary mechanism via which stress leads to disease (McEwen, 1998). More specifically, the allostatic load model states that adaptation in response to stressful circumstances involves activation of neural, neuroendocrine, and immune systems (e.g., the SAM and HPA axes) so as to allow the body to cope with challenges that it may not survive otherwise. For example, secretion of primary stress mediators leads to increased blood pressure, blood sugar levels, and analgesia, thereby promoting survival functions (e.g., fighting off a predator; Everly & Lating, 2013). However, under certain circumstances (e.g., repeated activation), the SAM and HPA axes may produce functional changes in organs regulated by hormones of the SAM and HPA axes (e.g., thickening of blood vessels) and in term contribute to disease. For example, primary stress mediators also influence the functioning of immune cells (e.g., cytotoxic T cells), and, when prolonged, this effect can increase the risk of developing infectious diseases (Glaser & Kiecolt-Glaser, 2005). Similarly, repeated cardiovascular activation is thought to cause hypertension via structural adaptation of blood vessels (i.e., vessel wall thickening; Johsson & Hansson, 1977). In sum, the cumulative effects of stressors on biological systems are thought to lead to disease susceptibility via excessive exposure to primary stress mediators.

However, the cumulative effect of stressors on health as outlined by the allostatic load model does not apply uniformly to all stressors or all individuals. Consistent with Lazarus and Folkman's approach to understanding stress, individual appraisals can influence this relationship. For example, perceptions of high status inhibit responses to stress (Adler et al., 2000; Akinola & Mendes, 2014) and, thus, limit the rate of stress-related disease in high-status populations (Adler et al., 1994; Sapolsky, 2005). In addition, stressor characteristics can influence biological response patterns (e.g., Dickerson and Kemeny, 2004, and Segerstrom and Miller, 2004). For example, pain-inducing physical stressors (e.g., submerging ones that had in ice water) reliably lead to the secretion of epinephrine and norepinephrine (Biondi & Picardi, 1999), whereas psychological stressors characterized by social evaluative threat, or events that challenge self-esteem and social status (e.g., a hostile job interview), reliably lead to increased cortisol secretion (Dickerson & Kemeny, 2004; Dickerson et al., 2008). In sum, some stressors are more potent activators of certain physiological stress responses than others and, thus, may contribute more significantly to physical "wear and tear" and disease.

Why Is Chronic Stress Toxic?

Yet, a single short-lived stressor is unlikely to lead to disease. Instead, excessive exposure to primary stress mediators is a lengthy process, whereby recurrent or ongoing stressors (or mental representation of stressors; Brosschot et al., 2006) repeatedly activate the SAM and HPA axes. Consistent with this view, mounting evidence suggests that the *duration* of a stressor is an important determinant of its eventual health effects (Dhabhar, 2014; Juster et al., 2010; Mcewen, 2004; Segerstrom & Miller, 2004; Taylor et al., 2008). Acute stressors are a relative short-term, non-recurring, or low-frequency events (e.g., a single argument with a spouse) and typically do not continue to be appraised as overwhelming long after they have ended (Cohen et al., 2007a, b). In contrast, chronic stressors include persistent events (e.g., caring for a sick spouse), frequent or recurring forms of acute stressors (e.g., daily arguments with a spouse), and non-recurring events that continue to be experienced as overwhelming long after they have ended (e.g., sexual assault).

The effect of acute stressors on health is commonly studied in humans using laboratory tasks like the Trier Social Stress Test (Kirschbaum et al., 1993), which combines public speaking with mental arithmetic or short naturalistic stressors like academic examinations. In contrast, the study of chronic stress is restricted to naturalistic studies among humans. Some work on chronic stress has focused on measuring self-reported major life events and daily hassles over a specific period of time (Dimsdale et al., 1994; Miller et al., 2004). Additional work has focused on traumatic stressors like natural disasters (Solomon et al., 1997). Most notably, chronic stress has been measured by examining populations facing ongoing challenges like individuals caring for a spouse suffering from dementia (Kiecolt-Glaser et al., 1987), mothers of pre-term very-low-birthweight infants (Gennaro et al., 1997b), professional soldiers (Lauc et al., 1998), prisoners of war (Dekaris, 1993), unemployed adults (Ockenfels et al., 1995), bereaved individuals (Kemeny et al., 1994), and victims of child abuse (Felitti et al., 1998). Among these chronically stressed populations, spousal dementia caregivers are among the most extensively studied. Caring for a spouse with dementia is thought to be particularly challenging because dementia caregivers find themselves providing care for over 10 h per day and may do so for over 5 years (Donelan et al., 2002; Wimo et al., 2002). Moreover, dementia caregivers face significant distress as they see their spouse slowly lose their personality and intellect (Kiecolt-Glaser et al., 1991).

The distinction between acute and chronic stresses is particularly important in the context of immune function. Among the work examining the effect of acute and chronic stressors on immunity, some of the most compelling evidence comes from rodent studies that manipulated both acute and chronic stresses. In a study by Dhabhar and McEwen (1997), rats were restrained, shook, or both restrained and shook (a stressor nicknamed the "New York City Subway Stress") for 2 h to manipulate varying intensities of acute stress. Chronic stress was manipulated by applying a random sequence of restraint, shaking, or restraint and shaking each hour, 6 h per day, for 3 weeks. Rats which were acutely stressed showed improvement in some aspects of their immune response, whereas chronically stressed rats showed widespread immunodeficiency. This work is consistent with reviews of human studies suggesting that chronic and acute stresses produce different immune effects (Segerstrom & Miller, 2004).

Why Focus on the Immune System?

Maintaining a functional immune system is essential to survival. The immune system serves to protect organisms against external threats like virus and bacteria as well as internal threats like cancerous cells. When one or more components of an organism's immune system is no longer active, that organism is said to be immunodeficient and is subject to increased frequency or complications of common infections (Chinen & Shearer, 2010). In extreme cases, immunodeficiency can lead to death. For example, the majority individuals infected by human immunodeficiency virus (HIV) are expected to die within 2 years of the onset of acquired immunodeficiency syndrome (AIDS) if they remain untreated (Poorolajal et al., 2016). It is also possible for some components of the immune system to become too active and damage cells of the body (i.e., autoimmune disease, chronic inflammation). For example, in rheumatoid arthritis, the immune system is overactive and wrongfully attacks the body by degrading cartilage and bone tissue in articulations. Like AIDS, some diseases resulting from immune overactivity can lead to death (e.g., multiple sclerosis; Brønnum-Hansen et al., 2004). Thus, a balance in immune activity is necessary to maintain health. Indeed, dysregulated immune function implicates the development of many major and potentially fatal diseases, including cardiovascular disease (Hansson & Hermansson, 2011), cancer (de Visser et al., 2006), and diabetes (Pickup & Crook, 1998). In summary, to the extent that chronic stress can influence the immune system, it yields a powerful influence on health and well-being.

The Immune System

Overview

Understanding the effects of stressors on immune functioning relies to some degree on understanding how the immune system functions under normal conditions. The present chapter only briefly reviews major components of the immune system to provide a general frame of reference for studies linking chronic stress to immunity. However, the human immune system is remarkably complex, and those wishing to fully appreciate the complexity of the immune system may benefit from reviewing textbooks which more closely focus on the subject (e.g., Abbas et al., 2014, and Daruna, 2012).

The immune system is a complex array of cells, proteins, and physical barriers aimed at protecting the body against pathogens or foreign proteins, viruses, bacteria, parasites, and fungi that can cause illness and result in death. Over many generations, the human body has acquired numerous mechanisms to keep pathogens out of the body or prevent them from causing harm. The skin, gastrointestinal tract, respiratory tract, nasopharynx, cilia, eyelashes, and body hair all serve to physically prevent pathogens from entering or infecting the body. In addition, the body secretes a variety of substances to hinder pathogen entry. For example, the stomach tends to produce an acidic chemical environment which is hostile to foreign bacteria. If a foreign pathogen manages to penetrate physical barriers, cells and proteins of the innate immune system promptly mobilize to neutralize it. Finally, pathogens are also targeted by the adaptive immune system which mobilizes cell-mediated and antibody-mediated responses to neutralize the pathogen on initial exposure and improve the body's defense against future exposure to the same pathogen. The immune system can therefore be divided in two parts: the innate and adaptive immune system. These branches of the immune system rely on distinct strategies to protect the body against pathogens and are considered complementary.

The Innate Immune System

The innate immune system is often called non-specific because the proteins and cells that comprise the innate immune system respond to a broad class of foreign pathogens but do not typically attempt to recognize or adapt their response to a specific type of pathogen. This is in contrast to the adaptive immune response which develops a form of cellular memory aimed at protecting the body against future exposure to a unique pathogen. Albeit non-specific, the innate immune system is a powerful first line of defense. Within hours of infection, cells and proteins of the innate immune system will migrate to the site of infection, ingest pathogens, and coordinate a wide array of cell-to-cell signaling.

Cells of the innate immune system include macrophages, mast cells, neutrophils, eosinophils, basophils, natural killer cells, and dendritic cells. Accumulation of these cells at the site of infection along with increased vascular dilation and permeability is part of inflammation (Abbas et al., 2014). Many of these cells (i.e., neutrophils, monocytes, and dendritic cells) travel to the site of infection to ingest and kill pathogens (via a process known as phagocytosis). For example, neutrophils are mass produced in the bone marrow following infection and promptly move to the site of infection where they ingest microbes for intracellular killing. In addition, damaged cells and some cells of the innate immune system (e.g., macrophages) produce soluble proteins known as cytokines. Cytokines like interleukin-1 and tumor necrosis factor alpha (TNF-α) circulate in the blood and attract other macrophages to the site of infection, stimulating inflammation. In addition, mast cells produce other signaling molecules known as anaphylatoxins (e.g., histamine, serotonin, and prostaglandins) which increase vascular dilation and permeability to ease the movement of immune cells to the site of infection. Thus, the innate immune response could be summarized as a form of cellular redeployment or migration to the site of infection aided by a variety of chemical alarm signals.

The innate immune response is not exclusively carried out by cells. Another notable component of the innate immune system is the complement cascade. The complement cascade is an array of proteins which are activated following infection and circulate in the blood to supplement other parts of the immune response. The complement cascade is able to mark pathogens and infected cells for ingestion, recruit neutrophils to the site of infection, break down the membrane of some pathogens (e.g., bacteria), and cluster pathogens together to ease the action of other immune cells.

Finally, the innate immune system also plays an important role in initiating the adaptive immune response via a process known as antigen presentation. After ingesting and breaking down the ingested pathogen into small protein fragments, cells of the innate immune system (e.g., dendritic cells and macrophages) will present these fragments on the surface of their membrane. The protein fragment then becomes visible to some cells of the adaptive immune system, including T cells. Numerous T cells are then able to bind to the protein fragments and check if they produce a matching protein or the one that physically fits the pathogen fragment. If they do, they will proliferate, thus initiating the cell-mediated component of the adaptive immune response.

The Adaptive Immune System

The hallmark of adaptive immunity is the ability to identify and neutralize specific pathogens. Humoral, or antibody-mediated, responses rely on the ability of B cells to identify specific pathogens using antibodies, or Y-shaped proteins, which bind to part of a pathogen (e.g., a foreign surface protein). The part of a pathogen which binds to an antibody is called an antigen. Much like a lock and a key, a given antibody can only bind to a relatively unique protein structure and thus is considered specific to an antigen. The variable part of antibodies can take on many shapes (10^{10} possible combinations) because the portion of DNA which codes for the variable portion of antibodies is shuffled during the development of B cells. As a result, the body produces a diverse population of B cells which each can only react with a unique protein structure. B-cell activation therefore relies on a potentially lengthy trial-and-error process whereby various antibody-producing B cells try to bind with an antigen until a match is produced. Once a naïve B cell has found an antigen matching the membrane-bound antibody it produces, that B cell will activate (with the help of other immune cells) and rapidly multiply in the form of plasma cells and memory B cells. Plasma cells secrete large quantities of antibodies which travel through blood and lymph to neutralize target antigens. Memory B cells remain dormant until the pathogen presents itself again. Memory B cells can therefore accelerate the antibody-mediated immune response as soon as the second exposure to the same pathogen. The cellular memory of B cells is the principle mechanisms underlying the use of vaccine where a weakened or inactive virus is injected in an organism to activate its "matching" B cell and protect the organism against future exposure to the active virus. Circulating antibodies protect against infections by blocking parts of a pathogen needed for it to function (neutralization), clustering pathogens together (thus facilitating the ingestion of pathogens by other immune cells through the process called phagocytosis), and activating the complement cascade.

Another important function of B cells (along with dendritic cells and macrophages) is antigen presentation. The cell-mediated adaptive immune response relies on antigen-presenting cells to interact with the pathogen. More specifically, helper T cells and cytotoxic T cells produce T-cell receptor which acts much like antibodies in that they express a highly variable structure and thus can bind to a specific antigen. However, the receptors of helper T cells and cytotoxic T cells cannot bind to free (i.e., non-ingested) pathogens; they can only bind to pathogen fragments processed by professional antigen-presenting cells (e.g., B cells, dendritic cells, or macrophages). Once again, through a process of trial and error, numerous naïve T cells will attempt to bind to the antigen displayed by antigen-presenting cells, and once a match is found, these T cells will activate. Activated T cells will proliferate in the form of effector T cells and memory T cells for current and future action.

Summary and Implications

In summary, the innate immune system protects the body from a broad range of pathogens via inflammation, whereas the adaptive immune system relies on humoral and cell-mediated responses to neutralize a unique pathogen on initial exposure and improve immune resistance to future encounters with the same pathogen. This process relies on the interaction of numerous cells and highlights the degree of interconnectivity present in the human immune system. However, this interconnectivity is not limited to the immune system. The central nervous system and neuroendocrine system also participate in the chemical cross talk carried out by the immune system. Psychoneuroimmunology is the study of this cross talk or the interdisciplinary research field that addresses the interactions of the central nervous system, the neuroendocrine system, and the immune system

(Glaser & Kiecolt-Glaser, 2005). A major focus of psychoneuroimmunology research is on the topic of chronic stress and immune function.

Chronic Stress and Immune Function

Chronic stress may influence immune function by increasing the production of primary stress mediators (e.g., epinephrine, norepinephrine, and cortisol) over long periods of time. The complexity of the immune system provides researchers with numerous opportunities to assess the effect of chronic stress on immune function. For example, measuring the quantity of antibodies secreted in response to a vaccine can serve as a valuable indicator of the antibody-mediated response. Similarly, antibody secretion to latent infections (e.g., herpesvirus), immune cell proliferation (e.g., NK cells), and cytokine production (e.g., TNF-α) have all been used to infer the activity of distinct components of the immune system (e.g., Kiecolt-Glaser et al., 1987; Nakano et al., 1998; and Vedhara et al., 1999). Although animal researchers are able to manipulate stressors' intensity and duration to examine the effects of chronic stress on immune function (e.g., Dhabhar and Mcewen, 1997), researchers examining the association between chronic stress and immunity in humans have had to rely on different methodology. For example, prior work has examined the association between self-reported stress and immune function (e.g., Miller et al., 2004) or compared groups of people facing chronic stressors to groups of individuals facing no chronic stress (e.g., Kiecolt-Glaser et al., 1991). Fewer have followed chronically stressed individuals and longitudinally examined immune outcomes (e.g., comparing immune function of unemployed adults before and after they obtained a new job; Cohen et al., 2007a, b). As a result, the current state of psychoneuroimmunology literature suggests a fairly reliable association between chronic stress and immune function in humans (rather than causation per se). To illustrate the association between chronic stress and immune function, we review evidence linking chronic stress to vaccine response, immune control over latent infections, lymphocyte proliferation, natural killer cell activity, systemic inflammation, wound healing, and cellular aging of immune cells.

Vaccine Response

Vaccines serve to protect individuals/organisms by eliciting humoral and cell-mediated immune responses to a weakened or inactive virus, which generate memory cells that allow for faster and more robust immune responses to future infections. As such, measuring the degree to which individuals mount an immune response to vaccination can serve as a valuable indicator of immune functioning, whereby larger responses indicate stronger immunity. Multiple studies have linked chronic stress to immune responses to vaccination. For example, a study of 32 individuals caring for a spouse with dementia showed that dementia caregivers had diminished responses to an influenza vaccine relative to 32 control (non-caregiving) participants of similar age, sex, and socioeconomic status (Kiecolt-Glaser et al., 1996). More specifically, participants in this study were administered an influenza vaccine and provided periodic blood draws for up to 6 months post-vaccination to assess immune responses. Results showed that dementia caregivers showed significantly lower increases in interleukin-1β and interleukin-2 relative to the control (non-caregiving) group, suggesting that chronically stressed older adults may suffer from diminished natural and specific immune responses to vaccines. Similarly, some work indicates that spousal dementia caregivers have a reduced antibody response to an influenza vaccine relative to non-caregiving controls (Vedhara et al., 1999). Additional work implies that dementia caregivers do not continue to produce viral antibodies following vaccination for as long as non-chronically stressed (control) participants (Glaser et al., 2000), suggesting that the protective effect of vaccination may subside more rapidly in chronically stressed older adults compared to their non-stressed counterparts. Chronic stress also influences vaccine responses in young adults (Miller et al., 2004). In one such study, healthy young adults (N = 83) were followed for a 13-day period, during which they were vaccinated for influenza (on day 3) and reported the degree to which they felt stressed or overwhelmed daily. Young adults who reported being more stressed across the entire 13-day observation period were found to produce fewer antibodies for the influenza virus. In summary, chronic stress is associated with reduced immune responses to vaccination.

Control over Latent Viral Infections

Herpesvirus, including the Epstein-Barr virus, the herpes simplex virus type I, and varicella-zoster virus, are common (Glaser & Kiecolt-Glaser, 1994; Nahmias & Roizman, 1973) and are able to permanently infect a host by copying their viral sequence into the host's DNA. However, latent viruses are not always active; the immune system generally keeps such viruses in a dormant state by inhibiting viral replication. When immunity is compromised, the latent virus can reactivate; in turn, the body increases its production of antibodies to neutralize the active virus. As such, monitoring the degree to which individuals produce antibodies specific to latent viruses can serve as a valuable indicator of immune function, whereby greater antibody production to latent viruses implies compromised immunity for the host.

Chronic stress predicts antibody production to latent herpesvirus (i.e., compromised immunity) in multiple studies (Kiecolt-Glaser et al., 1991; Kiecolt-Glaser et al., 1987; McKinnon et al., 1989). In one cross-sectional study, individuals caregiving for a family member with dementia displayed elevated Epstein-Barr virus antibody production relative to sociodemographically matched non-caregiving controls (Kiecolt-Glaser et al., 1987). Similarly, a longitudinal study assessing Epstein-Barr virus antibody production over two measurements (13 months apart on average) found that dementia caregivers displayed elevated viral antibody production during the second measurement relative to non-caregiving control participants who were matched on the basis of age, education, and income (Kiecolt-Glaser et al., 1991). Finally, residents living within 5 miles of the damaged Three Mile Island nuclear power plant (a population thought to be chronically stressed due to their proximity with a damaged nuclear plant; Baum et al. , 1983) exhibited increased antibody production to herpes simplex virus type I relative to control participants living more than 80 miles away (McKinnon et al., 1989). Taken together, these studies show that the immune systems of chronically stressed individuals may be impaired in their ability to maintain herpesviruses in a dormant state.

Lymphocyte Proliferation

In response to immune challenge, such as exposure to bacterial toxins, numerous cells of the innate and adaptive immune system must undergo rapid cell division (i.e., cell proliferation). As a result, researchers have measured the rate of immune cell proliferation to broadly infer immune function among individuals, with greater lymphocyte proliferation indicating better immune functioning. Among this line of work, some have observed lower lymphocyte (i.e., T cells, B cells, and NK cells) proliferation in chronically stressed populations. For example, one study examined lymphocyte proliferation among Japanese taxi drivers under conditions of low economic strain as well as under conditions of high economic strain and healthy control participants (Nakano et al., 1998). Results implied that lymphocyte proliferation was comparable for control participants and taxi drivers under low economic strain. However, taxi drivers operating under high economic strain displayed reduced lymphocyte proliferation relative to controls. Similarly, multiple other studies have found that dementia caregivers tend to display lower lymphocyte proliferation than non-caregiving controls (Bauer et al., 2000; Cacioppo et al., 1998; Fonareva et al., 2011; Kiecolt-Glaser et al., 1991). Finally, lower lymphocyte proliferation is also observed among chronically stressed populations including mothers of pre-term low-birthweight infants (Gennaro et al., 1997a) and women who were forced out of their home in a war-ravaged region (Sabioncello et al., 2000). In summary, chronically stressed individuals may be less able to mass produce immune cells through lymphocyte proliferation during acute infection.

Natural Killer Cell Activity

Natural killer (NK) cells owe their name to their ability to bind to cancerous and virus-infected cells without antigen stimulation to induce cell death. Nearly all cells of the body present fragments of the proteins they produce on their surface such that most cells of the immune system can identify them as foreign or non-foreign (self) cells. For example, when an infected cell produces viral protein fragments, these fragments are presented on the cell's surface and allow cytotoxic T cells to recognize that the cell is infected. Related NK cells rely on a similar mechanism to identify infected cells but can trigger cell death even when cells stop presenting protein fragments on their surface. This ability allows NK cells to identify and kill some cancerous cells. NK cells also secrete antimicrobial molecules which kill bacteria by disrupting their cell wall (Iannello et al., 2008). NK cells are therefore an important part of the innate immune system because they can protect the body immediately and with no prior exposure to pathogen against a wide variety of foreign or infected cells as well as cancerous cells.

As such, the ability of NK cells to destroy foreign cells (i.e., NK cell activity or NK cell cytotoxicity) serves as a valuable indicator to the innate immune response. In the context of chronic stress, NK cell activity has been examined in a variety of chronically stressed populations (Dekaris, 1993; Lutgendorf et al., 1999; Vitaliano et al., 1998). For example, war prisoners suffering from prolonged stress and malnutrition were found to show decreased NK cell cytotoxicity and decreased phagocytic functions of ingestion and digestion (Dekaris, 1993), suggesting that their NK cells were less able to damage, ingest, and digest foreign/cancerous cells. Similar links between chronic stress and decreased NK cell activity have been documented in the context of older adults undergoing voluntarily housing relocation (Lutgendorf et al., 1999) and among dementia caregivers with a history of cancer (Vitaliano et al., 1998). Finally, a study that compared mothers of pre-term low-birthweight infants to mothers of healthy full-term infants showed that pre-term mothers exhibited decreased NK cell activity relative to mothers of healthy infants after delivery (Gennaro et al., 1997a). Women who give birth pre-term to low-birthweight infant may experience ongoing stress during the perinatal period, and this study suggests that this type of chronic stress may reduce NK cell activity.

NK cells work in concert with other immune cells to increase their activity during infection. For example, inflammatory cytokines can render NK cells more active as well as promote NK cell proliferation. As such, NK cells that are less responsive to cytokines may indicate a less coordinated or impoverished innate immune response. Therefore, examining NK cell responses to cytokine stimulation is another useful functional immune measure to consider in the context of stress. In one study, current and former dementia caregivers were found to have NK cells which were less responsive to interferon gamma (IFN-γ) and interleukin-2 stimulation compared to non-caregiving controls who were matched on the basis of age, sex, education, and race (Esterling et al., 1994). This finding suggests that NK cells of chronically stressed older adults are less responsive to chemical signaling and that such deficits may endure even after the chronic stressor has ended. However, it should be noted that former caregivers may still experience some chronic stress post-caregiving in the form of bereavement. Additional work shows that unemployed adults displayed reduced NK cell cytotoxicity relative to employed adults, but that NK cell cytotoxicity also substantially increased for unemployed adults who eventually obtained a job (Cohen et al., 2007a, b). Taken together, these results indicate that NK cells of chronically stressed individuals may be less active and less well coordinated with other immune responses, but these stress-related effects may recover post-stressor.

Systemic Inflammation

Much like other stress responses (e.g., activation of the HPA and SAM axes), inflammation is an adaptive process in the short term. Inflammation is driven by a redistribution of immune cells at the site of infection aided by cytokine and ana-phylatoxins release. However, when inflammation endures for too long, it can also damage the body. For example, cytokines also influence metabolism, and chronic elevation of cytokine levels is thought to increase risk for cardiovascular disease by promoting the formation of fatty plaques in arteries (Sattar et al., 2003). Given that cells of the innate immune system rely on chemical messages to move to the site of infection, measuring the amount of pro-inflammatory cytokines and other inflam-matory markers (e.g., C-reactive protein or CRP) can serve as a valuable indicator of inflammation. Related, chronically stressed individuals tend to display elevated levels of inflammatory markers like CRP and TNF-α. For example, in a 3-year lon-gitudinal study, dementia caregivers exhibited greater levels of TNF-α than matched controls, and longer duration of caregiving was associated with elevated CRP levels (von Känel et al., 2012). Furthermore, Känel et al. (2012) found that circulating CRP levels decreased in caregivers 3 months after the death of the spouse with Alzheimer's disease, suggesting that CRP production was temporally tied to care-giving duties. Consistent with this finding, dementia caregivers tend to display greater levels of CRP than non-caregiving controls (Fonareva et al., 2011; Gouin et al., 2012), and lymphocytes of dementia caregivers tend to produce more TNF-α than non-caregiving controls (Damjanovic et al., 2007). In summary, chronic stress may promote varied and impactful disease states (e.g., cardiovascular disease) by inducing systemic inflammation.

Wound Healing

Mounting research suggests that stressors can influence wound healing. In humans, numerous studies show that stressors of varied duration influence the healing of wounds, which are commonly inflicted experimentally by researchers in a standard-ized way (see Gouin & Kiecolt-Glaser, 2011, for a review). For example, women caring for a relative suffering from dementia showed slower wound healing than controls matched for age and income (Kiecolt-Glaser et al., 1995). In this study, a small wound was applied to the forearm of all participants, and wound size was monitored every 2–8 days thereafter. On average, dementia caregivers needed 9 more days to heal fully. In addition, dementia caregivers produced less IL-1β (a cytokine implicated in wound healing via its effect on tissue remodeling; Barbul,

1990). Similar effects of stressors on wound healing have been documented in dental students facing examinations (Marucha et al., 1998), as well as couples discussing marital disagreements (Kiecolt-Glaser et al., 2005). In summary, prior work suggests that chronic as well as acute forms of stress slow wound healing across varied populations.

Cellular Aging

The innate and adaptive immune responses often rely on rapid cell division to protect the body against infections. For example, neutrophils tend to be mass produced in the bone marrow following infection. Yet, cell division is finite because DNA polymerase (the protein complex responsible for copying DNA during cell division) cannot produce full copies of a chromosomal DNA strand. Instead DNA polymerase always produces a slightly shorter copy with each replication. As a result, the protective ends of chromosomes (i.e., telomeres) progressively shorten and can serve as a useful indicator of cellular aging. Moreover, if stress is able to activate the immune system, then chronic stress may accelerate the rate of cellular aging in immune cells (which in turn may impair immune function). To investigate this claim, researchers collected blood samples from mothers of chronically ill children to assess the rate of cellular aging of white blood cells (Epel et al., 2004). Mothers of chronically ill children who had been caregiving for longer (i.e., were more chronically stressed) showed shortened white blood cell telomere length, suggesting accelerated cellular aging. Consistent with this finding, others have found that the peripheral blood mononuclear cells (i.e., T cells, B cells, NK cells, macrophages, and dendritic cells) of dementia caregivers also display shorter telomeres than non-caregiving controls (Damjanovic et al., 2007). Since telomere shortening is expected to continue at a relatively steady rate with normal aging, this marker allows one to compare telomere shortening expected under normal conditions to stress-induced telomere shortening. In this fashion, Epel et al. (2004) estimated that white blood cells of women who reported high levels of stress could be considered to be approximately 10 years older than those of women with low stress. In summary, chronic stressors may accelerate the rate of cellular aging of immune cells and thus impair immunity.

Concluding Remarks

Summary

Chronic stressors are thought to over activate physiological stress responses and thus may promote disease via their prolonged effect on the nervous, neuroendocrine, and immune systems. The present review provides support for the links between chronic stressors and diverse components of the immune system among humans. Relative to healthy controls, chronically stressed individuals tend to exhibit impoverished immune responses to vaccines, reduced control over latent infections, diminished lymphocyte proliferation, reduced NK cell activity and cytotoxicity, chronically elevated cytokine levels (i.e., systemic inflammation), slowed wound healing, and accelerated cellular aging of immune cells. Altogether, this review implies that chronic stressors may lead to widespread dysregulation of the immune system and thus yield a strong influence on health and well-being.

Additional Considerations

It is important to note that the present review of evidence linking chronic stress to immune outcomes is far from exhaustive. For example, other work has examined the association between stressor chronicity and oxidative stress (e.g., Aschbacher et al., 2013), platelet activity (e.g., Aschbacher et al., 2009), mucosal immunity (e.g., Bristow et al., 2008), β2-adrenergic receptor sensitivity (e.g., Mausbach et al., 2007), markers of blood coagulation (e.g., Aschbacher et al., 2006), and other markers of inflammation (e.g., von Känel et al., 2006). The literature linking chronic stress to immune functioning therefore includes numerous outcomes beyond those reviewed in this chapter. It is also worth mentioning that some work has linked chronic *appraisals* of threat to immune function. For example, HIV-infected men who report greater rejection sensitivity also display increase CD4+ T-cell decline, earlier AIDS onset, and mortality (over 9 years) than less sensitive individuals (Cole et al., 1997), thus suggesting that chronic stressor appraisals could also contribute to poor immune function. In addition, the reader should be aware that the effects of chronic stressors on immune functioning are not unequivocally supported in the literature. For example, prior work reports divergent findings with regard to the association between chronic stress and some immune parameters, including CRP (Vitaliano et al., 2007) and NK cell activity (Irwin et al., 1991, 1997). As such, it will be important for future work to continue to investigate the factors that may contribute to such mixed findings.

In addition, few studies have examined the ability of the human immune system to recover after a chronic stressor has ended. This may be in part because much of the previous research on chronic stress has focused spousal caregivers. For this population, the end of caregiving duties is typically followed by another stressor: bereavement. We reviewed one exception by Cohen et al. (2007a, b), who followed unemployed adults pre- and post-employment and found some evidence for immune recovery. To move forward in examining immunological recovery from chronic stress, it will be necessary to use novel methodology and not rely solely on spousal caregivers. In addition to examining unemployment, researchers might consider chronic interpersonal or environmental stressors such as relationship difficulties, workplace stress, or discrimination—all of which may end with positive and/or negative outcomes. Further complicating the issue of examining post-stressor recovery is the need to examine pre-stressor functioning. In other words, future study designs may need to follow participants before chronic stress begins and after it has ended to thoroughly assess the extent to which individuals return to pre-stressor levels of immune functioning. In summary, the logistics to following individuals before and after they experience a chronic stressor remains a major barrier to the study of immune recovery in humans.

Future Directions

A promising venue for future research lies in identifying potential mechanisms and moderators of the relationship between chronic stress and immunity and leveraging the influence of such factors to promote health. Indeed, a variety of past studies support the notion that improving psychological outcomes can contribute to improved immune status. For example, a review of the effects of psychological interventions on neuroendocrine and immune functioning among HIV-infected individuals indicates that interventions which improve psychological outcomes (e.g., cognitive behavioral stress management) tend to have salutary effects on immunity (Carrico & Antoni, 2008). Similarly, stressed individuals who reported more active coping strategies also tended to display greater leukocyte (i.e., white blood cells) proliferation than stressed individuals who report more passive coping strategies (Stowell et al. 2001). In closing, although chronic stressors are often overwhelming, the experience of a chronic stressor is unique to each individual, and future research has the potential to leverage individual variability in coping and stress appraisals to minimize the effect of chronic stressors on health and well-being.

References

Abbas, A. K., Lichtman, A. H., & Pillai, S. (2014). *Basic immunology: Functions and disorders of the immune system*. Elsevier.

Adler, N. E., Boyce, T., Chesney, M. A., Cohen, S., Folkman, S., Kahn, R. L., & Leonard, S. (1994). Socioeconomic status and health: The challenge of the gradient. *American Psychologist, 49*(1), 15–24.

Adler, N. E., Epel, E. S., Castellazzo, G., & Ickovics, J. R. (2000). Relationship of subjective and objective social status with psychological and physiological functioning: Preliminary data in healthy, white women. *Health Psychology, 19*(6), 586–592.

Akinola, M., & Mendes, W. B. (2014). It's good to be the king: Neurobiological benefits of higher social standing. *Social Psychological and Personality Science, 5*(1), 43–51.

Aschbacher, K., O'Donovan, A., Wolkowitz, O. M., Dhabhar, F. S., Su, Y., & Epel, E. (2013). Good stress, bad stress and oxidative stress: Insights from anticipatory cortisol reactivity. *Psychoneuroendocrinology, 38*(9), 1698–1708.

Aschbacher, K., Roepke, S. K., von Känel, R., Mills, P. J., Mausbach, B. T., Patterson, T. L., … Grant, I. (2009). Persistent versus transient depressive symptoms in relation to platelet hyperactivation: A longitudinal analysis of dementia caregivers. *Journal of Affective Disorders, 116*(1–2), 80–87.

Aschbacher, K., von Känel, R., Dimsdale, J. E., Patterson, T. L., Mills, P. J., Mausbach, B. T., … Grant, I. (2006). Dementia severity of the care receiver predicts procoagulant response in Alzheimer caregivers. *The American Journal of Geriatric Psychiatry, 14*(8), 694–703.

Barbul, A. (1990). Immune aspects of wound repair. *Clinics in Plastic Surgery, 17*(3), 433–442.

Bauer, M. E., Vedhara, K., Perks, P., Wilcock, G. K., Lightman, S. L., & Shanks, N. (2000). Chronic stress in caregivers of dementia patients is associated with reduced lymphocyte sensitivity to glucocorticoids. *Journal of Neuroimmunology, 103*(1), 84–92.

Baum, A., Gatchel, R. J., & Schaeffer, M. A. (1983). Emotional, behavioral, and physiological effects of chronic stress at Three Mile Island. *Journal of Consulting and Clinical Psychology, 51*(4), 565–572.

Biondi, M., & Picardi, A. (1999). Psychological stress and neuroendocrine function in humans: The last two decades of research. *Psychotherapy and Psychosomatics, 68*(3), 114–150.

Black, P. H. (2003). The inflammatory response is an integral part of the stress response: Implications for atherosclerosis, insulin resistance, type II diabetes and metabolic syndrome X. *Brain, Behavior, and Immunity, 17*(5), 350–364.

Black, P. H., & Garbutt, L. D. (2002). Stress, inflammation and cardiovascular disease. *Journal of Psychosomatic Research, 52*(1), 1–23.

Bloom, B. L., Asher, S. J., & White, S. W. (1978). Marital disruption as a stressor: A review and analysis. *Psychological Bulletin, 85*(4), 867–894.

Bloomberg, G. R., & Chen, E. (2005). The relationship of psychologic stress with childhood asthma. *Immunology and Allergy Clinics, 25*(1), 83–105.

Bristow, M., Cook, R., Erzinclioglu, S., & Hodges, J. (2008). Stress, distress and mucosal immunity in carers of a partner with fronto-temporal dementia. *Aging & Mental Health, 12*(5), 595–604.

Brønnum-Hansen, H., Koch-Henriksen, N., & Stenager, E. (2004). Trends in survival and cause of death in Danish patients with multiple sclerosis. *Brain, 127*(4), 844–850.

Brosschot, J. F., Gerin, W., & Thayer, J. F. (2006). The perseverative cognition hypothesis: A review of worry, prolonged stress-related physiological activation, and health. *Journal of Psychosomatic Research, 60*(2), 113–124.

Cacioppo, J. T., Poehlmann, K. M., Kiecolt-Glaser, J. K., Malarkey, W. B., & Burleson, M. H. (1998). Cellular immune responses to acute stress in female caregivers of dementia patients and matched controls. *Health Psychology, 17*(2), 182–189.

Carrico, A. W., & Antoni, M. H. (2008). Effects of psychological interventions on neuroendocrine hormone regulation and immune status in HIV-positive persons: A review of randomized controlled trials. *Psychosomatic Medicine, 70*(5), 575–584.

Chinen, J., & Shearer, W. T. (2010). Secondary immunodeficiencies, including HIV infection. *Journal of Allergy and Clinical Immunology, 125*(2), S195–S203.

Cohen, F., Kemeny, M. E., Zegans, L. S., Johnson, P., Kearney, K. A., & Stites, D. P. (2007b). Immune function declines with unemployment and recovers after stressor termination. *Psychosomatic Medicine, 69*(3), 225–234.

Cohen, S., Tyrrell, D. A., & Smith, A. P. (1993). Negative life events, perceived stress, negative affect, and susceptibility to the common cold. *Journal of Personality and Social Psychology, 64*(1), 131–140.

Cohen, S., Gianaros, P. J., & Manuck, S. B. (2016). A stage model of stress and disease. *Perspectives on Psychological Science, 11*(4), 456–463.

Cohen, S., Janicki-Deverts, D., & Miller, G. E. (2007a). Psychological stress and disease. *JAMA, 298*(14), 1685–1687.

Cohen, S., Kessler, R., & Gordon, L. (1995). Strategies for measuring stress in studies of psychiatric and physical disorders. In *Measuring stress: A guide for health and social scientists* (pp. 3–26).

Cole, S. W., Kemeny, M. E., & Taylor, S. E. (1997). Social identity and physical health: Accelerated HIV progression in rejection-sensitive gay men. *Journal of Personality and Social Psychology, 72*(2), 320–335.

Damjanovic, A. K., Yang, Y., Glaser, R., Kiecolt-Glaser, J. K., Nguyen, H., Laskowski, B., … Weng, N. -p. (2007). Accelerated telomere erosion is associated with a declining immune function of caregivers of Alzheimer's disease patients. *The Journal of Immunology, 179*(6), 4249–4254.

Daruna, J. H. (2012). *Introduction to psychoneuroimmunology.* Academic Press.

de Visser, K. E., Eichten, A., & Coussens, L. M. (2006). Paradoxical roles of the immune system during cancer development. *Nature Reviews Cancer, 6*(1), 24–37.

Dekaris, D. (1993). Multiple changes of immunologic parameters in prisoners of war: Assessments after release from a camp in Manjača, Bosnia. *JAMA, 270*(5), 595.

Dhabhar, F. S. (2014). Effects of stress on immune function: The good, the bad, and the beautiful. *Immunologic Research, 58*(2–3), 193–210.

Dhabhar, F. S., & Mcewen, B. S. (1997). Acute stress enhances while chronic stress suppresses cell-mediated immunity in vivo: A potential role for leukocyte trafficking. *Brain, Behavior, and Immunity, 11*(4), 286–306.

Dickerson, S. S., & Kemeny, M. E. (2004). Acute stressors and cortisol responses: A theoretical integration and synthesis of laboratory research. *Psychological Bulletin, 130*(3), 355–391.

Dickerson, S. S., Mycek, P. J., & Zaldivar, F. (2008). Negative social evaluation, but not mere social presence, elicits cortisol responses to a laboratory stressor task. *Health Psychology, 27*(1), 116–121.

Dimsdale, J. E., Mills, P., Patterson, T., Ziegler, M., & Dillon, E. (1994). Effects of chronic stress on beta-adrenergic receptors in the homeless. *Psychosomatic Medicine, 56*(4), 290–295.

Donelan, K., Hill, C. A., Hoffman, C., Scoles, K., Feldman, P. H., Levine, C., & Gould, D. (2002). Challenged to care: Informal caregivers in a changing health system. *Health Affairs, 21*(4), 222–231.

Elenkov, I. J., & Chrousos, G. P. (2002). Stress hormones, proinflammatory and antiinflammatory cytokines, and autoimmunity. *Annals of the New York Academy of Sciences, 966*(1), 290–303.

Epel, E. S., Blackburn, E. H., Lin, J., Dhabhar, F. S., Adler, N. E., Morrow, J. D., & Cawthon, R. M. (2004). Accelerated telomere shortening in response to life stress. *Proceedings of the National Academy of Sciences, 101*(49), 17312–17315.

Esterling, B. A., Kiecolt-Glaser, J. K., Bodnar, J. C., & Glaser, R. (1994). Chronic stress, social support, and persistent alterations in the natural killer cell response to cytokines in older adults. *Health Psychology, 13*(4), 291–298.

Evans, D. L., Leserman, J., Perkins, D. O., Stern, R. A., Murphy, C., Zheng, B., … Petitto, J. M. (1997). Severe life stress as a predictor of early disease progression in HIV infection. *The American Journal of Psychiatry, 154*(5), 630–634.

Everly, G. S., & Lating, J. M. (2013). The anatomy and physiology of the human stress response. In G. S. Everly & J. M. Lating (Eds.), *A clinical guide to the treatment of the human stress response* (pp. 17–51). Springer.

Felitti, V. J., Anda, R. F., Nordenberg, D., Williamson, D. F., Spitz, A. M., Edwards, V., … Marks, J. S. (1998). Relationship of childhood abuse and household dysfunction to many of the leading causes of death in adults. *American Journal of Preventive Medicine, 14*(4), 245–258.

Fonareva, I., Amen, D. M., Zajdel, D. P., Ellingson, R. M., & Oken, B. S. (2011). Assessing sleep architecture in dementia caregivers at home using an ambulatory polysomnographic system. *Journal of Geriatric Psychiatry and Neurology, 24*(1), 50–59.

Gennaro, S., Fehder, W., Nuamah, I. F., Campbell, D. E., & Douglas, S. D. (1997b). Caregiving to very low birthweight infants: A model of stress and immune response. *Brain, Behavior, and Immunity, 11*(3), 201–215.

Gennaro, S., Fehder, W. P., Cnaan, A., York, R., Campbell, D. E., Gallagher, P. R., & Douglas, S. D. (1997a). Immune responses in mothers of term and preterm very-low-birth-weight infants. *Clinical and Diagnostic Laboratory Immunology, 4*, 7.

Glaser, R., & Kiecolt-Glaser, J. K. (1994). Stress-associated immune modulation and its implications for reactivation of latent herpesviruses. In *Human Herpesvirus infections* (pp. 245–270).

Glaser, R., & Kiecolt-Glaser, J. K. (2005). Stress-induced immune dysfunction: Implications for health. *Nature Reviews Immunology, 5*(3), 243.

Glaser, R., Sheridan, J., Malarkey, W. B., MacCallum, R. C., & Kiecolt-Glaser, J. K. (2000). Chronic stress modulates the immune response to a pneumococcal pneumonia vaccine. *Psychosomatic Medicine, 62*(6), 804–807.

Godbout, J. P., & Glaser, R. (2006). Stress-induced immune dysregulation: Implications for wound healing, infectious disease and cancer. *Journal of Neuroimmune Pharmacology, 1*(4), 421–427.

Gouin, J.-P., Glaser, R., Malarkey, W. B., Beversdorf, D., & Kiecolt-Glaser, J. (2012). Chronic stress, daily stressors, and circulating inflammatory markers. *Health Psychology, 31*(2), 264–268.

Gouin, J.-P., Hantsoo, L., & Kiecolt-Glaser, J. K. (2008). Immune dysregulation and chronic stress among older adults: A review. *Neuroimmunomodulation, 15*(4–6), 251–259.

Gouin, J.-P., & Kiecolt-Glaser, J. K. (2011). The impact of psychological stress on wound healing: Methods and mechanisms. *Immunology and Allergy Clinics of North America, 31*(1), 81–93.

Hammen, C. (2005). Stress and depression. *Annual Review of Clinical Psychology, 1*(1), 293–319.

Hansson, G. K., & Hermansson, A. (2011). The immune system in atherosclerosis. *Nature Immunology, 12*(3), 204–212.

Holmes, T. H., & Rahe, R. H. (1967). The social readjustment rating scale. *Journal of Psychosomatic Research, 11*(2), 213–218.

Iannello, A., Debbeche, O., Samarani, S., & Ahmad, A. (2008). Antiviral NK cell responses in HIV infection: I. NK cell receptor genes as determinants of HIV resistance and progression to AIDS. *Journal of Leukocyte Biology, 84*(1), 1–26.

Irwin, M., Brown, M., Patterson, T., Hauger, R., Mascovich, A., & Grant, I. (1991). Neuropeptide Y and natural killer cell activity: Findings in depression and Alzheimer caregiver stress. *The FASEB Journal, 5*(15), 3100–3107.

Irwin, M., Hauger, R., Patterson, T. L., Semple, S., Ziegler, M., & Grant, I. (1997). Alzheimer caregiver stress: Basal natural killer cell activity, pituitary-adrenal cortical function, and sympathetic tone. *Annals of Behavioral Medicine, 19*(2), 83–90.

Johsson, A., & Hansson, L. (1977). Prolonged exposure to a stressful stimulus (noise) as a cause of raised blood-pressure in man. *Lancet, 1*(8002), 86–87.

Juster, R.-P., McEwen, B. S., & Lupien, S. J. (2010). Allostatic load biomarkers of chronic stress and impact on health and cognition. *Neuroscience & Biobehavioral Reviews, 35*(1), 2–16.

Kemeny, M. E., Weiner, H., Taylor, S. E., Schneider, S., Visscher, B., & Fahey, J. L. (1994). Repeated bereavement, depressed mood, and immune parameters in HIV seropositive and seronegative gay men. *Health Psychology, 13*(1), 14–24.

Kiecolt-Glaser, J. K., Dura, J. R., Speicher, C. E., Trask, O. J., & Glaser, R. (1991). Spousal caregivers of dementia victims: Longitudinal changes in immunity and health. *Psychosomatic Medicine, 53*(4), 345–362.

Kiecolt-Glaser, J. K., Glaser, R., Gravenstein, S., Malarkey, W. B., & Sheridan, J. (1996). Chronic stress alters the immune response to influenza virus vaccine in older adults. *Proceedings of the National Academy of Sciences, 93*(7), 3043–3047.

Kiecolt-Glaser, J. K., Loving, T. J., Stowell, J. R., Malarkey, W. B., Lemeshow, S., Dickinson, S. L., & Glaser, R. (2005). Hostile marital interactions, proinflammatory cytokine production, and wound healing. *Archives of General Psychiatry, 62*(12), 1377–1384.

Kiecolt-Glaser, J. K., Shuttleworth, E. C., Dyer, C. S., Ogrocki, P., & Speicher, C. E. (1987). Chronic stress and immunity in family caregivers of Alzheimer's disease victims. *Psychosomatic Medicine, 13*.

Kiecolt-Glaser, J. K., Marucha, P. T., Mercado, A. M., Malarkey, W. B., & Glaser, R. (1995). Slowing of wound healing by psychological stress. *The Lancet, 346*(8984), 1194–1196.

Kirschbaum, C., Pirke, K.-M., & Hellhammer, D. H. (1993). The "Trier social stress test": A tool for investigating psychobiological stress responses in a laboratory setting. *Neuropsychobiology, 28*(1–2), 76–81.

Lang, A. J., Rodgers, C. S., Laffaye, C., Satz, L. E., Dresselhaus, T. R., & Stein, M. B. (2003). Sexual trauma, posttraumatic stress disorder, and health behavior. *Behavioral Medicine, 28*(4), 150–158.

Lauc, G., Dabelic, S., Dumic, J., & Flogel, M. (1998). Stressin and natural killer cell activity in professional soldiers. *Annals of the New York Academy of Sciences, 851*(1), 526–530.

Lazarus, R., & Folkman, S. (1984). Stress, appraisal, and coping. .

Lutgendorf, S. K., Vitaliano, P. P., Tripp-Reimer, T., Harvey, J. H., & Lubaroff, D. M. (1999). Sense of coherence moderates the relationship between life stress and natural killer cell activity in healthy older adults. *Psychology and Aging, 14*(4), 12.

Marucha, P. T., Kiecolt-Glaser, J. K., & Favagehi, M. (1998). Mucosal wound healing is impaired by examination stress. *Psychosomatic Medicine, 60*(3), 362–365.

Mausbach, B. T., Mills, P. J., Patterson, T. L., Aschbacher, K., Dimsdale, J. E., Ancoli-Israel, S., ... Grant, I. (2007). Stress-related reduction in personal mastery is associated with reduced immune cell β2-adrenergic receptor sensitivity. *International Psychogeriatrics, 19*(05), 935.

McEwen, B. S. (1998). Stress, adaptation, and disease: Allostasis and allostatic load. *Annals of the New York Academy of Sciences, 840*(1), 33–44.

Mcewen, B. S. (2004). Protection and damage from acute and chronic stress: Allostasis and allostatic overload and relevance to the pathophysiology of psychiatric disorders. *Annals of the New York Academy of Sciences, 1032*(1), 1–7.

McKinnon, W., Weisse, C. S., Reynolds, C. P., Bowles, C. A., & Baum, A. (1989). Chronic stress, leukocyte subpopulations, and humoral response to latent viruses. *Health Psychology, 8*(4), 389–402.

Miller, G. E., Cohen, S., Pressman, S., Barkin, A., Rabin, B. S., & Treanor, J. J. (2004). Psychological stress and antibody response to influenza vaccination: When is the critical period for stress, and how does it get inside the body? *Psychosomatic Medicine, 66*(2), 215–223.

Nahmias, A. J., & Roizman, B. (1973). Infection with herpes-simplex viruses 1 and 2. *The New England Journal of Medicine, 289*(13), 667–674.

Nakano, Y., Nakamura, S., Hirata, M., Harada, K., Ando, K., Tabuchi, T., ... Oda, H. (1998). Immune function and lifestyle of taxi drivers in Japan. *Industrial Health, 36*(1), 32–39.

Ockenfels, M. C., Porter, L., Smyth, J., Kirschbaum, C., Hellhammer, D. H., & Stone, A. A. (1995). Effect of chronic stress associated with unemployment on salivary cortisol: Overall cortisol levels, diurnal rhythm, and acute stress reactivity. *Psychosomatic Medicine, 57*(5), 460.

Park, C. L., & Iacocca, M. O. (2014). A stress and coping perspective on health behaviors: Theoretical and methodological considerations. *Anxiety, Stress, & Coping, 27*(2), 123–137.

Pickup, J. C., & Crook, M. A. (1998). Is type II diabetes mellitus a disease of the innate immune system? *Diabetologia, 41*(10), 1241–1248.

Poorolajal, J., Hooshmand, E., Mahjub, H., Esmailnasab, N., & Jenabi, E. (2016). Survival rate of AIDS disease and mortality in HIV-infected patients: A meta-analysis. *Public Health, 139*, 3–12.

Sabioncello, A., Kocijan-Hercigonja, D., Rabatić, S., Tomašić, J., Jeren, T., Matijević, L., ... Dekaris, D. (2000). Immune, endocrine, and psychological responses in civilians displaced by war. *Psychosomatic Medicine, 62*(4), 502–508.

Sapolsky, R. M. (2005). The influence of social hierarchy on primate health. *Science, 308*(5722), 648–652.

Sarafino, E. P., Caltabiano, M. L., & Byrne, D. (2008). *Health psychology: Biopsychosocial interactions [second Australasian edition]*. Wiley. Retrieved from http://au.wiley.com/WileyCDA/WileyTitle/productCd-EHEP002216.html

Sattar, N., McCarey, D. W., Capell, H., & McInnes, I. B. (2003). Explaining how "high-grade" systemic inflammation accelerates vascular risk in rheumatoid arthritis. *Circulation, 108*(24), 2957–2963.

Segerstrom, S. C., & Miller, G. E. (2004). Psychological stress and the human immune system: A meta-analytic study of 30 years of inquiry. *Psychological Bulletin, 130*(4), 601–630.

Solomon, G. F., Segerstrom, S. C., Grohr, P., Kemeny, M., & Fahey, J. (1997). Shaking up immunity: Psychological and immunologic changes after a natural disaster. *Psychosomatic Medicine, 59*(2), 114–127.

Stowell, J. R., Kiecolt-Glaser, J. K., & Glaser, R. (2001). Perceived stress and cellular immunity: When coping counts. *Journal of Behavioral Medicine, 24*(4), 323–339.

Taylor, D. H., Ezell, M., Kuchibhatla, M., Østbye, T., & Clipp, E. C. (2008). Identifying trajectories of depressive symptoms for women caring for their husbands with dementia. *Journal of the American Geriatrics Society, 56*(2), 322–327.

Vedhara, K., Cox, N. K., Wilcock, G. K., Perks, P., Hunt, M., Anderson, S., … Shanks, N. M. (1999). Chronic stress in elderly carers of dementia patients and antibody response to influenza vaccination. *The Lancet, 353*(9153), 627–631.

Vitaliano, P., Echeverria, D., Shelkey, M., Zhang, J., & Scanlan, J. (2007). A cognitive psychophysiological model to predict functional decline in chronically stressed older adults. *Journal of Clinical Psychology in Medical Settings, 14*(3), 177–190.

Vitaliano, P., Scanlan, J. M., Ochs, H. D., Syrjala, K., Siegler, I. C., & Snyder, E. A. (1998). Psychosocial stress moderates the relationship of cancer history with natural killer cell activity. *Annals of Behavioral Medicine, 20*(3), 199–208.

von Känel, R., Dimsdale, J. E., Mills, P. J., Ancoli-Israel, S., Patterson, T. L., Mausbach, B. T., & Grant, I. (2006). Effect of Alzheimer caregiving stress and age on frailty markers interleukin-6, C-reactive protein, and D-dimer. *The Journals of Gerontology Series A: Biological Sciences and Medical Sciences, 61*(9), 963–969.

von Känel, R., Mills, P. J., Mausbach, B. T., Dimsdale, J. E., Patterson, T. L., Ziegler, M. G., … Grant, I. (2012). Effect of Alzheimer caregiving on circulating levels of C-reactive protein and other biomarkers relevant to cardiovascular disease risk: A longitudinal study. *Gerontology, 58*(4), 354–365.

Wimo, A., von Strauss, E., Nordberg, G., Sassi, F., & Johansson, L. (2002). Time spent on informal and formal care giving for persons with dementia in Sweden. *Health Policy, 61*(3), 255–268.

Chapter 5
The Roles of Appraisal and Perception in Stress Responses, and Leveraging Appraisals and Mindsets to Improve Stress Responses

Jeremy P. Jamieson and Emily J. Hangen

Introduction

Stress is too frequently a "dirty word" in modern society. Mass media outlets highlight studies that focus on the deleterious effects of stress on health, and myriad wellness and health improvement programs tout the benefits of stress reduction and relaxation. The result is that individuals believe avoiding and/or reducing stress in their daily lives is the best route to attain positive health outcomes, academic achievement, and vocational success. To illustrate, when prompted to choose the optimal method for coping with an evaluative stressor, laypeople overwhelmingly (91%) indicated that remaining calm and relaxed is the optimal method (Brooks, 2014). Even researchers equate stress with distress. For instance, self-report scales that purport to assess "perceptions of stress" are frequently constructed entirely of negative items (or reverse-scored positive items), such as "how often have you been upset because of something that happened unexpectedly?" (e.g., Cohen et al., 1983). As a result, people often devote considerable time and energy engaging in stress reduction methods, including taking vacations, completing relaxation retreats, or even stopping at the local pub after work for a calming drink.

Although maladaptive acute stress responses or chronic stressors can predict poor health decisions and health outcomes (e.g., Jamieson & Mendes, 2016; Jefferson et al., 2010; Juster et al., 2010), stress is not unilaterally negative. In fact, early stress theories emphasized the nondiagnostic nature of stress responses. For instance, Hans Selye (1936) labeled stress as the body's *nonspecific* response for any demand for change. Based on this conceptualization, "a painful blow and a

J. P. Jamieson (✉)
University of Rochester, Rochester, NY, USA
e-mail: jeremy.jamieson@rochester.edu

E. J. Hangen
Harvard University, Cambridge, MA, USA

© Springer Nature Switzerland AG 2021   105
H. Hazlett-Stevens (ed.), *Biopsychosocial Factors of Stress, and Mindfulness for Stress Reduction*, https://doi.org/10.1007/978-3-030-81245-4_5

passionate kiss can be equally stressful" (Selye, 1974). Thus, it is not possible to completely avoid stress in daily life without cutting oneself off from the social environment. Stress, however, is not an experience one should necessarily seek to avoid. In fact, in the absence of stress or adversity, people would miss growth opportunities, be less resilient, and even suffer negative health outcomes (e.g., Crum et al., 2013; Seery et al., 2010a, b). Along these lines, the research and theories reviewed here are rooted in the idea that stress is a multifaceted psychobiological state that can result in myriad adaptive or maladaptive outcomes depending on contextual, temporal, and psychological factors. First, we review meta-level beliefs about stress, mindsets about stress, and contextually grounded biopsychosocial processes. Then, we offer suggestions for intervening to promote active coping with stressors based on mindset and appraisal theories. Finally, we offer suggestions for integrating extant models, with an emphasis on extensions to emotion regulation and interpersonal processes.

Stress Mindsets and Implicit Theories Models

Although the full breadth of the concept of "stress" encompasses complex, multidimensional biobehavioral response patterns, beliefs and mindsets about stress serve as perceptual lenses through which individuals orient their expectations and motivations when anticipating or responding in stressful situations. Classic research on stress hardiness embraced the idea that beliefs about stressors and stress responses are integrated in determining coping responses (e.g., Kobasa et al., 1982). For instance, a "hardy" personality style has been associated with active, adaptive coping strategies in the face of stressors, as well as resistance to the damaging effects of negative stressful experiences (Eschleman et al., 2010). Stress hardiness beliefs are rooted in three dimensions: commitment, control, and "challenge" (Kobasa et al., 1982). Commitment beliefs determine the extent of engagement/disengagement with the social environment, control beliefs inform individuals whether they can influence events in their life, and challenge beliefs shape whether difficult situations are generally perceived as positive challenges or negative threats. The "stress hardy" typology, specifically, is thus characterized by high commitment, control, and challenge beliefs. To illustrate how hardiness might manifest, consider a student being teased or bullied at school. Even in the face of the negative social feedback, a hardy student would not shy away from engaging with future social contexts, believe that they retain control over the trajectory of their academic and personal development, and view bullying situations as challenges to be overcome.

Along similar lines, early research on stress and coherence suggests a person's general worldview can shape how they respond to stressors. More specifically, one's sense of coherence (SOC) is a dispositional lens through which the world is perceived as more or less meaningful, comprehensible, and manageable (Antonovsky, 1998). That is, beliefs that internal and external experiences are comprehensible, can be managed with available coping resources, and are positive challenges in which one can find meaning signal "healthy" or adaptive beliefs about stress. Importantly for understanding SOC as a belief system that is trait-like, rather than contextually bound such as appraisal processes, research demonstrates that SOC is stable across stressful contexts (Schnyder et al., 2000). More downstream, using a salutogenic framework of health, individuals reporting high levels of SOC have been shown to orient closer to the "ease" end of the ease-disease continuum, thus promoting subjective health and resilience (Eriksson & Lindström, 2006).

Stress Mindsets

As reviewed above, early lines of research on beliefs about stress and stress processes focused on understanding multiple dimensions of beliefs to understand (and promote) positive stress coping and resilience. More recent work has sought to consolidate multifaceted belief systems studied previously into a unified theory using the mindsets and implicit theories literatures as organizing frameworks. *Stress mindsets* are beliefs about the nature of stress in general (Crum et al., 2013). That is, whether the experience of stress is enhancing or debilitating. Research on stress mindsets emerged from the large corpus of research on implicit theories and mindsets (for a review, see Yeager & Dweck, 2012). Implicit theories are labeled as "implicit" because meta-level belief systems need not be explicitly activated or consciously accessible to exert effects on psychological and behavioral outcomes downstream. Also, like any other scientific theory, beliefs and mindsets are "theories" because they create a perceptual lens through which hypotheses are constructed and attributes of the world are interpreted. Thus, experiential events are interpreted as evidence that confirms or disconfirms hypotheses. Mindsets, implicit theories, and naïve/"lay" theories are often equated. Labeling these belief systems as naïve or "lay" theories touches on the "commonsense" nature of hypotheses derived from subjective beliefs. That is, unlike scientific theories based on scientific research, implicit theories refer to explanations grounded in daily experiences and social communications (Dweck, 2000; Molden & Dweck, 2006; Yeager & Dweck, 2012).

As the literature on implicit or naïve/lay theories has progressed, however, researchers have primarily utilized the term "mindset" to refer to meta-level belief systems that create perceptual lenses, rather than "implicit theories." This terminological preference is particularly relevant for intervention research that seeks to change beliefs to encourage individuals to endorse one implicit theory over another so as to enact positive change (e.g., Dweck, 2006; Paunesku et al., 2015; Yeager & Walton, 2011). In other words, mindset interventions make implicit theories explicitly accessible. The first wave of intervention research on mindset processes examined intraindividual factors such as intelligence (Blackwell et al., 2007) or personality (Yeager et al., 2013). More recently, mindsets research has expanded beyond individual-level factors to beliefs whether groups possess the potential for change (Goldenberg et al., 2018), the nature of failure (Haimovitz & Dweck, 2016), health behaviors (e.g., Crum & Langer, 2007), and as outlined in this chapter, the utility of stress.

A growing body of evidence suggests benefits of endorsing a "stress-is-enhancing" mindset. For instance, students holding a stress-is-enhancing mindset are more open to evaluative feedback from peers and teachers, and exhibit healthier physiological responses during evaluative episodes (Crum et al., 2013). Moreover, other research from educational settings demonstrates that a stress-is-enhancing mindset weakens associations between negative life events and distress responses (Park et al., 2017). Furthermore, employees reporting a stress-is-enhancing mindset were more approach motivated, and demonstrated better task engagement (Casper et al., 2017). Particularly relevant for mapping links between mindsets and mindfulness processes, individual differences in stress mindsets predicted employees' life satisfaction, resilience, and mindfulness (Crum et al., 2013). That is, the more individuals endorsed a stress-is-enhancing mindset, the more they perceived value in present consciousness without judgment (Walach et al., 2006).

As reviewed above, stress mindsets provide a lens that focuses attentional engagement and response patterns in anticipation of stressors and in the face of stressful situations. Mindsets, however, are "situation general," meaning that they are diffuse perceptual processes that tap into meta-level beliefs about the nature of stress. However, mindsets do not necessarily provide situation-specific tools for implementing adaptive stress responses. Thus, recent theoretical advances have sought to understand the interplay between mindsets and other cognitive processes in stressful situations. Notably, situation-specific stress appraisal processes have emerged a promising focus for integration with mindsets to optimize responses in stressful situations (Crum et al., 2017; Jamieson et al., 2018; Yeager et al., 2016). In the following section, we review stress appraisal processes and then focus on leveraging mindset and appraisal processes to promote active coping in stressful contexts.

Stress Appraisals and Biopsychosocial (BPS) Models

Appraisal Theory of Emotion

Psychologists have long considered stress appraisals to be contextually grounded. For instance, in seminal research appraisals of internal states directly influenced emotional responses (Schachter & Singer, 1962). One prominent model, the *appraisal theory of emotion*, introduced notions of "challenge and threat" states derived from cognitive appraisal processes (see Lazarus, 1991, for a review). This model argued multiple (and diffuse) processes informed stress appraisals and downstream emotional responses, including bodily states, episodic memories, and situational factors (e.g., Lazarus et al., 1985). More specifically, the appraisal theory of emotion specifies two "levels" of appraisals: primary and secondary. Primary appraisals address relevance: whether situations are irrelevant, benign, or stressful.

Stressful primary appraisals are subdivided into "threat" and "challenge." Threatening situations involve the possibility of harm (physical or social), whereas challenging situations refer to growth opportunities (Lazarus, 1991). Primary appraisals, however, are not sufficient to determine stress responses. Secondary appraisals then appraise what coping resources are available. In other words, secondary appraisals indicate how individuals can address/cope with the stressors that they face (e.g., Folkman & Lazarus, 1985).

Importantly for understanding how appraisals inform stress responses, primary and secondary appraisals can act synergistically or independently (e.g., Folkman & Lazarus, 1980). For instance, primary appraisals could suggest a "challenge" situation that includes growth potential, such as learning a new language, but secondary appraisals could indicate that one does not possess resources to attain growth, such as low competence. So, the "challenge" situation could be threatening if resources cannot meet perceived demands. In sum, appraisal theory of emotion solidified the notion that appraisals directly inform affective responses in stressful situations via appraisals. Building on this earlier model, subsequent research sought to refine how appraisals operate to inform stress responses (e.g., Blascovich, 1992; Tomaka et al., 1993), which led to the development of the biopsychosocial (BPS) model of challenge and threat (e.g., Blascovich & Tomaka, 1996).

The Biopsychosocial Model of Challenge and Threat

Appraisal Processes . Appraisals are central to the BPS model of challenge and threat. Specifically, appraisals of situational demands interact with appraisals of available coping resources (at the same level) to produce challenge- and threat-type stress responses (see Mendes & Park, 2014, for a review). In the appraisal model of emotion reviewed above, "challenge" and "threat" refer to primary appraisals rooted in the potential for gain versus loss. Secondary appraisals then assess coping ability. However, appraisal processes are integrated into a single "level" in the BPS model to produce challenge or threat responses. Another advance of the BPS model is that challenge/threat represents anchors along a continuum of stress responses (e.g., Jamieson et al., 2013a), which is represented by how BPS researchers often utilize an index to assess challenge/threat responses (e.g., Hangen et al., 2016; Seery et al., 2004, 2009).

As in the appraisal theory of emotion, *challenge* and *threat* responses manifest in "motivated-performance" situations that involve situational demands, but differ in antecedent appraisals. Challenge is experienced when appraisals of coping exceed appraisals of demands, whereas threat responses occur when perceived demands exceed perceived resources. To demonstrate, consider an entrepreneur pitching an idea for a new product to a team of executives offering potential sources of funding (e.g., *Shark Tank*). Regardless of the quality of the product or the entrepreneur's mindset, this situation is stressful. There is an immediate demand (evaluation of the product) that requires instrumental responding (delivering the presentation). A well-trained and well-practiced presenter might appraise the situation as challenging, believing that their skills, training, and experience (i.e., resources) allow them to handle the demands of the difficult evaluative situation, whereas presenters may experience threat if the demands of the product pitch context are appraised as outweighing their (potentially low) skills and experience level. Thus, the stress arousal felt by presenters in the above example is semantically and psychologically imprecise (Blascovich, 1992) – stress arousal levels are elevated as a consequence of engagement with the situation. The *form* the arousal takes – threat vs. challenge – depends on how appraisal processes unfold.

Importantly for understanding stress appraisals in the context of the BPS model, "demands" and "resources" are considered to be multidimensional. For instance, demands may consist of perceptions of uncertainty, danger, and/or effort, and these facets can be independent or intertwined. For instance, consider an unfamiliar situation in which there is potential for evaluation, such as a student taking an important placement exam. Here, the test-taker may not know the layout of the exam and would certainly not know exact questions (i.e., uncertainty appraisals), which would require the test-taker to devote effort to parsing the format/instructions (i.e., effort appraisals). Thus, appraisals of uncertainty and effort are intertwined in this example. However, placement test situations also include the potential for negative evaluation and social harm (i.e., "danger" appraisals), which are not necessarily tied to appraisals of uncertainty or perceptions of effort needed to reduce uncertainty.

In addition to appraisals being multifaceted *within* resource and demand categories, appraisals of resources can be independent from appraisals of demands or a single appraisal could inform both categories. That is, resource and demand appraisals may be ontologically distinct. Perceptions of knowledge, ability, or skills (resource appraisals) are typically distinct from perceptions of danger, difficulty, or effort (demand appraisals). However, resource and demand appraisals can (and often do) index bipolar factors with relevance for both. For example, it is often difficult to dissociate appraisals of familiarity (resource) from appraisals of uncertainty (demand), or appraisals of safety (resource) from appraisals of danger (demand). Thus, these facets of appraisal typically represent dimensions that simultaneously impact both resources and demands (e.g., as familiarity increases relative to uncertainty, resource appraisals increase relative to demand appraisals, Blascovich, 2008).

To demonstrate how appraisals can be bipolar and intertwined, consider a firefighter about to charge into a burning building to locate a person in distress. The firefighter may or may not be familiar with the layout of the building, which has direct consequences for her/his perceptions of danger. That is, appraisals of familiarity (e.g., knowledge of the location of stairs, corridors, beams, etc.) are resources that would attenuate appraisals of danger because the firefighter can better predict sources of potential harm. Alternatively, an unfamiliar building presents demands (uncertainty about the structure) that increase appraisals of the potential for danger. Here, appraisals of familiarity/uncertainty are intertwined – one necessarily affects the other – but can also influence another facet of demand/resource appraisals: perceptions of danger/safety.

Because appraisals that produce challenge and threat responses are derived from multiple components that vary across situations, BPS researchers have preferred experimental methods. That is, manipulating situational factors and/or distinct facets of appraisals can help isolate mechanisms (e.g., Oveis et al., n.d.; Karnilowicz et al., 2018; Peters et al., 2018). For instance, studies have manipulated dyadic interaction context to induce uncertainty in couples (Peters et al., 2018), or have manipulated the valence (positive vs. negative) of feedback during social evaluation (Kassam et al., 2009). Manipulating situational factors or facets of resource/demand appraisals also helps to avoid limitations of self-reports of appraisals, although explicit measures are often used to elucidate appraisal content (e.g., Jamieson et al., 2016). Focusing on experimental methods is important because individuals process information in stressful situations consciously *and* unconsciously, and may or may not have access to the full breadth of appraisal processes when completing explicit reports.

Physiological Processes. Importantly for mapping how appraisals impact downstream health and behavior, challenge and threat responses reflect differential patterns of physiological responses (see Mendes & Park, 2014, for a review). Informed by models of physiological toughness (Dienstbier, 1989), challenge and threat focus on two core systems: the sympathetic-adrenal-medullary (SAM) and hypothalamic-pituitary-adrenal (HPA) (aka, pituitary-adrenocortical (PAC)) axes. The SAM axis may be loosely conceived of as reflecting general sympathetic activation, whereas the HPA system is more conservative, responding after long(er) exposures to (typically negative) stressors.

Upstream, SAM activation stimulates release of epinephrine from the adrenal medulla, and increased epinephrine levels lead to elevated heart rate, dilated blood vessels, and release of glucose from the liver. HPA activation results in the release of cortisol from the *zona fasciculata* of the adrenal gland. Given the chemical signaling sequence – hypothalamus releases corticotropin, which triggers the pituitary to release adrenocorticotropin, which then travels through bloodstream to the adrenal glands to stimulate release of cortisol – cortisol levels typically peak 15–20 min after stress onset (Dickerson & Kemeny, 2004). Both challenge- *and* threat-type appraisal patterns are theorized to stimulate the SAM axes, but threat only is accompanied by HPA activation (see Blascovich, 2013, for a review).

More downstream, physiological patterns of challenge and threat appraisals can be observed in cardiovascular (CV) responses. In fact, CV responses are commonly used to index challenge and threat responses in vivo during acute stress (e.g., Blascovich et al., 1999; Hangen et al., 2019; Jamieson & Mendes, 2016). The most common measures derived from the CV system used to index engagement with stressful situations are heart rate (HR) and pre-ejection period (PEP). HR is the rate of left ventricle contraction. Engagement produces increases in HR primarily through increased sympathetic tone, but vagal withdrawal (decrease parasympathetic tone) may also contribute to increases in HR in situations involving cognitive effort (e.g., Appelhans & Luecken, 2006). PEP measures the time from left ventricle contraction to the opening of the aortic valve, and thus, is an index of ventricular contractility (VC), with shorter PEP intervals thus indexing increased arousal.

After individuals are engaged with a stressful situation (i.e., when individuals experience sympathetic arousal), to differentiate challenge from threat, researchers often examine changes in cardiac output (CO) and total peripheral resistance (TPR) (see Seery, 2011, for a review). CO indexes the amount of blood pumped through the CV system per minute (usually in liters) and is calculated by assessing stroke volume (SV) – the amount of blood ejected from the heart at each beat – and multiplying that by HR. TPR indexes resistance in peripheral vasculature and is often calculated using the following validated formula: TPR = (mean arterial pressure/CO) *80 (see Sherwood et al., 1990; for examples, see Jamieson et al., 2012; Hangen et al., 2019; Oveis et al., n.d.). Challenge is marked by an increase in cardiac efficiency (i.e., increased SV or CO) combined with reduced resistance in the peripheral vasculature. This response pattern helps to deliver oxygen to peripheral sites (e.g., the brain) to facilitate active coping with stressors. Threat, on the other hand, elicits declines or little change in CO because although cardiac activity increases, this increase is not accompanied by dilation of the vasculature. That is, when threatened, vascular resistance increases so as to limit blood flow to the periphery in anticipation of damage or defeat. See Fig. 5.1 for a diagram for the sequence whereby challenge and threat responses unfold in the context of the BPS model.

As noted above, although challenge and threat responses elicit differential patterns of physiological responses, they are at their core, psychological states. That is, appraisals of coping resources and situational demands interact to produce

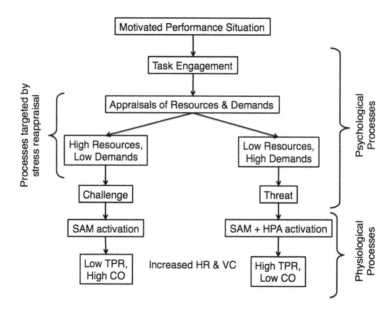

Fig. 5.1 Overview of the psychological and physiological processes of the biopsychosocial model of challenge and threat. *SAM* sympathetic-adrenal-medullary axis, *HPA* hypothalamic-pituitary-adrenal axis, *HR* heart rate, *VC* ventricular contractility, *TPR* total peripheral resistance, *CO* cardiac output

downstream responses. In the following section, we review how appraisals and mindsets can be harnessed to improve health, performance, and well-being outcomes in stressful situations.

Regulating Stress Responses via Mindsets and Appraisals

Stress Mindset Interventions

Intervention programs that seek to alter mindsets about stress have taken two forms. Seminal research on stress mindset interventions presented information conveyed in videos that described the enhancing or debilitating nature of stress (Crum et al., 2013). Notably, although the content communicated in the intervention materials was accurate in both the stress-is-enhancing and stress-is-debilitating conditions, content was overwhelmingly unbalanced, which does not reflect the "true" multifaceted nature of stress. That is, the first iterations of stress mindset manipulations presented a unilateral position about the *general* nature of stress not tied to any particular stressful situation. These "unilateral" manipulations were shown to be efficacious in experimental paradigms (e.g., Crum et al., 2013). For instance, compared to communicating a stress-is-debilitating mindset (i.e., what people typically believe), teaching individuals a stress-is-enhancing mindset led to higher levels of health-supportive anabolic hormones (Crum et al., 2017; see https://mbl.stanford.edu/instruments/stress-mindset-manipulation-videos for intervention materials).

Presenting unbalanced information can pose concerns. First, presenting unilaterally enhancing or debilitating information about stress can be viewed as ethically questionable given the "true" nature of stress is more nuanced and multifaceted: Stress can be debilitating and enhancing, and variable from situation-to-situation. Moreover, the durability and effectiveness of mindset-based interventions for stress may be limited. For instance, individuals informed about a unilateral mindset about stress (enhancing or debilitating) will likely encounter oppositional evidence at some point, thus potentially compromising the efficacy of the information communicated prior. To more accurately reflect the "true" nature of stress, more recent research has developed materials for regulating stress that present more balanced information on both the enhancing and debilitating aspects of stress while also communicating the positive benefits of a stress-is-enhancing mindset (Crum et al., 2018a, b).

This more complex, and less biased, interaction approach (referred to as stress mindset training) has been delivered both in a 2-h training and in online modules. This shift to presenting more multifaceted information has demonstrated that stress mindset interventions can be durable beyond delivery points. For instance, employees at a financial company who attended one, 2-h stress mindset training exhibited improvements in health and well-being up to a month after the intervention was delivered (Crum et al., 2018b). In addition, a field experiment with college students found that an intervention educating students about a stress-is-enhancing mindset delivered in the summer prior to beginning college led to more positive affect during the *spring*-semester exams over freshman *and* sophomore years (Goyer et al., 2018, in preparation).

Importantly for understanding the interplay of cognitive processes underlying stress responses, stress mindset interventions can interact with challenge and threat processes as delineated by the BPS model. For instance, stressful evaluative situations involving instrumental responding that are presented as positive (i.e., "challenging") or negative ("threatening") via valenced feedback interact with mindsets about stress (Crum et al., 2017). In positive, challenging situations, a stress-is-enhancing mindset led to greater cognitive flexibility, reduced attentional bias, and more positive affect relative to a stress-is-debilitating mindset (Crum et al., 2017). However, the challenge and threat processes manipulated in the aforementioned study were not stress appraisal processes, but rather attributes of the context. Other intervention approaches reviewed in the following have more tightly focused on shifting appraisals of resources and demands in stressful situations.

Stress Reappraisal Interventions

In acute performance situations, experiencing more challenge-type stress responses predicts improved performance, cognitive flexibility, and short-term health benefits. As noted in the prior section on stress appraisals, patterns of challenge and threat responding are derived from appraisals of resources and demands. So, manipulating appraisals is an efficient way to improve stress responses. Along these lines, stress reappraisal focuses on manipulating appraisal processes – primarily resource appraisals – to improve stress responses (Beltzer et al., 2014; Brady et al., 2017; Hangen et al., 2019; Jamieson et al., 2010, 2012, 2013b, 2013c, 2016; John-Henderson et al., 2015; Moore et al., 2015; Rozek et al., 2019; Sammy et al., 2017). In this growing body of research, the arousal that individuals experience during stressful situations is conceptualized as a functional *resource* that can benefit outcomes. To date, the literature on stress reappraisal has primarily utilized two types of manipulations: a) an ~10-min reading/Q&A exercise comprised of summaries of scientific articles on the adaptive benefits of stress responses, similar to implicit theories of personality interventions (e.g., Yeager et al., 2016; materials available at http://socialstresslab.wixsite.com/urochester/research) and b) a "short form," paragraph length instruction (see Jamieson et al., 2010).

Notably, stress reappraisal acknowledges the demanding nature of stressful situations, and performance contexts in particular. That is, these manipulations do not focus on reducing perceptions of demands; the aim is not to eliminate or dampen sympathetic arousal or engagement. Rather, stress reappraisal focuses on boosting appraisals of resources so as to change the *type* of stress response experienced (threat → challenge). Maintaining, not decreasing, adaptive levels of arousal is needed to optimize performance and coping.

As touched on above, the focal mechanism of stress reappraisal is the resource component of stress appraisals within the context of challenge and threat theory. In fact, the stress response *itself* is presented as a coping resource (e.g., Jamieson, 2017). The emphasis on altering appraisals of resources is an important mechanistic distinction when individuals encounter stressful situations that cannot be avoided. For example, educational systems across the world require students to take evaluative exams that have direct relevance for outcomes such as grades, placements, and even one's future vocation. Students cannot avoid the demands of testing situations. However, students who reframe the stress they experience in these situations as functional can experience stress as a "skill" that then facilitates their performance (e.g., Jamieson et al., 2010, 2016).

In laboratory settings, stress reappraisal has demonstrated efficacy in improving performance, physiological functioning, attentional processes, and affective displays (e.g., Beltzer et al., 2014; Jamieson et al., 2012; John-Henderson et al., 2015). For instance, in one study (Jamieson et al., 2012), participants completed the commonly used stress induction procedure, the Trier Social Stress Test (Kirschbaum et al., 1993). Prior to speaking, participants were randomly assigned to reappraise stress, "ignore" stress (expectancy control), or received no instructions. Relative to the other conditions, reappraisal participants exhibited cardiovascular responses indicative of more challenge/less threat. Moreover, following the Trier task, attention for emotionally negative cues was assessed (Emotional Stroop: Williams et al., 1996), and reappraisal participants exhibited less vigilance for threatening cues than those assigned to the other conditions.

Research has also tested for applied effects of stress reappraisal (e.g., Brady et al., 2017; Jamieson et al., 2016; Rozek et al., 2019). For instance, in a double-blind field experiment (Jamieson et al., 2016), community college students were randomly assigned to receive stress reappraisal or expectancy control ("ignore stress") instructions prior to an in-class exam. Reappraisal participants reported less math evaluation anxiety and outperformed controls. An extension of this work tracked the temporal dynamics of stress reappraisal (Jamieson et al., n.d.). Relative to controls, reappraisal students exhibited a healthier pattern of neuroendocrine functioning (lower cortisol and higher testosterone), improved exam performance, and better psychological outcomes. Importantly for understanding temporal dynamics, reappraisal students reported more adaptive stress appraisals and higher performance approach goals compared to controls outside the classroom, and these psychological processes fed-forward to improve neuroendocrine functioning and academic performance during a subsequent exam situation.

Although research on stress mindsets and reappraisal has yielded promising findings, these are not the *only* cognitive operations that can be modified to improve stress responses. In fact, a central aim of current and future work is integrating these approaches with other processes to develop interventions. To illustrate, a combined stress reappraisal and expressive writing intervention program was implemented and delivered to ninth grade high school students to reduce socioeconomic (SES) disparities in Science, Technology, Engineering, and Math (STEM) education (Rozek et al., 2019). One notable highlight from this innovative integration was a

50% reduction in failure rate for low SES students assigned to the interventions. Thus, achievement outcomes can be improved by targeting stress arousal processes (the focus of stress reappraisal) and cognitive-affective processes, such as worries (the focus of expressive writing). In another potentially fruitful line, researchers have started bridging stress mindset and stress reappraisal interventions. The following section highlights a hypothesized integrated model.

Integrating Stress Mindsets and Stress Appraisals

At first glance, mindsets about the enhancing/debilitating nature of stress and appraisals of whether sufficient coping resources are available to meet demands in stressful contexts appear to be similar (or at least highly related) cognitive processes. Although these processes likely interact in important ways, they are psychologically distinct. One critical distinction between stress mindsets and stress appraisal processes is that mindsets do not focus on *specific* stressful situations. Rather, the foci of stress mindsets are more general and tap into the nature of stress (i.e., whether stressful experiences are enhancing or debilitating). That is, stress mindsets are meta-cognitive processes that shape responses in stressful contexts in potentially independent ways from how demands and resources are appraised. Moreover, although research on stress mindsets and stress reappraisal demonstrate that people can be active agents in constructing their stress responses, neither is a "magic bullet" for improving stress.

Initial empirical work has laid the groundwork for integrating mindset research with processes derived from the BPS model of challenge and threat – albeit no research to date has directly integrated intervention approaches derived from these research traditions. For instance, a set of laboratory and field experiments manipulated implicit theories of personality (not stress) and examined adolescents' responses to a social evaluative stressor (Yeager et al., 2016). High school students were assigned to receive incremental theory of personality – the belief that people have the capacity for change – or educational ("areas of the brain") control instructions. Notably, the incremental theory instructions attenuated associations among daily negative stressful events, threat-type stress appraisals, and cortisol levels (a catabolic stress hormone). Thus, teaching students an incremental theory of personality mindset altered how they appraised stressful events in their daily lives. With the exception of the preceding research and follow-up studies that also examined effects of implicit theories of personality on stress appraisals (Lee et al., 2018; Lee et al., in press), regulation approaches focusing on mindset and appraisal processes have preceded independently in the stress coping literature.

More pertinent for integrating mindsets about the nature of *stress* and challenge-threat processes, an extant experimental paradigm sought to test the effects of stress mindsets in situations designed to elicit differential patterns of stress appraisals – though appraisals were not directly assessed nor manipulated (Crum, Santoro, et al., 2018b). In that research, the valence of social feedback was manipulated in an evaluative performance task. When participants received negative evaluation, a stress-is-enhancing mindset improved outcomes. Optimal outcomes manifested, however, in positive evaluation contexts when participants endorsed a stress-is-enhancing mindset. However, stress appraisal processes and mindset concepts were not integrated in that prior research. In fact, to date, no empirical research has directly integrated stress mindsets and appraisals.

Future integration efforts may seek to focus on combining concepts from each approach (i.e., stress mindsets and stress appraisals) to promote stress optimization. A nexus for integration of mindset and appraisal processes is the idea that stressful experiences have the potential to promote physiological and psychological thriving (i.e., stress is not only negative). More specifically, stress mindsets seek to encourage individuals to see opportunities inherent in stress, and stress reappraisal instructs individuals to perceive stress arousal itself as functional and adaptive. Although mindset- and appraisal-based interventions target domain general and domain-specific processes, respectively, an underlying theme of both is the notion that altering interpretations of psychological and situational factors can promote coping and lead to thriving in demanding conditions.

Any integrated intervention incorporating mindset and appraisal concepts should, thus, not deny the negative realities of stress, but rather elucidate how stress can help optimize responses in stressful situations. That is, just as it is important not to unilaterally conceptualize and categorize stress (and accompanying responses) as negative, presenting stress as *only* positive or adaptive does not match reality. Stress is multifaceted and including content acknowledging potentially negative aspects of stress adds flexibility to intervention approaches. However, stress reappraisal approaches overwhelmingly focus on the adaptive, functional aspects of stress responses. This shortcoming could reduce the effectiveness of reappraising stress for some individuals. To illustrate the utility of maintaining flexibility, consider students taught to reappraise stress as functional and adaptive prior to taking an exam. If they receive a low score after appraising their stress as a resource, students may not reappraise stress in future exam situations because they believe doing so confers no benefits based on the prior situation. Increasing the flexibility of stress processes in an integrated mindset-appraisal intervention would help prevent people from overgeneralizing negative experiences in one particular stress context to every stress context.

Moreover, stress-is-enhancing mindset approaches seek to change meta-level beliefs about the *nature* of stress. Its focus is not on providing situation-specific tools (appraisals) to assist individuals in regulating their stress responses in particular contexts. For instance, one could adopt a meta-level mindset that the experience of stress is enhancing, but be at a loss for how to efficiently implement the belief system in a specific stress context (i.e., demands appraised as outweighing resources). Integrating processes from stress reappraisal can potentially help people *apply* their mindsets by elucidating the adaptive benefits of stress responses and promoting resource appraisal. That is, meta-level mindsets can orient individuals to engage with stressful situations, and appraisal processes can be leveraged to direct contextually specific responses (see also Yeager et al., 2016). The multi-level focus of a stress mindset and stress reappraisal integration is relevant for myriad other psychological processes too. Notably, recent advances in emotion regulation highlight the importance of considering how different psychological systems interact to regulate affective responses (Gross, 2015a). Toward this end, the following section delves into how mindsets and appraisals about stress can inform the development of models of emotion regulation.

Emotion Regulation

Future lines of research on mindsets and appraisals of stress should seek to incorporate concepts from the emotion regulation literature. In particular, stress mindsets and appraisals have direct relevance for the *extended process model of emotion regulation* (EPM; Gross, 2015a, b). The EPM posits that affective responses (such as stress responses) involve valuations. For example, whether a stressor or stress response is valued as "good for me" versus "bad for me" is central to the regulation of stress responses. Thus, a defining feature of the EPM regulation is the activation of valuation systems to influence trajectories of affective responses (Gross, 2015a, b). Myriad valuation systems operate to make evaluative distinctions, but each system focuses on different inputs, promotes different outputs, and unfolds differentially over time (e.g., Ochsner & Gross, 2014). Importantly, valuation systems operate at multiple levels and can interact such that one valuation system can target another valuation system (for more extensive reviews, see Gross, 2015a, b).

Emotion or stress generation valuation systems may be conceptualized as first-level systems, whereas emotion or stress regulation valuation systems operate as second-level systems. Thus, demand and resource appraisals (as defined by the BPS model) can be understood as second-level valuations that inform the cyclical emotion regulation-emotion generation chain (Ochsner & Gross, 2014). Stress mindsets can also function as second-level valuation systems interfacing with emotion generation systems at a more general level than situation-specific appraisals. To engage these second-level valuation/regulation systems, one must first evaluate whether representations of the environment (i.e., the "state of the world") match representations of the desired environment (e.g., "Is the way I'm feeling now how I want to feel?"). Then, if desired and current states do not align, one engages regulatory "second-level" valuation systems to produce action outputs to address the gap.

When regulating responses to stress, relaxation and stress attenuation approaches begin with the initial valuation that "stress is bad for me," and thus, regulatory goals focus on the reduction, avoidance, or elimination of stressful experiences altogether. However, stress mindset and stress reappraisal approaches engage second-level valuation systems to optimize stress responses based on the idea that stress responses can be helpful and/or enhancing. In other words, these approaches do not encourage relaxation, but rather trigger second-level valuations to generate beneficial high arousal challenge states. As noted previously, within the context of stressful performance situations, stress arousal is needed to optimize outcomes (Dienstbier, 1989).

Stress mindsets and reappraisal seek to appropriately match the emotion regulation strategy to the stressful context.

Valuation systems as conceptualized by the EPM, however, are not restricted to *only* mindset and appraisal processes. Moreover, stress appraisals, alone, are not even singular neural processes (Mende-Siedlecki et al., 2015). Beliefs about whether stress is enhancing/debilitating, or appraisals about whether sufficient resources are available to address situational demands, involve multiple, distinct cognitive operations, including (but not limited to) assessments of the meaning of stressful contexts for the self, attentional deployment to demanding aspects of the situation, episodic memories of prior stressful experiences, and/or executive processes to select and implement regulatory strategies (for reviews, see Ochsner et al., 2012; Schmitz & Johnson, 2007; Silvers et al., 2013).

The flexibility inherent in conceptualizations of valuation systems allows for the dynamic regulation of affect in stressful contexts, as well as a model to inform how unique processes interact. Importantly, dynamic processes have the potential to provide important insights into how (brief) interventions based on mindset and appraisal processes can exert long(er)-term effects on behavior and performance (and potentially interact in an additive sequence). For instance, modifying stress appraisals at one time-point can impact proximal physiological and performance outcomes, and distal goal processes, which can subsequently feed-forward to change appraisals and outcomes in future situations.

Dynamics that unfold between psychological processes are also important to consider. For instance, manipulating beliefs about stress has the potential to directly impact situation-specific stress appraisal relevant for challenge/threat responses (for a review, see Jamieson et al., 2018). Interpreted in the vernacular of the EPM, mindsets can be conceptualized as a "lens" through which situation-specific stress appraisals are implemented (i.e., two valuation systems interacting) to impact stress responding. For instance, if one believes stress is debilitating, stress responses cannot function as resource. For such individuals, situation-specific stress appraisals and responses will be particularly sensitive to perceptions of demands. That is, action outputs will be rooted in the demand component of BPS-derived appraisal processes. Thus, the EPM provides an organizing framework through which to understand recursive, temporal dynamics processes, and how stress mindsets and appraisals interact. Another important aspect of the dynamics of stress is how stress processes communicate between individuals and groups. The following section explores the potential for stress mindsets and appraisals to inform how stress is regulated in interpersonal or social contexts.

Interpersonal Processes

The extant research on stress mindsets and appraisals has focused on elucidating how individuals regulate their *own* affective responses in stressful situations. Stress regulation, however, does not occur in a social vacuum. Rich interpersonal and social processes inherent to naturalistic stress contexts should be fully considered so as to best understand the effects of stress regulation on downstream outcomes. That is, stress regulation processes can be transmitted between people, and these interpersonal processes have relevance for one's own stress responses *and* the stress responses of those with whom one interacts. However, relatively less is known about how stress mindsets and appraisals strategies enacted by "regulators" impact outcomes in interaction partners. This section highlights the interpersonal dynamics of stress mindset and appraisal processes and offers suggestions for areas of future research along these lines.

Models of interpersonal emotion regulation (IER) and physiological linkage can guide ongoing and future research on interpersonal stress regulation and the transmission between individuals. IER models (e.g., Zaki & Williams, 2013) specify how social resources can be leveraged to regulate emotional responses. The underlying premise of these models is that the regulation of affective responses is frequently interpersonal, and individuals recruit social coping resources when regulating affect (including stress responses, Uchino et al., 1996). Although one could posit that *any* social interaction influences emotional responding, IER processes are relevant only when individuals engage in *goal-directed* actions to alter affective states (similar to the goal-directed focus of the extended process model), though IER efforts can be *intrinsic* – outputs aimed at engaging the social environment to regulate one's own emotions – or *extrinsic*, outputs aimed at regulating others' emotions (Zaki & Williams, 2013). For example, in stressful interpersonal situations, an individual could leverage their social coping resources (e.g., seeking assistance to complete a task) in an effort to improve their own stress responses, or they could seek to help improve stress responding in others (e.g., providing unconditional positive regard). Each would constitute a goal-directed pursuit of affect change, but the former would be intrinsic, whereas the latter is extrinsic.

Physiological linkage models focus on the interdependence in interaction partners' physiological responses (often autonomic stress indexes), though we note there are myriad terms that include linkage effects, such as "contagion," "coupling," and "synchrony," to name a few (e.g., Bachrach et al., 2015; Levenson & Gottman, 1983; Waters et al., 2014). Although these terms can differ in their specific theoretical and practical implications (see Thorson et al., 2018), the perspective espoused in this chapter favors an integrated approach that focuses on processes that explain shared physiological states in stressful contexts. As highlighted by Thorson et al. (2018), understanding linkage effects is predicated on several underlying assumptions, the first of which is that communication can manifest within and across specific physiological measures. For instance, one dyad member's pre-ejection period (PEP, a measure of sympathetic arousal) may predict their interaction partner's respiratory sinus arrhythmia (RSA, a parasympathetic processes).

Related to the above, the upstream psychological processes that direct physiological responses can (and often are) distinct across interaction partners. To illustrate, consider an individual seeking to suppress their stress arousal so as to appear "calm, cool, and collected" to their interaction partner(s). If she/he engages emotion suppression processes to do so, this can potentially increase perceptions of uncertainty in their partners (e.g., *Why is my partner acting 'odd' or 'flat?'*). Thus, each interaction partner may be expected to experience physiological threat responses (Peters & Jamieson, 2016) and may even link up (West et al., 2017), but the physiological responses of the agent trying to *appear* calm are driven by response-focused regulation psychological processes, whereas the physiological responses of the target partner are driven by uncertainty reduction processes. The above example highlights the importance of understanding linkage processes by not only studying "transmitters" or agents of stress regulation, but also attributes of "receivers" or the targets of regulatory efforts. The subsections below highlight areas of research that have sought to map how stress regulation unfolds in interpersonal contexts rooted in principles from IER and linkage models.

Teams

Working with others is stressful and ubiquitous. From group projects in educational settings to joint projects in vocational settings, people work together to meet deadlines under evaluative pressure. In other words, groups marshal resources to address situational demands. Stress responses are inherently tied to the group context, and individuals often take actions to regulate their stress response in team settings. However, to date, little is known about how efforts to regulate their one's *own* stress responses impacts *others* they work with, even though a large corpus of research focuses on team design optimization (e.g., Hackman, 1987). The dearth of research along these lines is notable given the impact of teams' stress responses for organizational and educational outcomes, and the health and well-being of individual team members engaged in collaborative work.

Along these lines, initial research has started to map interpersonal dynamics of stress regulation in team performance settings. For instance, one study examined how stress reappraisal impacted teammates' physiological responses (Oveis et al., under review). This research is rooted in the idea that intrapersonal stress regulation efforts that impact one's own responses should impact social partners' responses. For example, if two teammates are developing a marketing plan, and one experiences a maladaptive threat response, this could "spill over" and negatively impact team members. However, if one teammate regulates stress responses and experiences an adaptive challenge response, other team members may contagiously benefit. To test these ideas, two strangers were brought to the lab to complete a collaborative task. Upon arrival, participants were separated for a resting baseline physiological measurement and task instructions. In addition, during their time alone, emotion regulation strategy was manipulated in the manipulated teammate, who was instructed to reappraise stress (reappraisal), suppress emotional displays (suppression), or received no instructions (control). The non-manipulated teammate received no instructions. After task instructions, participants were brought together to perform a collaborative task (develop a marketing plan for a product). Stress reappraisal benefited *both* teammates during collaborative work, eliciting challenge-like physiological responses relative to the suppression and control conditions. Thus, the stress regulation strategies enacted by the manipulated teammate spilled over to improve stress responses in their work partner who was unaware of the manipulation.

After collaborative work, participants pitched their product plan to evaluators trained to provide neutral-negative nonverbal feedback similar to a standard Trier Social Stress Test paradigm (e.g., Kirschbaum et al., 1993). During product pitch, the manipulated and non-manipulated teammates took turns presenting different aspects of the pitch (order was counterbalanced). Notably, the same physiological benefits experienced by both teammates during collaborative work also manifested during the product pitch epoch. That is, even though participants were not directly interacting during their pitches, non-manipulated teammates maintained benefits exhibited during collaborative work. Moreover, a mediation model suggested face-to-face interpersonal effects of stress reappraisal fed-forward to improve non-manipulated teammates' stress responses during individual performance. In sum, these data suggest that regulating stress responses with mindset- or appraisal-based methods has the potential to improve responses not only in regulators but also in others with whom they work.

Close Relationships

A common thread across all close relationships is that intrapersonal processes need to be interpreted within the relational context (e.g., West et al., 2008), and stress processes are important to examine in relational contexts. For instance, emotions are regulated in partner dyads (Butler & Randall, 2013), and partners can be sources of stress (i.e., create demands) and/or coping (e.g., social coping resources) (e.g., Major et al., 1997; Schoebi & Randall, 2015). To demonstrate how relational factors motivate and direct stress regulation, consider a situation in which a couple is discussing a relationship conflict, such as one member of a marital couple airing grievances about their partner which their partner must then respond to. In this case, the stressful conflict situation was generated by relational factors, and necessitates regulatory efforts by both partners to attenuate or optimize stress responses in such situations. If couples are dissatisfied and/or chronically initiate conflict discussions, negative stress process will map these relationship dynamics. For example, in a classic study, Levenson and Gottman (1983) found that dissatisfied partners were more likely to exhibit similar (negative) physiological responses during conflict discussions, but not during neutral discussions devoid of conflict.

Emotion co-regulation models are prominent in the close relationships literature and provide insights into interpersonal dynamics of stress processes (see Butler, 2015; Butler & Randall, 2013, for reviews). The organizing principle of co-regulation (and co-dysregulation) models is that dyadic affective processes are regulated in "self-regulating systems" that take into account experiential, behavioral, and physiological channels that vary across time and between partners (Butler & Randall, 2013). Thus, adaptive stress responses can be instantiated at the individual *and* interpersonal levels using myriad channels in which feedback between the levels informs downstream regulatory outcomes. Applying concepts from the stress mindsets and reappraisal literatures, there may be several lines to improve coping in close relationship dyads from teaching both members a stress-is-enhancing mindset and mapping how those beliefs feed-forward to impact how couples interact in future stressful situations, or educating individual members of a couple about reappraising stress when they find themselves embroiled in stressful relationship contexts to approach resolution goals. Or, as highlighted in the previous section, simply helping one member of a couple regulate their stress responses has the potential to spillover to positively impact their partner.

Studying stress mindset and appraisal processes in relationship contexts has much promise as highlighted by recent research on dyadic stress responses in noncorrespondent situations – those in which self-oriented interests do not align with partner's interests (Peters et al., 2018). Specifically, research examined how news that is positive for one relationship partner but that conflicts with the other partner's interests produced maladaptive stress responses in dyad members by asking couples to engage in a discussion in which one person (the *discloser*) revealed she/he had just gotten her/his dream job and the other person (the *responder*) reacted to the news. Couples were randomly assigned to discuss the discloser's positive news in a situation where they would live apart (misaligned self and relationship interests) or live together (aligned self and relationship interests). Responders *and* disclosers who discussed the noncorrespondent long-distance relationship scenario *and* exhibited threat-type stress responses were then behaviorally less responsive to their partners (Peters et al., 2018). Thus, if members of a relationship dyad appraised the demands of living apart as exceeding their resources to cope (i.e., a threat stress appraisal pattern), they were less able to be responsive to their partners, potentially exacerbating negative stress responses by removing social coping resources.

Summary and Conclusion

Stress is typically conceptualized as having negative effects on performance, well-being, and health. However, stress is a multifaceted psychobiological process, and is not inherently negative for downstream outcomes. Along these lines, the research reviewed here is predicated on the idea that stress may not only be "not negative," but rather beneficial for achievement, personal growth, and even mental and physical health. To highlight the many sides of stress, we focused on two psychological processes, stress mindsets and stress appraisals. A growing body of research suggests that optimizing mindset and appraisal processes has the potential to facilitate stress coping and promote thriving.

Although empirical evidence suggests mindset- and reappraisal-based stress interventions benefit individuals in *acute* stress contexts, chronic stress (i.e., prolonged stress exposure) can create allostatic load and cause negative health outcomes (e.g., Juster et al., 2010). However, as highlighted in the emotion regulation section above, it is possible that future stress mindset and stress appraisal research may target chronic stress processes through recursive and reciprocal processes. When mapping longitudinal processes, however, the social context and nature of the stressors being studied are important to consider. For instance, if one lives in a dangerous environment with high levels of violent crime, engaging reappraisal processes to attenuate the experience of threat may be unadvisable because attentional vigilance for threatening cues can be adaptive when navigating dangerous situations. In such chronically stressful environments, research may seek to target and understand resilience processes to optimize outcomes (McEwen, 2016), rather than implement stress reappraisal/mindset techniques.

Recently, researchers have started to formally integrate stress mindsets and stress appraisals into a unified framework (Jamieson et al., 2018), but this synthesis should also include other pertinent psychological processes, such as mindfulness. Particularly relevant for stress optimization, Eastern mindfulness traditions emphasize *awareness* in the present moment (i.e., "base awareness") and (often) nonjudgmental appraisals, acceptance, and/or a "quiet brain" (Brown et al., 2007). Present moment awareness is an integral factor for both stress mindset and stress reappraisal interventions. From the stress mindsets perspective, momentary awareness can orient attention to one's current mindsets – which typically operate nonconsciously – to allow for the modification of mindset content (e.g., Crum & Lyddy, 2014). In fact, the first step in stress mindset interventions is teaching individuals to acknowledge (i.e., notice or be aware of) their stress (Crum et al., 2013). Mindfulness processes are also pertinent to stress reappraisal. The content of stress reappraisal interventions specifically draws an individual's attentional focus to bodily signs of stress arousal (e.g., racing heart, sweaty palms, etc.). Rather than noticing and promoting acceptance of stress responses, though, reappraisal-based interventions encourage individuals to actively appraise their stress as a coping resource to help them succeed. However, without the initial awareness of and attention to internal stress processes, the effectiveness of reappraisal is limited. Future research on stress mindsets and stress appraisals – and especially interventions based on those processes – should seek to more closely integrate and study mindfulness processes to maximize impact.

Finally, it is important to emphasize that improving stress responses can be achieved through multiple means, not *only* via mindset- and reappraisal-based approaches. That is, mindsets and appraisals are not unique hubs for *all* psychological interventions to improve (or reduce, if that is the goal) stress responses. The research reviewed above, instead, highlights how bridging processes and theories that share underlying assumptions and goals can be conducted to help facilitate stress coping in active contexts. More broadly, it is our hope that this chapter can serve as an example for how science can be advanced not only by creating new theories or models (i.e., creating islands) but also by bridging existing theories or models (i.e., building bridges).

References

Antonovsky, A. (1998). *Stress, coping, and health in families: Sense of coherence and resiliency.* Sage Publications.

Appelhans, B. M., & Luecken, L. J. (2006). Heart rate variability as an index of regulated emotional responding. *Review of General Psychology, 10*(3), 229–240.

Bachrach, A., Fontbonne, Y., Joufflineau, C., & Ulloa, J. L. (2015). Audience entrainment during live contemporary dance performance: Physiological and cognitive measures. *Frontiers in Human Neuroscience, 9*, 179.

Beltzer, M. L., Nock, M. K., Peters, B. J., & Jamieson, J. P. (2014). Rethinking butterflies: The affective, physiological, and performance effects of reappraising arousal during social evaluation. *Emotion, 14*, 761–768.

Blackwell, L. S., Trzesniewski, K. H., & Dweck, C. S. (2007). Implicit theories of intelligence predict achievement across an adolescent transition: A longitudinal study and an intervention. *Child Development, 78*(1), 246–263.

Blascovich, J. (1992). A biopsychosocial approach to arousal regulation. *Journal of Social and Clinical Psychology, 11*, 213–237.

Blascovich, J. (2008). Challenge, threat, and health. In J. Y. Shah, Gardner, & W. L. Gardner (Eds.), *Handbook of motivation science* (pp. 481–493). Guilford Press.

Blascovich, J. (2013). Challenge and threat. In A. J. Elliot (Ed.), *Handbook of approach and avoidance motivation* (pp. 431–446). Psychology Press.

Blascovich, J., & Tomaka, J. (1996). The biopsychosocial model of arousal regulation. *Advances in Experimental Social Psychology, 28*, 1–52.

Blascovich, J., Mendes, W. B., Hunter, S. B., & Salomon, K. (1999). Social "facilitation" as challenge and threat. *Journal of Personality and Social Psychology, 77*(1), 68–77.

Brady, S. T., Hard, B. M., & Gross, J. J. (2017). Reappraising test anxiety increases academic performance of first-year college students. *Journal of Educational Psychology, 110*(3), 395–406.

Brooks, A. W. (2014). Get excited: Reappraising pre-performance anxiety as excitement. *Journal of Experimental Psychology: General, 143*(3), 1144–1158.

Brown, K. W., Ryan, R. M., & Creswell, J. D. (2007). Mindfulness: Theoretical foundations and evidence for its salutary effects. *Psychological Inquiry, 18*(4), 211–237.

Butler, E. A. (2015). Interpersonal affect dynamics: It takes two (and time) to tango. *Emotion Review, 7*(4), 336–341.

Butler, E. A., & Randall, A. K. (2013). Emotional coregulation in close relationships. *Emotion Review, 5*(2), 202–210.

Casper, A., Sonnentag, S., & Tremmel, S. (2017). Mindset matters: The role of employees' stress mindset for day-specific reactions to workload anticipation. *European Journal of Work and Organizational Psychology, 26*(6), 798–810.

Cohen, S., Kamarck, T., & Mermelstein, R. (1983). A global measure of perceived stress. *Journal of Health and Social Behavior, 24*, 385–396.

Crum, A. J., & Langer, E. J. (2007). Mind-set matters: Exercise and the placebo effect. *Psychological Science, 18*(2), 165–171.

Crum, A., & Lyddy, C. (2014). De-stressing stress: The power of mindsets and the art of stressing mindfully. In A. Ie, C. T. Ngnoumen, & E. J. Langer (Eds.), *The Wiley Blackwell handbook of mindfulness*. Wiley-Blackwell.

Crum, A. J., Salovey, P., & Achor, S. (2013). Rethinking stress: The role of mindsets in determining the stress response. *Journal of Personality and Social Psychology, 104*(4), 716.

Crum, A. J., Akinola, M., Martin, A., & Fath, S. (2017). The role of stress mindset in shaping cognitive, emotional, and physiological responses to challenging and threatening stress. *Anxiety, Stress, & Coping, 30*(4), 379–395.

Crum, A. J., Akinola, M., Turnwald, B. P., Kaptchuk, T. J., & Hall, K. T. (2018a). Catechol-O-methyltransferase moderates effect of stress mindset on affect and cognition. *PLoS One, 13*(4), e0195883.

Crum, A., Santoro, E., Smith, E. N., Salovey, P., Achor, S., & Mooraveji, N. (2018b). *Rethinking stress: Changing mindsets to harness the enhancing effects of stress.* Manuscript submitted for publication (copy on file with author).

Dickerson, S. S., & Kemeny, M. E. (2004). Acute stressors and cortisol responses: A theoretical integration and synthesis of laboratory research. *Psychological Bulletin, 130*(3), 355–391.

Dienstbier, R. A. (1989). Arousal and physiological toughness: Implications for mental and physical health. *Psychological Review, 96*(1), 84–100.

Dweck, C. S. (2000). *Self-theories: Their role in motivation, personality and development.* Taylor & Francis.

Eriksson, M., & Lindström, B. (2006). Antonovsky's sense of coherence scale and the relation with health: A systematic review. *Journal of Epidemiology & Community Health, 60*(5), 376–381.

Eschleman, K. J., Bowling, N. A., & Alarcon, G. M. (2010). A meta-analytic examination of hardiness. *International Journal of Stress Management, 17*(4), 277.

Folkman, S., & Lazarus, R. S. (1980). An analysis of coping in a middle-aged community sample. *Journal of Health and Social Behavior, 21*, 219–239.

Folkman, S., & Lazarus, R. S. (1985). If it changes it must be a process: Study of emotion and coping during three stages of a college examination. *Journal of Personality and Social Psychology, 48*(1), 150–170.

Goldenberg, A., Cohen-Chen, S., Goyer, J. P., Dweck, C. S., Gross, J. J., & Halperin, E. (2018). Testing the impact and durability of a group malleability intervention in the context of the Israeli–Palestinian conflict. *Proceedings of the National Academy of Sciences, 115*(4), 696–701.

Goyer, J. P., Akinola, M., Grunberg, R., & Crum, A. J. (2018). *Evaluation of a stress mindset intervention to improve performance and wellbeing in underrepresented minority college students at a selective institution.* Manuscript in preparation.

Gross, J. J. (2015a). Emotion regulation: Current status and future prospects. *Psychological Inquiry, 26*(1), 1–26.

Gross, J. J. (2015b). The extended process model of emotion regulation: Elaborations, applications, and future directions. *Psychological Inquiry, 26*(1), 130–137.

Hackman, J. R. (1987). The design of work teams. In J. Lorsch (Ed.), *Handbook of organizational behavior* (pp. 315–342). Prentice-Hall.

Haimovitz, K., & Dweck, C. S. (2016). What predicts children's fixed and growth intelligence mind-sets? Not their parents' views of intelligence but their parents' views of failure. *Psychological Science, 27*(6), 859–869.

Hangen, E. J., Elliot, A. J., & Jamieson, J. P. (2016). The opposing processes model of competition: Elucidating the effects of competition on risk-taking. *Motivation Science, 2*(3), 157–170.

Hangen, E. J., Elliot, A. J., & Jamieson, J. P. (2019). Stress reappraisal during a mathematics competition: Testing effects on cardiovascular approach-oriented states and exploring the moderating role of gender. *Anxiety, Stress, & Coping, 32*(1), 95–108.

Jamieson, J. P. (2017). Challenge and threat appraisals. In A. Elliot, C. Dweck, & D. Yeager (Eds.), *Handbook of motivation and cognition* (2nd ed., pp. 175–191). Guilford Press.

Jamieson, J. P., & Mendes, W. B. (2016). Social stress facilitates risk in youths. *Journal of Experimental Psychology: General, 145*(4), 467–485.

Jamieson, J. P., Mendes, W. B., Blackstock, E., & Schmader, T. (2010). Turning the knots in your stomach into bows: Reappraising arousal improves performance on the GRE. *Journal of Experimental Social Psychology, 46*(1), 208–212.

Jamieson, J. P., Nock, M. K., & Mendes, W. B. (2012). Mind over matter: Reappraising arousal improves cardiovascular and cognitive responses to stress. *Journal of Experimental Psychology: General, 141*(3), 417–422.

Jamieson, J. P., Koslov, K., Nock, M. K., & Mendes, W. B. (2013a). Experiencing discrimination increases risk taking. *Psychological Science, 24*, 131–139.

Jamieson, J. P., Mendes, W. B., & Nock, M. K. (2013b). Improving acute stress responses the power of reappraisal. *Current Directions in Psychological Science, 22*(1), 51–56.

Jamieson, J. P., Nock, M. K., & Mendes, W. B. (2013c). Changing the conceptualization of stress in social anxiety disorder affective and physiological consequences. *Clinical Psychological Science, 1*(4), 363–374.

Jamieson, J. P., Peters, B. J., Greenwood, E. J., & Altose, A. J. (2016). Reappraising stress arousal improves performance and reduces evaluation anxiety in classroom exam situations. *Social Psychological and Personality Science, 7*(6), 579–587.

Jamieson, J. P., Crum, A. J., Goyer, J. P., Marotta, M. E., & Akinola, M. (2018). Optimizing stress responses with reappraisal and mindset interventions: An integrated model. *Anxiety, Stress, & Coping, 31*(3), 245–261.

Jamieson, J. P., Black, A., Pelaia, L., Altose, A. J., & Reis, H. T. (manuscript in preparation). *The effects of stress reappraisal on neuroendocrine functioning, academic performance, and temporal dynamics in a naturalistic academic context.*

Jefferson, A. L., Himali, J. J., Beiser, A. S., Au, R., Massaro, J. M., Seshadri, S., ... Manning, W. J. (2010). Cardiac index is associated with brain aging: The Framingham heart study. *Circulation, 122*, 690–697.

John-Henderson, N. A., Rheinschmidt, M. L., & Mendoza-Denton, R. (2015). Cytokine responses and math performance: The role of stereotype threat and anxiety reappraisals. *Journal of Experimental Social Psychology, 56*, 203–206.

Juster, R. P., McEwen, B. S., & Lupien, S. J. (2010). Allostatic load biomarkers of chronic stress and impact on health and cognition. *Neuroscience & Biobehavioral Reviews, 35*(1), 2–16.

Karnilowicz, H. R., Waters, S. F., & Mendes, W. B. (2018). Not in front of the kids: Effects of parental suppression on socialization behaviors during cooperative parent–child interactions. *Emotion.* Advanced online publication.

Kassam, K. S., Koslov, K., & Mendes, W. B. (2009). Decisions under distress: Stress profiles influence anchoring and adjustment. *Psychological Science, 20*(11), 1394–1399.

Kirschbaum, C., Pirke, K. M., & Hellhammer, D. H. (1993). The 'Trier social stress test'–a tool for investigating psychobiological stress responses in a laboratory setting. *Neuropsychobiology, 28*(1–2), 76–81.

Kobasa, S. C., Maddi, S. R., & Kahn, S. (1982). Hardiness and health: A prospective study. *Journal of Personality and Social Psychology, 42*, 168–177.

Lazarus, R. S. (1991). Progress on a cognitive-motivational-relational theory of emotion. *American Psychologist, 46*(8), 819–834.

Lazarus, R. S., DeLongis, A., Folkman, S., & Gruen, R. (1985). Stress and adaptational outcomes: The problem of confounded measures. *American Psychologist, 40*(7), 770–779.

Lee, H. Y., Jamieson, J. P., Miu, A. S., Josephs, R. A., & Yeager, D. S. (2018). An entity theory of intelligence predicts higher cortisol levels when high school grades are declining. *Child Development, 90*, e849–e867.

Lee, H. Y., Jamieson, J. P., Reis, H. T., Beevers, C. G., Josephs, R. A., Mullarkey, M., O'Brien, J., & Yeager, D. S. (in press). Getting few likes on social media can make adolescents feel rejected. *Child Development.*

Levenson, R. W., & Gottman, J. M. (1983). Marital interaction: Physiological linkage and affective exchange. *Journal of Personality and Social Psychology, 45*(3), 587–597.

Major, B., Zubek, J. M., Cooper, M. L., Cozzarelli, C., & Richards, C. (1997). Mixed messages: Implications of social conflict and social support within close relationships for adjustment to a stressful life event. *Journal of Personality and Social Psychology, 72*(6), 1349–1363.

McEwen, B. S. (2016). In pursuit of resilience: Stress, epigenetics, and brain plasticity. *Annals of the New York Academy of Sciences, 1373*(1), 56–64.

Mendes, W. B., & Park, J. (2014). Neurobiological concomitants of motivational states. *Advances in Motivation Science, 1*, 233–270.

Mende-Siedlecki, P., Kober, H., & Ochsner, K. N. (2015). Emotion regulation: Neural bases and beyond. In J. Decety & J. T. Cacioppo (Eds.), *The Oxford handbook of social neuroscience* (pp. 277–293). Oxford University Press.

Molden, D. C., & Dweck, C. S. (2006). Finding "meaning" in psychology: A lay theories approach to self-regulation, social perception, and social development. *American Psychologist, 61*(3), 192–203.

Moore, L. J., Vine, S. J., Wilson, M. R., & Freeman, P. (2015). Reappraising threat: How to optimize performance under pressure. *Journal of Sport and Exercise Psychology, 37*(3), 339–343.

Ochsner, K. N., & Gross, J. J. (2014). The neural bases of emotion and emotion regulation: A valuation perspective. In *Handbook of emotional regulation* (2nd ed., pp. 23–41). Guilford.

Ochsner, K. N., Silvers, J. A., & Buhle, J. T. (2012). Functional imaging studies of emotion regulation: A synthetic review and evolving model of the cognitive control of emotion. *Annals of the New York Academy of Sciences, 1251*(1), E1–E24.

Oveis, C., Gu, Y., Ocampo, J. M., Hangen, E. J., & Jamieson, J. P. (revision under review). Emotion regulation contagion: Stress reappraisal promotes challenge responses in teammates. *Journal of Experimental Psychology: General*

Park, C. L., Riley, K. E., Braun, T. D., Jung, J. Y., Suh, H. G., Pescatello, L. S., & Antoni, M. H. (2017). Yoga and cognitive-behavioral interventions to reduce stress in incoming college students: A pilot study. *Journal of Applied Biobehavioral Research, 22*(4), e12068.

Paunesku, D., Walton, G. M., Romero, C., Smith, E. N., Yeager, D. S., & Dweck, C. S. (2015). Mind-set interventions are a scalable treatment for academic underachievement. *Psychological Science, 26*(6), 784–793.

Peters, B. J., & Jamieson, J. P. (2016). The consequences of suppressing affective displays in romantic relationships: A challenge and threat perspective. *Emotion, 16*(7), 1050–1066.

Peters, B. J., Reis, H. T., & Jamieson, J. P. (2018). Cardiovascular indexes of threat impair responsiveness in situations of conflicting interests. *International Journal of Psychophysiology, 123*, 1–7.

Rozek, C. S., Ramirez, G., Fine, R. D., & Beilock, S. L. (2019). Reducing socioeconomic disparities in the STEM pipeline through student emotion regulation. *Proceedings of the National Academy of Sciences, 116*(5), 1553–1558.

Sammy, N., Anstiss, P. A., Moore, L. J., Freeman, P., Wilson, M. R., & Vine, S. J. (2017). The effects of arousal reappraisal on stress responses, performance and attention. *Anxiety, Stress, & Coping, 30*(6), 619–629.

Schachter, S., & Singer, J. (1962). Cognitive, social, and physiological determinants of emotional state. *Psychological Review, 69*(5), 379–399.

Schmitz, T. W., & Johnson, S. C. (2007). Relevance to self: A brief review and framework of neural systems underlying appraisal. *Neuroscience & Biobehavioral Reviews, 31*(4), 585–596.

Schnyder, U., Büchi, S., Sensky, T., & Klaghofer, R. (2000). Antonovsky's sense of coherence: Trait or state? *Psychotherapy and Psychosomatics, 69*(6), 296–302.

Schoebi, D., & Randall, A. K. (2015). Emotional dynamics in intimate relationships. *Emotion Review, 7*(4), 342–348.

Seery, M. D. (2011). Challenge or threat? Cardiovascular indexes of resilience and vulnerability to potential stress in humans. *Neuroscience & Biobehavioral Reviews, 35*(7), 1603–1610.

Seery, M. D., Blascovich, J., Weisbuch, M., & Vick, S. B. (2004). The relationship between self-esteem level, self-esteem stability, and cardiovascular reactions to performance feedback. *Journal of Personality and Social Psychology, 87*(1), 133–145.

Seery, M. D., Weisbuch, M., & Blascovich, J. (2009). Something to gain, something to lose: The cardiovascular consequences of outcome framing. *International Journal of Psychophysiology, 73*(3), 308–312.

Seery, M. D., Holman, E. A., & Silver, R. C. (2010a). Whatever does not kill us: Cumulative lifetime adversity, vulnerability, and resilience. *Journal of Personality and Social Psychology, 99*(6), 1025–1041.

Seery, M. D., Leo, R. J., Holman, E. A., & Silver, R. C. (2010b). Lifetime exposure to adversity predicts functional impairment and healthcare utilization among individuals with chronic back pain. *Pain, 150*(3), 507–515.

Selye, H. (1936). A syndrome produced by diverse nocuous agents. *Nature, 138*(3479), 32.

Sherwood, A., Allen, M. T., Fahrenberg, J., Kelsey, R. M., Lovallo, W. R., & van Doornen, L. J. (1990). Methodological guidelines for impedance cardiography. *Psychophysiology, 27*(1), 1–23.

Silvers, J. A., Buhle, J. T., Ochsner, K. N., & Silvers, J. (2013). The neuroscience of emotion regulation: Basic mechanisms and their role in development, aging and psychopathology. *The Handbook of Cognitive Neuroscience, 1*, 52–78.

Thorson, K. R., West, T. V., & Mendes, W. B. (2018). Measuring physiological influence in dyads: A guide to designing, implementing, and analyzing dyadic physiological studies. *Psychological Methods, 23*(4), 595–616.

Tomaka, J., Blascovich, J., Kelsey, R. M., & Leitten, C. L. (1993). Subjective, physiological, and behavioral effects of threat and challenge appraisal. *Journal of Personality and Social Psychology, 65*(2), 248–260.

Uchino, B. N., Cacioppo, J. T., & Kiecolt-Glaser, J. K. (1996). The relationship between social support and physiological processes: A review with emphasis on underlying mechanisms and implications for health. *Psychological Bulletin, 119*(3), 488–531.

Walach, H., Buchheld, N., Buttenmüller, V., Kleinknecht, N., & Schmidt, S. (2006). Measuring mindfulness—The Freiburg mindfulness inventory (FMI). *Personality and Individual Differences, 40*(8), 1543–1555.

Waters, S. F., West, T. V., & Mendes, W. B. (2014). Stress contagion: Physiological covariation between mothers and infants. *Psychological Science, 25*(4), 934–942.

West, T. V., Popp, D., & Kenny, D. A. (2008). A guide for the estimation of gender and sexual orientation effects in dyadic data: An actor-partner interdependence model approach. *Personality and Social Psychology Bulletin, 34*(3), 321–336.

West, T. V., Koslov, K., Page-Gould, E., Major, B., & Mendes, W. B. (2017). Contagious anxiety: Anxious European Americans can transmit their physiological reactivity to African Americans. *Psychological Science, 28*(12), 1796–1806.

Williams, J. M. G., Mathews, A., & MacLeod, C. (1996). The emotional Stroop task and psychopathology. *Psychological Bulletin, 120*(1), 3–24.

Yeager, D. S., & Dweck, C. S. (2012). Mindsets that promote resilience: When students believe that personal characteristics can be developed. *Educational Psychologist, 47*(4), 302–314.

Yeager, D. S., & Walton, G. M. (2011). Social-psychological interventions in education They're not magic. *Review of Educational Research, 81*(2), 267–301.

Yeager, D. S., Miu, A. S., Powers, J., & Dweck, C. S. (2013). Implicit theories of personality and attributions of hostile intent: A meta-analysis, an experiment, and a longitudinal intervention. *Child Development, 84*(5), 1651–1667.

Yeager, D. S., Lee, H. Y., & Jamieson, J. P. (2016). How to improve adolescent stress responses: Insights from integrating implicit theories of personality and biopsychosocial models. *Psychological Science, 27*(8), 1078–1091.

Zaki, J., & Williams, W. C. (2013). Interpersonal emotion regulation. *Emotion, 13*(5), 803–810.

Part II
Mindfulness, Stress Reduction, and Mechanisms of Change

Chapter 6
Historical Origins and Psychological Models of Mindfulness

Michael Gordon, Shauna Shapiro, and Selma A. Quist-Møller

Introduction

Thirty years ago, the construct of mindfulness lived on the fringes of public awareness somewhere between Eastern mysticism and alternative medicine. Today, mindfulness is over a $1.1 billion industry (IBISWorld, 2021). From corporate boardrooms to the cover of *Time Magazine*, interest in contemplative practice is booming. This explosion stems from the growing body of literature investigating the effectiveness of mindfulness interventions. In 1990, there were 500 peer-reviewed articles on the science of meditation; in 2018 there were over 4000 (US National Library of Medicine, Vieten et al., 2018). Much of this boom in publications may be attributed to the dissemination of Jon Kabat-Zinn's mindfulness-based stress reduction program or MBSR (Kabat-Zinn, 1990). His operational definition and standardized 8-week course provided a consistent, secular, empirically driven framework for investigating the salutogenic effects of mindfulness meditation. Researchers have demonstrated that mindfulness-based interventions (MBIs) contribute to a multitude of psychological, physical, and cognitive benefits including the decrease in negative emotions and an increase in positive emotions (Lutz et al., 2014), greater relaxation (Khanna & Greeson, 2013), increased empathy and compassion (Zeng et al., 2017), decreased stress and anxiety (Davis & Hayes, 2011), decreased addictive cravings and behaviors (Garland & Howard, 2018), increased emotional resilience (Meiklejohn et al., 2012), increased learning and memory capacity (Hölzel et al., 2010), and improved job performance (Shonin & Van Gordon, 2014).

The majority of research thus far has pursued the question: does mindfulness work? And to a large extent, the answer has been a resounding yes. As researchers

M. Gordon (✉) · S. Shapiro
Santa Clara University, Santa Clara, CA, USA

S. A. Quist-Møller
University of Copenhagen, Copenhagen, Denmark

© Springer Nature Switzerland AG 2021
H. Hazlett-Stevens (ed.), *Biopsychosocial Factors of Stress, and Mindfulness for Stress Reduction*, https://doi.org/10.1007/978-3-030-81245-4_6

continue to build a critical mass of empirical backing of the benefits of mindfulness, clinicians pursue methodologies for implementing these practices. These mindfulness-based therapies range from dialectical behavioral therapy (Linehan, 1993) and the Hakomi method (Kurtz, 1990; Weiss, 2009) to mindfulness-based cognitive therapy (Teasdale et al., 2000) and acceptance and commitment therapy (Hayes, 2004). As the application of MBIs continues to broaden, researches are raising a second wave of inquiry, expanding beyond *if* MBIs work to *how* do MBIs work? These explorations seek to unveil the black box panacea of mindfulness practices and explore the mechanisms of action underlying their efficacy. In order to better understand these practices, researchers must resolve two basic steps in the research method: conceptualization and operationalization.

To date, there remains a lack of consensus about a definition of mindfulness within the field of psychology. While the majority of researchers use Kabat-Zinn's MBSR framework, the nuances of their investigation demand more specific terms. The attempts to operationalize the disparate components of MBIs have, accordingly, yielded a nebulous, often discordant, picture of how these practices work. However, as the body of research continues to grow, searching for the root of both what mindfulness is and how it operates, the findings offer increasingly concrete and pragmatic tools for clinicians and the people they seek to help.

This chapter will present some of the leading psychological models attempting to define mindfulness and elucidate the mechanisms that explain how it affects positive change. We first review both the historical roots and contemporary translations of the practices to illustrate the evolution of MBIs. We then review contemporary models of mindfulness and explore the active components of MBIs that may be causing their effects. The concept and understanding of mindfulness are constantly evolving, and as it does, so too must the rigor and nuance of our study and practice of it.

Conceptualization of Mindfulness

Historical Context

The term mindfulness has its roots in Eastern contemplative traditions and is most often associated with the formal practice of meditation. It must be understood, however, that mindfulness is not meditation. Mindfulness is "inherently a state of consciousness" (Brown & Ryan, 2003), while meditation is a practice to develop and strengthen that state. Epistemologically, the Sanskrit term for meditation, *bävhana*, denotes "cultivation" or "causing to become," and the Tibetan word, *sgoms*, translates to "developing familiarity" (Vargo & Silbersweig, 2012; Thera, 1962; Rahula, 1974; Bodhi, 1999; Jinpa, 2009). Meditation can be seen, therefore, as the "scaffolding" for the state or skill of mindfulness (Kabat-Zinn, 2005). The conceptualization of this state, however, remains a point of contention. There is no authoritative and definitive definition of mindfulness.

One of the first articulations of the original Pali term for mindfulness in English comes from the British scholar Thomas Rhys Davids. He translated the word *sati* (Sanskrit: *smṛti*), as literally meaning "memory," closely related to the verb *sarati*, referring to the process of "remembering" (Davids, 1882). While there is considerable variance within traditional Buddhist scholarship on the exact concept of mindfulness (Dreyfus, 2011; Dunne, 2011), the prevailing consensus recognizes a practiced connection between the faculties of memory and attention (Thera, 1962). The *satipatthana sutta* (The Foundation of Mindfulness Discourse), one of the key discourses on mindfulness in the Theravada tradition of Buddhism, outlines four major frames of reference or objects of attention: the body (Pali: *kāyā*), sensations or feelings aroused by perception (Pali: *vedanā*), the mind or consciousness (Pali: *cittā*), and the elements of Buddhist teachings (Pali: *dhammas*) (Thanissarom Bhikkhu, 2008).

One then uses these four objects to develop five major skills or qualities through meditative practice: (1) a balanced intensity of effort and diligence (Pali: *ātāpi*), (2) wisdom of clear discernment or phenomenal clarity (Pali: *sampajaña*), (3) mindful awareness, (4) freedom from desire and discontent (Pali: *vineyya loke abhijjhā-domanassạm*), and (5) equanimity (Pali, *upekkhā*) translated as "on-looking" or "watching things as they arise" and is described to involve a balance of arousal without hyperexcitability or fatigue (Vago & Silbersweig, 2012; Buddhaghosa, 1991). By developing these skills, one practices mindfulness on a direct path to the "cessation of suffering" (Analayo, 2003).

It should be noted that the Buddhist conception of suffering extends well beyond acute physical or psychopathological ailments. The Pali term, *dukkha*, commonly translated as "suffering," also holds the connotation of "unsatisfactory" with the metaphoric reference to a poorly fitted axle causing the cart to continuously rattle and lurch. Some have even extended the translation to a more modern term, "stress" (Goldstein, 2013). Through mindfulness practice one becomes aware of habitual thought patterns and behaviors which lead to suffering. Accordingly, the Buddhist path of mindfulness engages not only with the present circumstances of suffering but also the underlying root causes of that pain, anxiety, and dissatisfaction.

In the *Milindapana* sutra, the oldest known attempt in Buddhist literature to define the term, the monk Nagasena states, "Sati, when it arises, calls to mind wholesome and unwholesome tendencies... Sati, when it arises, follows the courses of beneficial and unbeneficial tendencies: these tendencies are beneficial, these unbeneficial; these tendencies are helpful, these unhelpful" (Gethin, 2001). This ancient conception suggests both a "retrospective memory" and a "prospective" memory, by which the individual actively chooses the "most beneficial" course of action based on previous understandings and applies the said understandings to the present moment (Wallace, 2010).

One is not only observing their current circumstances but also actively connecting those phenomena to a framework of tendencies that predict "helpful and unhelpful" outcomes. In a Buddhist context, this ethical structure is called *sila* and is composed of (1) inner virtue (kindness, truthfulness, and patience), (2) virtuous actions of body and speech (not doing or causing harm to others), and (3) rules designed to keep the individual aligned with the ethical ideals (abstaining from killing, stealing, lying, sexual misconduct, and intoxicants) (Bhikkhu Bodhi, 1994). Within this context, the ethical structure is seen as a prerequisite for the development of mindfulness (Spiro, 1982; Bhikkhu Bodhi, 2010), whereas in a secular, scientific context, the ethical structure is seen as a result of developing mindfulness (Ruedy & Schweitzer, 2010; Olendzki, 2011; Shapiro et al., 2012). Under such an assumption, the ethical context for mindfulness becomes an auxiliary benefit rather than a fundamental component of cultivating the state.

Much of this discrepancy may be attributed to the way in which secular MBIs have treated mindfulness as a "technique for symptomatic relief" (Monteiro et al., 2014) and not a concept that only has a full meaning when it is incorporated into a system of practices and conduct that is essentially shared among the major Buddhist traditions (Grossman, 2014; Goldstein, 2002). Accordingly, many traditional Buddhist practitioners argue that a conception of mindfulness stripped of its ethical context can lead to negative outcomes (see Purser & Loy, 2013; Ricard, 2009; Senauke, 2013; Titmuss, 2013; Monteiro et al., 2014). Additionally, some scholars contend that the traditional context provides a nuanced language that is more psychologically accurate, informed, and relevant to contemporary work and life than modern translations (Maex, 2011; Monteiro et al., 2014).

We agree with Kabat-Zinn (2003, p. 147) that "different but complementary epistemologies" of Buddhism and the Western scientific models of mindfulness can mutually inform each other to reach a common goal: the alleviation of suffering.

Recognizing the Old, Realizing the New

The majority of cited definitions of mindfulness come from two primary sources: traditional Buddhist literature and contemporary secular models largely based on Jon Kabat-Zinn's mindfulness-based stress reduction (MBSR) course. The most commonly cited contemporary definition is the one offered by Kabat-Zinn (2003): "the awareness that emerges through paying attention on purpose, in the present moment, and nonjudgmentally to the unfolding of experience moment by moment." These three components of awareness (purposeful, present-centered, and nonjudgmental) function as broad placeholders for a continually evolving conceptualization of the term.

Together, the three components create the state of mindfulness, a state of awareness that is certainly more than the sum of its parts. In order to rigorously define and assess each component, a large body of research continues to investigate the parameters, efficacy, and neurobiological features of these mechanisms (see Baer, 2003; Shapiro et al., 2009; Hölzel et al., 2010; Quist Møller et al., 2019). It should be noted, however, that these mechanisms function in a non-linear cyclical process, each one continually building upon and affecting the others.

While MBSR does draw from specific Buddhist techniques intended for general stress reduction (Kabat-Zinn, 1990), its emphasis is on providing a "vehicle for the effective training of medical patients... free of the cultural, religious, and ideological factors associated with the Buddhist origins of mindfulness" (Kabat-Zinn, 2003). Accordingly, MBSR purposefully dissociates itself from much of the ethical and metaphysical context surrounding the traditional Buddhist roots of mindfulness. The tension between maintaining the complexity and authenticity of the ancient traditions while engaging in rigorous, empirical science forms the crux of establishing consensus on an unequivocal definition of mindfulness.

Some researchers have argued for a complete separation of the methods of mindfulness from their historical lineages. According to Hayes (2002), in order to properly integrate these methods into Western psychological practices, "the field must be free to interpret and transform them theoretically, without being limited by their religious and spiritual past." While this line of research has been crucial in establishing a rigorous scientific study of mindfulness and its benefits, it has also led to narrowly defined conceptions of the term and its corresponding practices. Below we will explore these in greater depth.

Researchers, such as Grossman (2014) and Grabovac et al. (2011), argue that the secularization of these practices has resulted in a neglect of the explicitly defined psychological and mechanistic models of mindfulness proposed by Buddhism, leading to "the unnecessary loss of the context that explains how these techniques work and why they are used" (Grabovac et al., 2011). The field is calling for increased specificity and depth in the definition of mindfulness, differentiating it as a trait, state, or practice (Davidson, 2010), as well as an examination of how component of mindfulness leads to specific outcomes (Coffey et al., 2010).

In response, this paper reviews six contemporary models of mindfulness: (1) the intention, attention, and attitude model (IAA) (Shapiro et al., 2006), (2) the Buddhist psychological model (BPM) (Grabovac et al., 2011), (3) the self-awareness, self-regulation, and self-transcendence model (S-ART) (Vago & Silbersweig, 2012), (4) two-component model of mindfulness (Bishop et al., 2004), (5) statistically derived model of mindfulness (Coffey et al., 2010), and (6) the attention regulation model (Carmody, 2009).

Contemporary Psychological Models of Mindfulness

The IAA Model (Shapiro et al., 2006)

Shapiro and her colleagues developed a model of mindfulness that posits three interwoven components of mindfulness: *intention, attention,* and *attitude*. These three elements correlate with Kabat-Zinn's definition of mindfulness:

1. "Paying attention on purpose" or intention
2. "In the present moment" or attention
3. "Non-judgmental/in a particular way" or attitude

The model is an attempt to break mindfulness down into a simple, yet nuanced construct that reflects the core components of the practice. Intention, attention, and attitude (IAA) are fundamental building blocks of mindfulness out of which four additional mechanisms emerge: (1) self-regulation and self-management; (2) emotional, cognitive, and behavioral flexibility, (3) values clarification; and (4) exposure.

The IAA model attempts to reground the conception of mindfulness in a sense of altruism and purpose. Shapiro (2009) writes: "when Western psychology attempted to extract the essence of mindfulness practice from its original religious/cultural roots, we lost, to some extent, the aspect of intention" (p. 557). In the Buddhist tradition, this is commonly referred to as *boddhichitta*, or the "spontaneous wish to attain enlightenment motivated by great compassion for all sentient beings, accompanied by a falling away of the attachment to the illusion of an inherently existing self" (Fischer, 2013, p. 11). While Shapiro and her colleagues developed a model free of religious and cultural context, they did attempt to preserve its fundamental characteristics of compassion and the will to alleviate suffering, first for one's self and, ultimately, for all others. This intentionality is seen as both dynamic and evolving, a direction rather than a destination.

In an earlier study, Deane Shapiro (1992) outlined a continuum of intention, whereby one moves from self-regulation to self-exploration and finally to self-transcendence. The study found that the outcomes of the mindfulness practice correlated with the participant's intentions: "Those whose goal was self-regulation and stress management attained self-regulation, those whose goal was self-exploration attained self-exploration, and those whose goal was self-liberation moved toward self-liberation and compassionate service" (Shapiro 1992, p. 26).

This is further supported by a later study by Ferguson and Sheldon (2013) who found that simply setting the intention to be happy elevated dopamine levels in the brain, creating positive mood. Thus, it is important to know that it is not just "doing" meditation that will create growth and change, but having some kind of personal vision is also necessary (Shapiro & Carlson, 2009; Kabat-Zinn, 1990). Intention puts us in touch with *why* we pay attention. It helps us connect with our unique, motivating force behind the practice.

This continuum parallels what Patrul Rinpoche (2010) describes as the three levels of loving kindness (*bodhichitta*): "The way of the King, who primarily seeks his own benefit but who recognizes that his benefit depends crucially on that of his kingdom and his subjects. The path of the boatman, who ferries his passengers across the river and simultaneously, of course, ferries himself as well, and finally that of the shepherd, who makes sure that all his sheep arrive safely ahead of him and places their welfare above his own" (p. 134). However, further research still needs to be conducted to properly validate the efficacy of intention.

The second element, attention, entails observing one's moment-moment internal and external experiences (e.g., thought, emotions, sensations) (Shapiro et al., 2006). The meditator learns to objectively examine the contents of consciousness, "suspending all the ways of interpreting experience and attending to experience itself, as it presents itself in the here and now" (p. 376). It involves a dynamic process of learning how to cultivate an attention that is discerning and nonreactive and sustained and concentrated, so that we can see clearly what is arising in the present moment (Shapiro et al., 2012). Shapiro et al. (2006) makes explicit connection to the way which cognitive psychology have studied the various components of attention, including the capacity to attend for long periods of time to one object (vigilance or sustained attention; Parasuraman, 1998; Posner & Rothbart, 1992), the ability to shift the focus of attention between objects and mental sets at will (switching; Posner et al., 1980), and the ability to inhibit secondary elaborative processing of thoughts, feelings, and sensations (cognitive inhibition; Williams et al., 1996).

The third element, attitude, relates to *how* one pays attention. The field of neuroplasticity demonstrates that our repeated experiences shape our brain. Thus, if we continually practice meditation with a cold, judgmental, and impatient attitude, these are the pathways that will grow stronger. Instead, Shapiro (2009, p. 558) outline that rather than slipping into negative habitual patterns, our intention instead is to practice with an attitude of kindness, curiosity, patience, compassionate, and non-striving attention.

From an operation level, "non-striving" and "accepting" may be treated as synonyms. Both refer to the capacity "not to continually strive for pleasant experiences, or to push aversive experiences away'" (Shapiro et al., 2006). Shapiro adds the "heart qualities," which she derives from the Japanese characters of mindfulness, composed of two interactive figures: one mind and the other heart (Santorelli, 1999). By adding patience, kindness, and compassion, Shapiro connects the concept of mindfulness to what some researchers are now referring to as compassion-based interventions (CBIs) (Gonzalo et al., 2018) or compassion-focused therapy (CFT) (Sommers-Spijkerman et al., 2018).

This effort to explicitly instruct patients on compassion building practices as part of mindfulness interventions has been shown to significantly increase empathetic concern and identification with all of humanity (Gonzalo et al., 2018), compounding positive emotions (Fredrickson & Cohn, 2008) and reducing psychological distress (Sommers-Spijkerman et al., 2018). It also addresses a return to what the Buddhist tradition refers to as *sila* or ethical conduct. By establishing compassion as a fundamental component of mindfulness, we might avoid the negative outcomes of mindfulness devoted solely to productivity and personal gain (see Purser & Loy, 2013; Ricard, 2009; Senauke, 2013; Titmuss, 2013; and Monteiro et al., 2014).

The theory behind IAA posits that through mindfulness practice of "intentionally (I) attending (A) with openness and non-judgmentalness (A), ... one becomes able to disidentify from the contents of consciousness and view one's moment-by-moment experience with great clarity and objectivity" (Shapiro & Carlson, 2009, p. 558). Shapiro and Carlson (2009) suggest that this leads to a fundamental shift in perspective, in which what was previously subject now becomes objective. They have termed this *reperceiving*. Essentially, reperceiving introduces a "space between one's perception and response" allowing the individual to disengage or "step outside" one's immediate experiences and simply be with the contents of consciousness instead of being defined by them or identifying with them. What was previously subject now becomes objective.

As Kabat-Zinn puts it (1990), this shift in perspective allows you to realize that, "your *awareness* of sensations, thoughts and feelings is different from the sensations, the thoughts and the feelings themselves" (p. 297). Thus, the emphasis is on changing one's relationship with thought rather than attempting to alter the content of thought itself. Practitioners may become able to observe their thoughts and emotions as temporary events in the mind, enabling them to experience thoughts and emotions with a sense of distance and objectivity, rather than identifying them with a sense of self (Teasdale, 1999).

Reperceiving is akin to the western psychological concepts of *decentering* (Safran & Segal, 1990; Fresco et al. 2007), *deautomatization* (Detloff & Deikman, 1982; Safran & Segal, 1990), *detachment* (Bohart, 1983) metacognitive awareness (Teasdale et al., 2002), *defusion* (Fletcher & Hayes, 2005), and *decreased rumination* (Deyo et al., 2009).

The IAA posits that the key mechanism through which mindfulness practice has its transformation effects is through the process of reperceiving. Preliminary evidence suggests that the ability to shift perspectives from an egocentric to a more objective view is salutary (Orzech et al., 2009, p. 111). For example, "if we are able to see *it,* then we are no longer merely *it,* that is, we must be more than *it*" (Shapiro & Carlson, 2009, p. 103). Meditators become able to rest in a witness consciousness where they develop the capacity to see their thoughts, emotions, and sensations, without becoming lost in them. As this shift in perspective continues to gain traction, the practitioner gradually begins a process of reorientation, whereby the recognition of something vaster than the individual leads to a natural decrease of self-conscious emotions. The practitioners' self-oriented sense of being, goals, and anxieties become less important as they experience a greater sense of connection to the larger world around them. Researchers have documented how this enhanced sense of self-diminishment and vastness, in which participants felt insignificant in the presence of something greater than the small self, can lead to less worry about day-to-day concerns (Shiota et al. 2007) and an increase in the collective dimensions of personal identity and prosocial behavior, such as generosity, helpfulness, and ethical conduct (Piff et al. 2015). This phenomenon parallels the traditional Buddhist teaching that through ethical practice (*sila*) and compassion for all beings (*metta* or *bodhichitta*) one realizes not only a disidentification but also a deconstruction of the individualized self, referred to as the not-self (*anatha*) (Grabovac et al., 2011) or self-transcendence (Vago & Silbersweig, 2012).

Shapiro et al. (2006) argue that this change in perspective may lead to additional mechanisms that in turn contribute to the positive outcomes produced by mindfulness practice. They highlight four additional mechanisms: (1) self-regulation and self-management; (2) values clarification; (3) cognitive, emotional, and behavioral flexibility; and (4) exposure. We will now explore each mechanism in greater depth.

Self-Regulation

Self-regulation is a process whereby systems maintain stability of functioning and adaptability to change by continually monitoring cognitive and emotional feedback loops. Similar to the attention regulation model (Carmody, 2009) presented later in this chapter, self-regulation interrupts the maladaptive habits of egocentric, depressive, anxious, etc. mental proliferation. Through the development of reperceiving, one comes to see that whatever arises, the conscious field of perception is, inherently, impermanent and will therefore inevitably pass away. Developing the mechanism of self-regulation through reperceiving also leads to what Hayes (2002) noted: "experiential avoidance becomes less automatic and less necessary" (p. 104). Rather than running from negative emotions, trauma, or pain, the individual learns to sit with them and observe their coming and going. According to Shapiro and Schwartz (1999, 2000), both intention and attention function to enhance these feedback loops and increase psychological wellness. This assertion is also supported by a study by Brown and Ryan (2003) in which they demonstrated that people who scored higher on a valid and reliable measure of mindfulness reported significantly greater self-regulated emotion and behavior.

Values Clarification

The IAA argues that reperceiving may also help people recognize what is meaningful for them and what they truly value. Often values have been conditioned by family, culture, and society, so that we may not realize whose values actually drive our choices in life. We become the value, instead of the one who observes the value (Shapiro et al., 2006). Frequently, we are pushed and pulled by what we believe is most important (based on cultural or familial conditioning) but fail to reflect upon whether it is truly important in the context of our own lives. However, when we are able to separate from (observe) our values and reflect upon them with greater objectivity, we have the opportunity to rediscover and choose values that may be truer for us. In other words, we become able to reflectively choose what has been previously reflexively adopted or conditioned. The literature suggests that automatic processing often limits considerations of options that would be more congruent with needs and values (Brown & Ryan, 2003; Ryan et al., 1997). However, an open, intentional awareness can help us choose behaviors that are congruent with our needs, interests, and values (Brown & Ryan, 2003; Ryan et al., 1997). A recent study found that when subjects are "acting mindfully," as assessed by the Mindful Attention Awareness Scale (MAAS) state measure, individuals act in ways that are more congruent with their actual values and interests (Brown & Ryan, 2003).

Cognitive, Emotional, and Behavioral Flexibility

Reperceiving may also facilitate more adaptive and flexible responding to the environment in contrast to the more rigid and reflexive patterns of reactivity that result from being overly identified with one's current experience. If we are able to see a situation and our own internal reactions to it with greater clarity, we will be able to respond with greater freedom of choice (i.e., in less conditioned, automatic ways) (Shapiro et al., 2006). As Borkovec (2002) points out, research from cognitive and social psychology demonstrates, "existing expectations or beliefs can distort the processing of newly available information" (p. 78).

Reperceiving facilitates this capacity to observe one's mental commentary about the experiences encountered in life. It enables us to see the present situation as it is in this moment and to respond with intentional purpose, instead of reacting with perfunctory thoughts, emotions, and behaviors triggered by prior habit, conditioning, and experience. Reperceiving affords a different place from which to view the present moment. For example, when we are caught on the surface of the ocean, and the waves are thrashing us about, it is difficult to see clearly. However when we drop down beneath the surface of the waves (which is analogous to observing and disidentifying from the movement of one's thoughts and emotional reactions), we enter a calmer, clearer space (Detloff & Deikman's (1982) "observing self" or what contemplative traditions refer to as "the Witness"). From this new vantage point, we can look up to the surface and see whatever is present more clearly—and therefore respond with greater consciousness and flexibility. Reperceiving enables the development of this capacity to observe our ever-changing inner experience and thereby see more clearly our mental-emotional content, which in turn fosters greater cognitive-behavioral flexibility and less automaticity or reactivity.

Exposure

The literature is replete with evidence of the efficacy of exposure in treating a variety of disorders (Barlow & Craske, 2000). Reperceiving—the capacity to dispassionately observe or witness the contents of one's consciousness—enables a person to experience even very strong emotions with greater objectivity and less reactivity. This capacity serves as a counter to the habitual tendency to avoid or deny difficult emotional states, thereby increasing exposure to such states. Through this direct exposure, one learns that his or her emotions, thoughts, or body sensations are not so overwhelming or frightening. Through mindfully attending to negative emotional states, one learns experientially and phenomenologically that such emotions need not be feared or avoided and that they eventually pass away (Segal et al., 2002). This experience eventually leads to the "extinction of fear responses and avoidance behaviors previously elicited by these stimuli" (Baer, 2003, p. 130). Goleman (1971) suggests that meditation provides a "global desensitization" as meditative awareness can be applied to all aspects of one's experience.

Baer (2003) provides an example of this process with chronic pain patients: "... prolonged exposure to the sensations of chronic pain, in the absence of catastrophic consequences, might lead to desensitization, with a reduction over time in the emotional responses elicited by the pain sensations. Thus the practice of mindfulness skills could lead to the ability to experience pain sensations without excessive emotional reactivity" (p. 128). Indeed, one of the first successful clinical applications of mindfulness was in the context of chronic pain (Kabat-Zinn, 1990). Another example of how facilitation of exposure to internal stimuli can help therapeutically comes from the literature on interoceptive exposure to physical sensations in panic disorder. Reperceiving allows one to explore and tolerate a broad range of thoughts, emotions, and sensations, which may in turn positively impact a number of debilitating conditions.

Buddhist Psychological Model (Grabovac et al., 2011)

The Buddhist psychological model (BPM), developed by Grabovac and her colleagues (Grabovac et al., 2011), utilizes the specificity and nuance of historical conceptions of mindfulness to provide a contemporary and holistic operationalization of the process. The authors draw primarily on the Abhidhamma Pitaka, a collection of Buddhist texts which contain the majority of what might be called Buddhist psychology, as a response to the lack of consensus about how contemporary mindfulness practices exert their physical, emotional, and psychological effects. Through this traditional lens, the model dissects the procedural components of mental activity leading to mental proliferation and pathological thought patterns.

The model focuses on the process by which a person reacts to the concordant feeling tone (pleasant, unpleasant, or neutral) associated with the perception of an object. The BPM then offers a definition of mindfulness in which the person observes three proposed characteristics of phenomena: impermanence, suffering, and not-self. By becoming aware of these three characteristics moment to moment, the individual develops a realization of equanimity or "viewing an object without attachment or aversion" (Grabovac et al., 2011, p. 3). Through the development of equanimity, the person experiences a "radical change in perception" and, in a clinical setting, pathological symptom reduction.

BPM's Components of Mental Activity

The BPM defines awareness of an object as, "when either a stimulus enters our field of perception and makes contact with a sense organ (i.e., sense impression) or when an object of cognition (a thought, memory, emotion) arises in the mind" (Grabovac et al., 2011, p. 2). This awareness is transient and usually only lasts for a few moments. Contrary to the majority of Western psychology, in the BPM, there is no distinction made between an awareness of sense impression (physical sensation) and cognitions (thoughts, memories, emotion, etc.). Both are treated equally in the mindfulness practices as arising (and passing) mental events.

Attentional resources are limited (Rogers & Monsell, 1995; Uncapher & Wagner, 2018), and an individual can only be aware of one object of awareness at a time.

Although we normally perceive a seemingly continuous "stream of consciousness," our experience is actually composed of a series of individual sense impressions (and cognitions). Grabovac et al. (2011) makes the analogy to the perceived movement of a film, which is actually composed of a series of still images. Like the passing of one still image to the next, our experience of sequential mental events occurs incredibly rapidly, with dozens of events occurring in a single second.

According to the BPM, each mental event has a concordal feeling tone: pleasant, unpleasant, or neutral. These feeling tones should not be confused with the complex physiocognitive states of emotions. Rather, they are "immediate and spontaneous affective experience of this awareness of a physical sensation or mental event" (Mendis, 2006, p. 20). And, as they occur rapidly and transiently, they often go unnoticed. However, these feeling tones are the fulcrum of our experience. Our habitual response is to pursue that which is pleasant and avoid that which is unpleasant. In a Buddhist context, this is known attachment and aversion, respectively. Through the process of learning, we group together these attachments and aversions into a web of memory, emotion, and cognitions to form a narrative around the object (Pali: *sankhara*). We say to ourselves, "I want the apple" (attachment) or "I am repulsed by the smell" (aversion). However, according to the Abhidhamma Pitaka and the BPM, attachment and aversion occur in reaction to the feeling tones rather than the object. When we say "I want the apple," what we are pursuing is the pleasant feeling we associate with the apple, not the apple itself.

In the early stages of meditation, Grabovac and her colleagues (2011) describe the cultivation of attention regulation. In order to sustain uninterrupted focus on a chosen object, attentional regulation entails attentional stability; the ability to let thoughts, emotions, and sensations go (similar to Shapiro's reperceiving); and the quality of acceptance. According to the BPM, mindfulness does not involve cognition. Accordingly, since these processes all require cognition, the BPM does not define them as active components of mindfulness. They are all components of attentional regulation.

BPM's Definition of Mindfulness

The BPM defines mindfulness as the active concentration on three foci of experience, which relate to the "three marks of existence" or "three characteristics" in Buddhist ontology

BPM foci	Buddhist "three characteristics"
1. Sense impressions and mental events are transient (they arise and pass away)	Impermanence (Pali: anicca)
2. Habitual reactions (i.e., attachment and aversion) to the feelings of a sense impression or mental event and a lack of awareness of this process, lead to suffering	Suffering (Pali: dukkha)
3. Sense impressions and mental events do not contain or constitute any lasting, separate entity that could be called a self	Not-self (Pali: anatta)

During any given mindfulness practice, the participant focuses on the way in which experiences arise and pass away (impermanence), elicit reactions of attachment and aversion which lead to a feeling of unsatisfactoriness (suffering), and do not contain any lasting, separate self-entity (not-self).

At first, the participant conceptually thinks about each of the three foci, noticing (and potentially narrating) the process described in Fig. 6.1 as it occurs. This process, known as insight, interrupts the habitual pattern of thought proliferation by recognizing both the sense impressions and mental events (as well as their concordal feeling tones and the habituated response of attachment and aversion) as transient happenings. By examining, naming, and observing each component of mental experience, one can begin to see the space between them (gaps between the still frames of the movie). Repeated practice then leads to a state of direct perception (Sanskrit: pratyakṣa) where cognitive reasoning is no longer necessary. As stated before, this state transcends all narrations of "name or genus…or [cognitive] expression," (Toru, 2005), and therefore the mental elaboration connected to attachment and aversion becomes superfluous. Grabovac et al. (2011) posits: "All sensory and mental events are allowed to naturally arise and fall away, without subsequent cognitive processing arising from either attachment or aversion. Sense impressions and mental events are still experienced as pleasant, unpleasant, or neutral; however, if there is no attachment, aversion, and thus no mental proliferation, adventitious suffering is not experienced" (p. 4).

It should be noted that the BPM's goal is the complete alleviation of suffering and it does not focus on symptom reduction (in a clinical sense). However, "reduction in symptoms resulting from practices such as mindfulness meditation is explainable as a reduction in these habitual reactions and resulting mental proliferation" (Grabovac et al., 2011, p. 4).

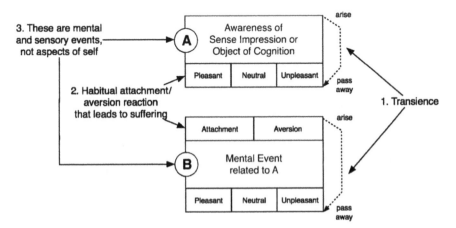

Fig. 6.1 Grabovac's schematic outlining the awareness of anicca (*1. transience*), dukkha (*2. habitual attachment/aversion*), and anatta (*3. not aspects of the self*)

Distinction Between BPM and Other Mindfulness Models

The authors of BPM make careful distinction between mindfulness or vipassana (an insight-oriented practice) and concentration or samatha (an attention regulation practice). While the latter seeks to stabilize attention on a single object, the former attempts to deconstruct the processes of mental activity and their subsequent consequences. In other words, mindfulness involves observing the three characteristics of experience, whereas concentration does not.

In the BPM, the notion of acceptance, critical to Kabat-Zinn's definition of mindfulness as well as most others (Bishop et al., 2004; Shapiro et al., 2006), is seen as a fundamental attitude, but not an active mechanism. Grabovac et al. (2011) argues that the attitude of acceptance is crucial to helping beginning meditators relax and avoid self-judgment. However, in the BPM, acceptance is a quality of awareness and does not involve cognition. Therefore, if one is thinking accepting thoughts to help stabilize their focus, this is a form of attention regulation, not mindfulness. As previously mentioned, the notions of metacognitive awareness (Teasdale et al., 2002), *decentering* (Fresco et al., 2007), and *reperceiving* (Shapiro et al., 2006) are all seen as cognitive processes which map onto the attention regulation component of BPM. Grabovac et al. (2011) posits, "in contrast [to attention regulation], insight and its side effects are non-conceptual and non-cognitive in their origin and result in reductions in attachment/aversion or mental proliferation without requiring any cognitive intervention or processing" (p. 9).

According to the BPM, without a non-conceptual awareness of the three characteristics (impermanence, suffering, and not-self), the models of mindfulness described are forms of attention regulation and not mindfulness itself. If patients over-emphasize attention regulation in MBIs, they may inadvertently strengthen their aversion to negative thoughts. If, for instance, a patient experiences a negative emotion, they may attempt to redirect their attention to the affect-neutral breath, rather than engaging with the transient nature of the emotion (impermanence), confronting their aversion to similarly challenging situations (suffering), or recognizing the futility of identifying with a passing emotion (not-self).

The deconstruction of the self (not-self) through equanimity is one of the major distinctions between the BPM and other models. While most models encourage the recognition of impermanence through one of the aforementioned cognitive mediators, and many engage with the general notion of suffering as an impetus for initiating MBIs (Kabat-Zinn, 1982; Carmody, 2009; Shapiro, 2009), the BPM asserts that the entire notion of a self is, in fact, a mental construction. The notion of the self can, therefore, be observed without attachment or aversion (with equanimity) the same way one observes sensations of breathing. As Grabovac et al. (2011) notes in a conversation with Steve Armstrong, "this is the point of equanimity, there is no reification of a sense of self." As we will discuss later in the chapter, this diminishing of a separate, isolated self may lead to increases in prosocial behavior (Piff et al. 2015), creativity (Sundararajan & Fatemi, 2016), and communal sharing (Fiske, 1991) as well as a feeling of unity (Barrett & Griffiths, 2017a, b).

Self-Awareness, Self-Regulation, and Self-Transcendence Model (Vago & Silbersweig, 2012)

The self-awareness, self-regulation, and self-transcendence (S-ART) model, created by Vago and Silbersweig (2012), provides a theoretical framework and system-based neurological model of mindfulness. In an effort to clarify the varying operationalizations of mindfulness, the model employs contemporary understandings of neural processes to map the ancient Buddhist descriptions of the mind. Vago and Silbersweig describes the way in which the Buddhist notion of suffering (Pali: *dukkha*) emerges from an "inflated sense of self-loathing and self-importance" (Thera, 1962; Vago & Silbersweig, 2012).

This process resonates with the way modern psychotherapists illustrate how habitual information processing biases reify a dysfunctional self-schema. Accordingly, the S-ART model outlines a systematic mental training that develops one's ability to recognize, observe, and remove these habitual distortions and biases. The training moves from meta-awareness (self-awareness) to the ability to effectively modulate one's behavior (self-regulation) and finally to a positive relationship between self and others that transcends self-focused needs and increased prosocial characteristics (self-transcendence).

Through meditation, one cultivates this transformation by engaging in seven specific neuropsychological mechanisms: intention and motivation, attention regulation, emotional regulation, extinction and reconsolidation, prosociality, non-attachment, and decentering. By examining the neurobiological structures correlated with each of these mechanisms, the S-ART framework also attempts to articulate both *how* mindfulness works and *what* the brain is doing to get there.

Following the path outlined by Buddhism, Vago and Silbersweig begin by confronting the causes of suffering. They propose that the distorted or biased sense of self colors one's perception of the world, provoking one to be hypervigilant toward certain stimuli and avoid other stimuli. Over time, this "sliver of reality" eventually crystallizes into a "sensory-affective-motor" script or scheme that dictates behavior. As Aaron Beck (2008) describes, the processing of external events or internal stimuli is biased and therefore systematically distorts the individual's construction of his or her experiences, leading to a variety of errors (e.g., overgeneralization, selective abstraction, personalization). In a psychopathological context, these negative views of the self become reified through a continuous feedback loop of affect-biased attention, rumination, and reinforcing behavior.

Similar to the BPM notion of equanimity (Grabovac et al., 2011), or the two-component model's "acceptance" (Bishop et al., 2004), Vago and Silbersweig (2012) argue, "one of the goals of mindfulness-based practice is to make no such distinction between positive, negative, or neutral valence and treat all incoming stimuli with impartiality and equipoise" (p. 5). By developing this sense of "impartiality and equipoise," the person learns to recognize the arising pathological feedback loop and let it fade away without getting wrapped up in the loop's emotional narrative.

To accomplish this goal, the S-ART model outlines seven mechanisms to culti-vate during meditation: intention and motivation, attention regulation, emotional regulation, extinction and reconsolidation, prosociality, non-attachment, and decen-tering. The first, as first described by Shapiro et al. (2006), spurs one to practice and, according to Vago and Silbersweig (2012), shifts from external motivation (goal oriented) to internal motivation (intrinsically validating) as the practice develops. The second, attention regulation, corresponds to the meta-awareness (self-awareness) component of S-ART, enabling the individual to detangle the contents of subjective experience from the "conditioned and consolidated schemas that dictate behavior" (Vago & Silbersweig, 2012). The third, emotional regulation, is synony-mous with the self-regulation component of S-ART and refers to the ability to shift focus of attention at will, rather than falling into elaborative emotional feedback loops, and actively choose one's behavior.

The extinction and reconsolidation mechanism refers to the end goal of mindful-ness or "stillness of the mind" (Sanskrit: *nirvana*) (Buddhaghosa, 1991). In Buddhism, the term "Nirvana" literally translates to "blowing out" or "extinguish-ing" the mental afflictions (Sanskrit: *klesha*) preventing happiness and well-being. According to the S-ART, these mental afflictions are rooted in "maladaptive habits, distorted perceptions, and biases accumulated through the conditioning or reifica-tion of the narrative-self" (Vago & Silbersweig, 2012). Mindfulness practices serve to break down the rigidity of the stories one tells about oneself, extinguish maladap-tive narratives, and rebuild more positive behavioral patterns. Through the process of memory retrieval, one can actively weaken (i.e., extinguish) the condition responses to a given stimuli (e.g., anger toward a difficult coworker) and habituate (i.e., consolidate) a more purposeful response (i.e., responding to the coworking with patience and openness).

The final two mechanisms are associated with the self-transcendence component of S-ART. Prosociality, the ability to dissolve the distinction between self and others and reflect loving-kindness toward both, leads to empathy, altruism, a greater sense of well-being (Lyubomirsky et al., 2005), and reliance (Champagne & Curley, 2008; Feder et al., 2009).

Non-attachment and decentering correspond to the Buddhist notion of "not-self" through a realization of impermanence. Vago and Silbersweig equate the process of decentering to "reperceiving" (Shapiro et al., 2006), whereby one finds "space between one's perception and response"; experiences one's thoughts, emotions, and feelings as "thing-like" (Varela et al., 1991); and, eventually, comes to see that there is no permanently existing self.

In addition to its theoretical contributions, the S-ART model outlines the specific neural networks associated with each of these mechanisms and their corresponding processes. While the explanation of these brain areas and their functions is beyond the scope of this chapter, it is of quintessential importance to developing the field of psychological mindfulness and warrants further rigor-ous investigation.

Two-Component Model of Mindfulness (Bishop et al., 2004)

One of the first attempts to establish an operational definition of mindfulness came from Scott Bishop and his team, resulting in the two-component model of mindfulness. In this model, mindfulness is a "metacognitive skill" rather than a state with two active components. The first component involves "self-regulation of attention so that it is maintained on immediate experience, thereby allowing for increased recognition of mental events in the present moment" (Bishop et al., 2004). Similar to the IAA's *attention*, this component focuses the observer's attention on a chosen object. Developing one's capacity to shift and hold one's attention is crucial for developing all further stages and components of mindfulness.

The second component calls for the adoption of a particular orientation toward those experiences characterized by "curiosity, openness and acceptance." This orientation to experience is similar to the third component of the IAA, *attitude*, which outlines the *qualities* one brings to attention. However, in addition to "openness and acceptance," Shapiro and her colleagues (2006) also stress the "heart-mindfulness" or a sense of "kindness."

The self-regulation of attention in this model stresses the need for *direct experience* of events in the mind and body rather than narrative descriptions, ruminations, or elaborations *about* those events. This language, though not explicitly stated, draws directly on the concept of *direct perception* (Sanskrit: pratyakṣa), which was defined by the Indian Buddhist Philosopher, Dharmakīrti, as "free from conceptual construction" (Toru, 2005). Regarded as one of the most crucial components of Buddhist philosophy, direct perception facilitates an experience of stimuli that is free from narrations of "name or genus...or [cognitive] expression." Thus, a given stimuli, be it physical or mental, can be perceived exactly as it is, without the filter of "beliefs, assumptions, expectations, and desires" (Bishop et al., 2004). This state of direct observation devoid of mental narration, seeing objects as if for the first time, is commonly known as "beginner's mind," (*Shoshin* in the Zen Tradition) a term popularized by Shunryu Suzuki in the 1970s. However, it should be noted that while the concept of direct perception applies to a much broader set of phenomena, including the pervasiveness of suffering (*dukkha*) construct of the self (*anatha*) and the impermanence of all things (*anicca*), Bishop and his colleagues (2004) limit their application of direct experience to the client's emotional and physical perception.

The second component, *orientation to experience*, forms the basis of what Shapiro (2006) would later call "attitude," what Kabat-Zinn refers to as "nonjudgmental awareness" (2003) and parallels the Buddhist concept of "equanimity." It combines an attitude of curiosity, non-striving, and acceptance. Bishop et al. define acceptance as "being experientially open to the reality of the present moment" (in Roemer & Orsillo, 2002).

In an effort to create a more descriptive explanation of mindfulness and its benefits, later models (Shapiro et al., 2006; Carmody, 2009; Grabovac et al., 2011; Vago & Silbersweig, 2012) dissect these two components into more precise, nuanced mechanisms.

Statistically Derived Model (Coffey et al., 2010)

Building upon the two-component model, Coffey and her colleagues (2010) developed a path analysis approach to clarify which components of mindfulness most influenced decreases in psychological distress and mental wellness. They worked with people who had little to no formal mindfulness training to assess whether present-centered attention or acceptance had a greater impact on managing negative emotions, rumination, and the extent to which one's happiness is independent of specific outcomes and events. The researchers found that acceptance of one's experience contributed more than present-centered attention to both decreasing psychological distress and increasing mental wellness.

The study also found a paradoxical relationship between present-centered attention and psychological distress. Coffey et al. (2010) writes: "Directly attending to one's present moment experience both decreased psychological distress, by beneficially impacting other constructs which then decreased distress, and increased psychological distress, via a direct association" (p. 248). This seeming contradiction may illuminate one of the major misunderstandings about mindfulness: it is not always easy. Given the growing body of evidence validating the benefits of mindfulness and the popular illustrations of meditation practices as "get calm now" remedies, many fail to see the difficulties inherent in the process. The inexperienced subjects in Coffey et al.'s (2010) studies illustrate that engaging with difficult emotions may temporarily increase psychological distress; however, by continuing to observe the feelings with acceptance, one builds the capacity and resilience to eventually outlast the pain.

The statistically derived model demonstrates a widespread effort to more precisely and systematically calculate the component processes of mindfulness. Coffey et al. (2010) argue for "the importance of treating these two dimensions of mindfulness [present centered attention and acceptance] as distinct constructs that may not be related when mindfulness is examined as a naturally varying individual difference" (p. 249). In order to study these components via the scientific method, they do, indeed, need to be treated as separate variables. However, both the Buddhist tradition and more contemporary models stress the interconnected non-linearity of the various mechanisms. While certain components may lead to more statistically evident outcomes, stressing isolated components over other may lead to negative outcomes. Mindfulness remains a complex state, process, and practice of interwoven mechanisms that is certainly more than the sum of its individual parts.

Attention Regulation Model (Carmody, 2009)

In an effort to find a testable model of mindfulness that was common to both the historical Buddhist definitions and modern clinical conceptualizations, James Carmody proposed the attention regulation model (2009). The objective of this model, like that of the ancient Buddhist scriptures and MBSR, is to most effectively alleviate the suffering of the patient. However, the author suggests that engaging with the complexities of the various interactive mechanisms, such as "present-centered awareness" or "non-judgmental awareness," may be redundant if not counterproductive. Instead, he argues that one should focus on attention. Carmody (2009) champions the development of the faculty of attention training as the central and common factor in salutogenic outcomes. He argues that the mind usually ruminates in an "established automaticity of constructed everyday experience." Without recognizing it, the individual learns to automatically link the thought, sensation, and feeling of a given experience and accordingly gets caught in the "association cycle" (see Fig. 6.1). In a clinical setting, a patient might be triggered by a memory (thought) which is associated with distress (sensation) and then leads to anxiety (feeling). So long as that patient continues to pay attention to that cycle, it will remain a close loop and the anxiety will increase (Figs. 6.2 and 6.3).

In order to "unlearn" the mind's habitual pattern of rumination, the attention regulation model utilizes mindfulness of breathing to break the cycle:

> In the deliberate relearning that is mindfulness practice, there is recognition of the individual components arising in the field of awareness and the construction process giving rise to everyday experience. (Carmody, 2009, p. 272)

This model does not reject notions of present-centered awareness or intention, but it does see them as "superfluous" (Carmody, 2009). The author argues that these terms, along with "present-centered" and "non-judgmental awareness," are implicitly learned in the process of bringing one's awareness back to an affect-neutral breath. Carmody also suggests that this model will provide a "parsimonious starting point for clinical research" as well as a more comprehensible definition of mindfulness for patients. Figures 6.4 and 6.5 illustrate this process.

Both the two-component model and the attention regulation model offer a high-level accessibility for clinicians and their patients. By emphasizing the faculty of

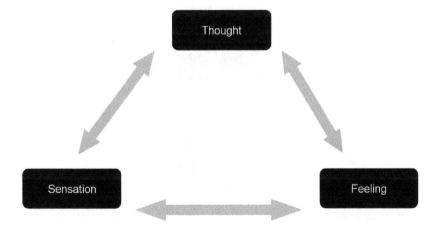

Fig. 6.2 Automatic association process when attention is undirected

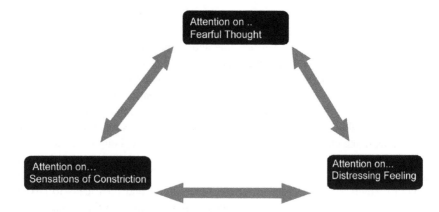

Fig. 6.3 Automatic maintenance of distress when attention is undirected

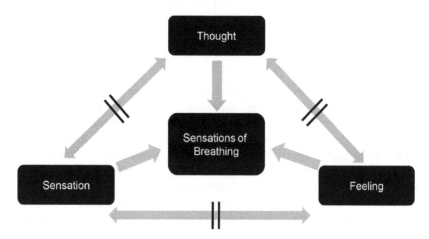

Fig. 6.4 Attention process in formal mindfulness concentration practice

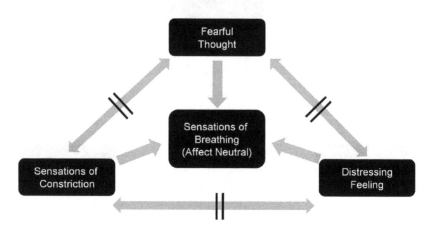

Fig. 6.5 Redirection of attention to the (affect neutral) breath

attention monitoring without explicitly engaging more of the contextual elements of mindfulness, these models provide a secular, direct, approachable conceptualization. However, as the field expands to include more neurological, biological, and systematic measurements of mindfulness, researchers require a more complex map to navigate the precision of their tools. Neither of these models outline the functional nuances of mental events. While it may make the models more approachable, they fail to engage with the root causes of what causes the "association cycle." In doing so, these models are often critiqued for distracting patients from negative thoughts or emotions (by returning to affect-neutral breath) rather than engaging with the source of that discomfort (Grabovac et al., 2011).

Conclusion

With the growing interest in the clinical applications of mindfulness and mindfulness-based interventions, a concomitant increase in rigorous research and exploration is needed to determine how best to understand and apply these practices to decrease suffering and increase well-being. Although we are encouraged by the development and publication of models of mindfulness in the literature and want to encourage similar future efforts, we also want to strongly impress the need for basic research investigating fundamental questions concerning mechanisms of action and common factors that exist across all MBIs. The field needs to find consensus on both the conceptual and operational aspects of mindfulness, and a reliable and valid instrument must be developed. Further, questions concerning mediating role and mechanisms of action must be investigated. Through these endeavors we have the opportunity to deepen our understanding and application of mindfulness to transform our individual and collective lives.

References

Analayo. (2003). *Satipatthana: The direct path to awakening*. Windhorse Publications.

Baer, R. A. (2003). Mindfulness training as a clinical intervention: A conceptual and empirical review. *Clinical Psychology: Science and Practice, 10*, 125–143.

Barlow, D. H., & Craske, M. G. (2000). *Mastery of your anxiety and panic* (3rd ed.). Graywind Publications Inc./The Psychological Corporation.

Barrett, F., & Griffiths, R. (2017a). Classic hallucinogens and mystical experiences: Phenomenology and neural correlates. *Current Topics in Behavioral Neuroscience*. https://doi.org/10.1007/7854_2017_474

Barrett, F. S., & Griffiths, R. R. (2017b). Classic hallucinogens and mystical experiences: Phenomenology and neural correlates. In A. L. Halberstadt, F. X. Vollenweider, & D. E. Nichols (Eds.), *Behavioral neurobiology of psychedelic drugs* (Vol. 36, pp. 393–430). Springer. https://doi.org/10.1007/7854_2017_474

Beck, A. T. (2008). The evolution of the cognitive model of depression and its neurobiological correlates. *The American Journal of Psychiatry, 165*, 969–977.

Bishop, S. R., Lau, M., Shapiro, S., Carlson, L., Anderson, N. D., Carmody, J., Segal, Z. V., Abbey, S., Speca, M., Velting, D., & Devins, G. (2004). Mindfulness: A proposed operational definition. *Clinical Psychology: Science and Practice, 11*, 230–241. https://doi.org/10.1093/clipsy.bph077

Bodhi, B. (1994). *Going for refuge & taking the precepts*. Retrieved from https://www.accesstoinsight.org/lib/authors/bodhi/wheel282.html

Bodhi, B. (1999). *A comprehensive manual of Abhidhamma: The philosophical psychology of Buddhism*. Buddhist Publication Society.

Bodhi, B. (2010). *The noble eightfold path: Way to the end of suffering*. Buddhist Publication Society.

Bohart, A. (1983). *Detachment: A variable common to many psychotherapies?* Paper presented at the 63rd annual convention of the Western Psychological Association, San Francisco, CA.

Borkovec, T. D. (2002). Life in the future versus life in the present. *Clinical Psychology: Science and Practice, 9*, 76–80.

Brown, K. W., & Ryan, R. M. (2003). The benefits of being present: Mindfulness and its role in psychological well-being. *Journal of Personality and Social Psychology, 84*(4), 822–848.

Buddhaghosa, B. (1991). *The path of purification (Visuddhimagga)*. Buddhist Publication Society Pariyatti Editions.

Carmody, J. (2009). Evolving conceptions of mindfulness in clinical settings. *Journal of Cognitive Psychotherapy, 23*, 270–280.

Champagne, F. A., & Curley, J. P. (2008). Maternal regulation of estrogen receptor alpha methylation. *Current Opinion in Pharmacology, 8*, 735–739.

Coffey, K., Hartman, M., & Fredrickson, B. (2010). Deconstructing mindfulness and constructing mental health: Understanding mindfulness and its mechanisms of action. *Mindfulness, 1*, 235–253. https://doi.org/10.1007/s12671-010-0033-2

Davids, R. (1882). *Buddhism: Being a sketch of the life and teachings of Gautama, the Buddha*. SPCK.

Davidson, R. J. (2010). Empirical explorations of mindfulness: Conceptual and methodological conundrums. *Emotion, 10*(1), 8–11.

Davis, D., & Hayes, J. (2011). What are the benefits of mindfulness? A practice review of psychotherapy-related research. *Psychotherapy (Chicago), 48*(2), 198–208.

Detloff, W., & Deikman, A. (1982). The observing self; Mysticism and psychotherapy. *The San Francisco Jung Institute Library Journal, 3*(4), 25–28. https://doi.org/10.1525/jung.1.1982.3.4.25

Deyo, M., Wilson, K., Ong, J., & Koopman, C. L. (2009). Mindfulness and rumination: Does mindfulness training lead to reductions in the ruminative thinking associated with depression? *Explore (New York, NY), 5*, 265–271. https://doi.org/10.1016/j.explore.2009.06.005

Dreyfus, G. (2011). Is mindfulness present-centred and non-judgmental? A discussion of the cognitive dimensions of mindfulness. *Contemporary Buddhism, 12*, 41–54.

Dunne, J. (2011). Toward an understanding of non-dual mindfulness. *Contemporary Buddhism, 12*(1), 71–88. https://doi.org/10.1080/14639947.2011.564820

Feder, A., Nestler, E. J., & Charney, D. S. (2009). Psychobiology and molecular genetics of resilience. *Nature Reviews Neuroscience, 10*, 446–457.

Ferguson, Y. & Sheldon, K. (2013) Trying to be happier really can work: Two experimental studies. *The Journal of Positive Psychology*, 8, 1, 23–33 https://doi.org/10.1080/17439760.2012.74700

Fredrickson, B. L., & Cohn, M. A. (2008). Positive emotions. In M. Lewis, J. M. Haviland-Jones, & L. F. Barrett (Eds.), *Handbook of Emotions*. 77–796. The Guilford Press.

Fischer, N. (2013). *Training in compassion: Zen teachings on the practice of Lojong* (p. 11). Shambhala Publications.

Fiske, A. P. (1991). *Structures of social life: The four elementary forms of human relations: Communal sharing, authority ranking, equality matching, market pricing*. Free Press.

Fletcher, L., & Hayes, S. (2005). Relational frame theory, acceptance and commitment therapy, and a functional analytic definition of mindfulness. *Journal of Rational-Emotive and Cognitive-Behavior Therapy, 23*, 315–336. https://doi.org/10.1007/s10942-005-0017-7

Fresco, D., Segal, Z., Buis, T., & Kennedy, S. (2007). Relationship of posttreatment decentering and cognitive reactivity to relapse in major depression. *Journal of Consulting and Clinical Psychology, 75*(3), 447–455.

Garland, E. L., & Howard, M. O. (2018, April 18). *Mindfulness-based treatment of addiction: Current state of the field and envisioning the next wave of research.* Retrieved from https://ascpjournal.biomedcentral.com/articles/10.1186/s13722-018-0115-3

Gethin, R. (2001). *The Buddhist path to awakening.* Oneworld.

Goldstein, J. (2002). *One dharma: The emerging Western Buddhism.* HarperCollins.

Goldstein, J. (2013). *Mindfulness: A practical guide to awakening.* Sounds True, Kindle Edition.

Goleman, D. (1971). Meditation as meta-therapy: Hypotheses toward a proposed fifth state of consciousness. *Journal of Transpersonal Psychology, 3*, 1–26.

Gonzalo, B., Campos, D., & Cebolla, A. (2018). Implicit or explicit compassion? Effects of compassion cultivation training and comparison with mindfulness-based stress reduction. *Mindfulness, 9.* https://doi.org/10.1007/s12671-018-0898-z

Grabovac, A. D., Lau, M. A., & Willett, B. R. (2011). Mechanisms of mindfulness: A Buddhist psychological model. *Mindfulness, 2*, 154. https://doi.org/10.1007/s12671-011-0054-5

Grossman, P. (2014, December 19). *Mindfulness: Awareness informed by an embodied ethic.* Retrieved from https://link.springer.com/article/10.1007/s12671-014-0372-5#citeas

Hayes, S. C. (2002). Acceptance, mindfulness, and science. *Clinical Psychology: Science and Practice, 9*(1), 101–106.

Hayes, S. C. (2004). Acceptance and commitment therapy and the new behavior therapies: Mindfulness, acceptance and relationship. In S. C. Hayes, V. M. Follette, & M. Linehan (Eds.), *Mindfulness and acceptance: Expanding the cognitive behavioral tradition* (pp. 1–29). Guilford.

Hölzel, B. K., Carmody, J., Vangel, M., Congleton, C., Yerramsetti, S. M., Gard, T., & Lazar, S. W. (2010). Mindfulness practice leads to increases in regional brain gray matter density. *Psychiatry Research, 191*(1), 36–43. https://doi.org/10.1016/j.pscychresns.2010.08.006

IBISWorld. (2021). *Alternative healthcare providers industry in the US – Market Research Report.* Available online at: https://www.ibisworld.com/industry-trends/market-research-reports/healthcare-social-assistance/ambulatory-health-care-services/alternative-healthcare-providers.html. Accessed 20 Dec 2018.

Jinpa, T. (2009). *Is meditation a means of knowing our mental world?* [Online]. Available online at: http://contemplativemind.wordpress.com/what-is-mindfulness/. Accessed 20 Dec 2018.

Kabat-Zinn, J. (1982). An outpatient program in behavioral medicine for chronic pain patients based on the practice of mindfulness meditation: Theoretical considerations and preliminary results. *General Hospital Psychiatry, 4*, 33–47.

Kabat-Zinn, J. (1990). *Full catastrophe living: Using the wisdom of your body and mind to face stress, pain and illness.* Delacorte.

Kabat-Zinn, J. (2003). Mindfulness-based interventions in context: Past, present, and future. *Clinical Psychology: Science and Practice, 10*(2), 144–156.

Kabat-Zinn, J. (2005). Bringing mindfulness to medicine: An interview with Jon Kabat-Zinn, PhD: Interview by Karolyn Gazella. *Advances in Mind-Body Medicine, 21*, 22–27.

Khanna, S., & Greeson, J. M. (2013). A narrative review of yoga and mindfulness as complementary therapies for addiction. *Complementary Therapies in Medicine, 21*, 244–252.

Kurtz, R. (1990). Body centered psychotherapy: The Hakomi method. ISBN:0-940795-03-5. Available from LifeRhythm, P.O. Box 806, Mendocino, CA 95460, Phone 707-937-1825. (1997). *Alternative Health Practitioner, 3*(2), 139–139.

Linehan, M. M. (1993). *Diagnosis and treatment of mental disorders. Cognitive-behavioral treatment of borderline personality disorder.* Guilford Press.

Lutz, J., Herwig, U., Opialla, S., Hittmeyer, A., Jäncke, L., Rufer, M., ..., Brühl, A. B. (2014, June). *Mindfulness and emotion regulation – An fMRI study*. Retrieved from https://www.ncbi.nlm.nih.gov/pubmed/23563850

Lyubomirsky, S., King, L., & Diener, E. (2005). The benefits of frequent positive affect: Does happiness lead to success? *Psychological Bulletin, 131*, 803–855. https://doi.org/10.1037/0033-2909.131.6.803

Maex, E. (2011). The Buddhist roots of mindfulness training: A practitioners view. *Contemporary Buddhism, 12*(1), 165–175. https://doi.org/10.1080/14639947.2011.564835

Meiklejohn, J., Phillips, C., Freedman, M. L., Griffin, M. L., Biegel, G., Roach, A., ..., Saltzman, A. (2012, March 14). *Integrating mindfulness training into K-12 education: Fostering the resilience of teachers and students*. Retrieved from https://link.springer.com/article/10.1007/s12671-012-0094-5#citeas

Mendis, N. K. G. (2006). *The Abhidhamma in practice*. Retrieved October 29, 2010, from http://www.accesstoinsight.org/lib/authors/mendis/wheel322.html

Monteiro, L. M., Musten, R., & Compson, J. (2014, April 29). *Traditional and contemporary mindfulness: Finding the middle path in the tangle of concerns*. Retrieved from https://doi.org/10.1007/s12671-014-0301

Olendzki, A. (2011). The construction of mindfulness. *Contemporary Buddhism, 12*, 55–70.

Orzech, K. M., Shapiro, S. L., Brown, K. W., & McKay, M. (2009). Intensive mindfulness training-related changes in cognitive and emotional experience. *The Journal of Positive Psychology, 4*(3), 212–222.

Parasuraman, R. (1998). The attentive brain: Issues and prospects. In R. Parasuraman (Ed.), *The Attentive Brain*. 3–15. MIT Press.

Patrul Rinpoche. (2010). *Words of my perfect teacher* (p. 134). Yale University Press.

Posner, M. I., Snyder, C. R., & Davidson, B. J. (1980). Attention and the detection of signals. *Journal of Experimental Psychology*: General, 109(2), 160–174. https://doi.org/10.1037/0096-3445.109.2.160

Posner, M. I., & Rothbart, M. K. (1992). Attentional mechanisms and conscious experience. *The Neuropsychology of Consciousness* 91–111. Academic press.

Piff, P. K., Dietze, P., Feinberg, M., Stancato, D. M., & Keltner, D. (2015). Awe, the small self, and prosocial behavior. *Journal of Personality and Social Psychology, 108*(6), 883–899.

Purser, R., & Loy, D. (2013). Beyond McMindfulness. *Huffington Post*. Retrieved from http://www.huffingtonpost.com/ron-purser/beyondmcmindfulness_b_3519289.html

Quist Møller, S. A., Sami, S., & Shapiro, S. L. (2019). Health benefits of (mindful) self-compassion meditation and the potential complementarity to mindfulness-based interventions: A review of randomized-controlled trials. *OBM Integrative and Complementary Medicine, 4*(1), 20. https://doi.org/10.21926/obm.icm.1901002

Rahula, W. (1974). *What the Buddha taught*. Grove Press.

Ricard, M. (2009). *A sniper's mindfulness*. Retrieved from http://www.matthieuricard.org/blog/posts/a-sniper-s-mindfulness

Roemer, L., & Orsillo, S. M. (2002). Expanding our conceptualization of and treatment for generalized anxiety disorder: Integrating mindfulness/acceptance-based approaches with existing cognitive-behavioral models. *Clinical Psychology: Science and Practice, 9*, 54–68.

Rogers, R., & Monsell, S. (1995). The costs of a predictable switch between simple cognitive tasks. *Journal of Experimental Psychology: General, 124*, 207–231.

Ruedy, N. E., & Schweitzer, M. E. (2010). In the moment: The effect of mindfulness on ethical decision making. *Journal of Business Ethics, 95*(1), 73–87.

Ryan, R. M., Kuhl, J., & Deci, E. L. (1997). Nature and autonomy: An organizational view of social and neurobiological aspects of self-regulation in behavior and development. *Development and Psychopathology, 9*, 701–728.

Safran, J. D., & Segal, Z. V. (1990). *Interpersonal process in cognitive therapy*. Rowman & Littlefield.

Santorelli, S. (1999). *Heal thy self: Lessons on mindfulness in medicine*. Random House.

Segal, Z. V., Williams, J. M. G., & Teasdale, J. D. (2002). *Mindfulness-based cognitive therapy for depression: A new approach to preventing relapse*. Guilford Press.

Senauke, A. (2013). *Wrong mindfulness: An interview with Hozan Alan Senauke*. Retrieved from http://www.tricycle.com/blog/wrongmindfulness

Shapiro, D. H. (1992). A preliminary study of long-term meditators: Goals, effects, religious orientation, cognitions. *Journal of Transpersonal Psychology, 24*(1), 23–39.

Shapiro, S. (2009). The integration of mindfulness and psychology. *Journal of Clinical Psychology, 65*(6), 555–560.

Shapiro, S., Jazaieri, H., & Goldin, P. R. (2012). Mindfulness-based stress reduction effects on moral reasoning and decision making. *The Journal of Positive Psychology, 7*(6), 504–515.

Shapiro, S. L., & Carlson, L. E. (2009). *The art and science of mindfulness: Integrating mindfulness into psychology and the helping professions*. American Psychological Association.

Shapiro, S. L., Carlson, L. E., Astin, J. A., & Freedman, B. S. (2006). Mechanisms of mindfulness. *Journal of Clinical Psychology, 62*(3), 373–386.

Shapiro, S. L., & Schwartz, G. E. (2000). The role of intention in self-regulation: Toward intentional systemic mindfulness. In M. Boekaerts, P. R. Pintrich, & M. Zeidner (Eds.), *Handbook of self-regulation* (pp. 253–273). Academic.

Shapiro, S. L., & Schwartz, G. E. R. (1999). Intentional systemic mindfulness: An integrative model for self-regulation and health. *Advances in Mind-Body Medicine, 15*, 128–134.

Shiota, M. N., Keltner, D., & Mossman, A. (2007). The nature of awe: Elicitors, appraisals, and effects on self-concept. *Cognition and Emotion, 21*. https://doi.org/10.1080/02699930600923668

Shonin, E., & Van Gordon, W. (2014). Managers' experiences of meditation awareness training. *Mindfulness, 6*. https://doi.org/10.1007/s12671-014-0334-y

Sommers-Spijkerman, M. P. J., Trompetter, H. R., Schreurs, K. M. G., & Bohlmeijer, E. T. (2018). Compassion-focused therapy as guided self-help for enhancing public mental health: A randomized controlled trial. *Journal of Consulting and Clinical Psychology, 86*(2), 101–115.

Spiro, M. E. (1982). *Buddhism and society: A great tradition and its Burmese vicissitudes*. University of California Press.

Sundararajan, L., & Fatemi, S. M. (2016). Creativity and symmetry restoration: Toward a cognitive account of mindfulness. *Journal of Theoretical and Philosophical Psychology, 36*(3), 131–141.

Suzuki, S. (1970). *Zen mind, beginner's mind*. Weatherhill.

Teasdale, J.D. (1999), Metacognition, mindfulness and the modification of mood disorders. *Clin. Psychol. Psychother.*, 6: 146–155. https://doi.org/10.1002/(SICI)1099-0879(199905)6:2<146::AID-CPP195>3.0.CO;2-E

Teasdale, J. D., Segal, Z. V., Williams, J. M. G., Ridgeway, V. A., Soulsby, J. M., & Lau, M. A. (2000). Prevention of relapse/recurrence in major depression by mindfulness-based cognitive therapy. *Journal of Consulting and Clinical Psychology, 68*, 615–623.

Teasdale, J. D., Moore, R. G., Hayhurst, H., Pope, M., Williams, S., & Segal, Z. V. (2002). Metacognitive awareness and prevention of relapse in depression: Empirical evidence. *Journal of Consulting and Clinical Psychology, 70*(2), 275–287.

Thanissarom, B. (2008). *Satipatthana Sutta: Frames of reference*. Retrieved from https://www.accesstoinsight.org/tipitaka/mn/mn.010.than.html

Thera, N. (1962). *The heart of Buddhist meditation: A handbook of mental training based on the Buddha's way of mindfulness*. Rider and Company.

Titmuss, C. (2013). *The Buddha of mindfulness. The politics of mindfulness*. Retrieved from http://christophertitmuss.org/blog/?p=1454

Toru, F. (2005). Perception, conceptual construction and yogic cognition according to Kamalaśīla's. *Epistemology. Chung-Hwa Buddhist Journal, 18*, 273–297.

Uncapher, M., & Wagner, A. (2018). Media multitasking, mind, and brain. *Proceedings of the National Academy of Sciences, 115*(40), 9889–9896. https://doi.org/10.1073/pnas.1611612115

Vago, D. R., & Silbersweig, D. A. (2012). Self-awareness, self-regulation, and self-transcendence (S-ART): A framework for understanding the neurobiological mechanisms of mindfulness. *Frontiers in Human Neuroscience, 6*, 296. https://doi.org/10.3389/fnhum.2012.00296

Varela, F. J., Thompson, E., & Rosch, E. (1991). *The embodied mind*. MIT Press.

Vieten, C., Wahbeh, H., Cahn, B. R., MacLean, K., Estrada, M., Mills, P., Murphy, M., Shapiro, S. L., & Delorme, A. (2018). Future directions in meditation research: Recommendations for expanding the field of contemplative science. *PLoS One*. https://doi.org/10.1371/journal.pone.0205740

Wallace, B. A. (2010). *The attention revolution: Unlocking the power of the focused mind*. Accessible Publishing Systems PTY.

Weiss, H. (2009). The use of mindfulness in psychodynamic and body oriented psychotherapy. *Body, Movement and Dance in Psychotherapy, 4*, 5–16. https://doi.org/10.1080/17432970801976305

Williams, J. M. G., Mathews, A., & MacLeod, C. (1996). The emotional Stroop task and psychopathology. *Psychological Bulletin, 120*(1), 3–24.

Zeng, X., Chio, F. H., Oei, T. P., Leung, F. Y., & Liu, X. (2017, February 6). *A systematic review of associations between amount of meditation practice and outcomes in interventions using the four immeasurables meditations*. Retrieved from https://www.ncbi.nlm.nih.gov/pmc/articles/PMC5292580/

Chapter 7
Mindfulness-Based Stress Reduction for Medical Conditions

Linda E. Carlson, Kirsti Toivonen, Michelle Flynn, Julie Deleemans, Katherine-Anne Piedalue, Utkarsh Subnis, Devesh Oberoi, Michaela Patton, Hassan Pirbhai, and Mohamad Baydoun

Stress and Health

In the first five chapters appearing in Part I of this book, the psychobiology of stress has been well documented. In summary, the psychological experience of distress and stress is most often accompanied by a biological stress response, which involves a cascade of neurological (through the sympathetic nervous system), hormonal (through stress hormones such as cortisol), immunological, and even gut-related effects. Overstimulation or chronic activation of these systems can affect health indicators both in the short and long term in many ways, and together their effects may be synergistic and affect a range of disease states.

In consideration of the effects of mindfulness on the psychological as well as biological experience of serious medical illness, it helps to first consider crucial elements of the illness experience that lead to the subjective experience of stress and the biological stress response. For example, in the case of cancer and other potentially life-threatening illnesses, there are existential, psychological, and physical effects. Existentially, people often confront their own pending mortality at a level that is, for the first time, personal rather than theoretical. While we are all aware that our lives will end in death, often we behave in a way that belies that reality, or we see it as some vague distant event. Being confronted with the direct possibility of death in the near rather than distant future may result in exacerbated fears of death, fear of pain and the process of dying, and confrontation with the lack of controllability over life that we often disregard. These existential concerns can cause anxiety, panic, depression, and other psychological symptoms.

L. E. Carlson (✉) · K. Toivonen · M. Flynn · J. Deleemans · K.-A. Piedalue · U. Subnis · D. Oberoi · M. Patton · H. Pirbhai · M. Baydoun
University of Calgary, Calgary, Canada
e-mail: lcarlso@ucalgary.ca

© Springer Nature Switzerland AG 2021
H. Hazlett-Stevens (ed.), *Biopsychosocial Factors of Stress, and Mindfulness for Stress Reduction*, https://doi.org/10.1007/978-3-030-81245-4_7

There is also a loss of perceived control and certainty in one's life, and practically schedules have to change and adapt around medical appointments and treatments, causing additional stress on the family and social support systems. There may be anger and symptoms of grief and loss as one has to adapt to changes in physical capabilities, changes in work schedules or abilities, and changes in relationships. There is inevitably a sense of grief, and often anger, when diagnosed with a serious medical condition. There may also be feelings of shame, guilt, and self-blame if the condition is related to one's lifestyle choices or other specific behaviors.

Psychological distress of this nature can contribute both directly and indirectly to many other physical symptoms. For example, common symptoms across conditions are fatigue, pain, sleep disturbances, weakness, discomfort, and loss of functional abilities. Chronic distress has a wide range of negative consequences now understood to be at a multisystem level, with bidirectional interactions between psychosocial and biological variables, including the brain, and the neuroendocrine and immune systems, involving stress hormones norepinephrine, epinephrine, and cortisol, for which almost all cells in the body have receptors (Bower et al., 2011).

These stress-induced immune dysregulations can affect people with chronic disease in three major ways: (1) the suppression of protective immunity, (2) the induction/exacerbation of chronic inflammation, and (3) the enhancement of immunosuppressive mechanisms (Antoni & Dhabhar, 2019), which may have implications for disease progression. For example, laboratory studies have demonstrated that stress activation of the sympathetic nervous system influences the entire cascade of the human immune response resulting in increased infiltration of invasive macrophage cells, formation of new blood vessels, and tumor invasion processes in cancer models (Cole et al., 2015; Sloan et al., 2010). This cellular response to sympathetic nervous system stress is corroborated by human studies which have demonstrated the role of chronic stress in promoting tumor growth and metastases in cancer, for example (Melhem-Bertrandt et al., 2011).

Hence, interventions which are capable of managing the psychological and physical symptoms associated with chronic disease will logically also have an impact on biomarkers of stress and health as well. Interventions which train participants in mindfulness meditation (mindfulness-based interventions or MBIs) are one such group of programs, the efficacy of which we will review across a range of medical conditions below.

Mindfulness for Medical Conditions

Mindfulness is often defined as paying attention, on purpose, in the present moment with a nonjudging and accepting attitude. It is a way of being in the world, and a practice, mindfulness meditation, which we use to train ourselves in this way of being. It has three components: intention (why we practice), attention (what we practice; attentional training), and attitude (how we practice; with nonjudgmental acceptance). Mindfulness training is most often offered and studied in Western medical settings through the mindfulness-based stress reduction (MBSR) program, or an adaptation thereof. MBSR as first developed by Jon Kabat-Zinn and colleagues is a group-based program which typically consists of eight weekly group classes in which principles of mindfulness and gentle Hatha yoga are taught and practiced. Participants also undertake home mindfulness practice of 30–45 min daily. Meditation practices include a lying body scan, sitting awareness of breath meditation, open awareness meditation, walking meditation, and loving kindness practice. Attitudes include non-striving, nonattachment, trust, patience, openness, curiosity, and self-compassion in addition to non-judgment and acceptance.

The question arises of how mindfulness training of this nature can address the range of difficult psychosocial as well as physical issues associated with chronic illness, as summarized above. The value of a mindfulness approach is that it is eminently adaptable to a wide array of circumstances. Absorbing the general understanding that the only certainty in life is change, and that sometimes the best thing to do to solve a problem is to just let it be, can be extremely relieving and even liberating to people who are desperately and often frantically trying to fix problems. Realizing that in fact they can slow down and see and accept their situation as it is, and learn ways to hold strong emotions and sensations that arise, can be transformative. The further realization that although specific symptoms may be unpleasant, they are tolerable and are constantly in flux can provide further liberation from suffering. Stepping back and seeing the racing thoughts, worries, and self-blame as just thoughts, and not necessarily the truth, provides yet another source of liberation. Hence, change occurs not only through training the mind in formal meditation practice, but via a shift in attitude and perspective that allows people to see their illness in a new light, without allowing fear to consume them and drive behavior.

In the following sections, we briefly summarize the latest research on the effects of various MBIs across a range of medical conditions, focusing on reviews and meta-analyses of both patient-reported outcomes and biomarkers associated with intervention participation, highlighting the latest rigorous clinical trials. While we focus on several common conditions with the largest evidence base, it's important to note that MBIs have been applied to an astonishingly wide array of medical conditions (summarized by Shapiro & Carlson, 2017). These include arthritis, asthma, chronic fatigue, diabetes, epilepsy, fibromyalgia, headache, hot flashes, multiple sclerosis, obesity, insomnia, psoriasis, solid organ transplant, tinnitus, and many other less common conditions. Typically, a small number of studies have been conducted in each of these areas resulting in weaker evidence of efficacy, but the pace of publication of MBI intervention research across diverse conditions continues unabated.

Mindfulness-Based Intervention Research Review

Cancer

Cancer patients and survivors frequently face significant, long-term side effects of treatment that can negatively affect their physical and mental health. Mind-body therapies such as MBSR have increasingly been used to help cancer patients

(Kabat-Zinn, 1990; Rouleau et al., 2015). MBSR has been reported effective in improving various psychosocial health outcomes in cancer patients such as mood disturbance (Carlson et al., 2003), anxiety and depression (Carlson et al., 2003; Kenne Sarenmalm et al., 2017; Lengacher et al., 2016), social support (Carlson et al., 2013), and overall quality of life (QoL) (Carlson et al., 2003, 2013). Some studies have also shown the beneficial effects of MBSR on other symptoms such as cancer-related fatigue (Carlson et al., 2003; Lengacher et al., 2016; Rahmani & Talepasand, 2015; Reich et al., 2014) and cognitive capacity such as concentration and attention (Rahmani et al., 2014).

Although effective in improving both mental and physical health outcomes in cancer patients, the benefits of MBSR on mental health outcomes outweigh those on physical health outcomes, although there is a dearth of studies which have included physical outcomes. In their meta-analysis investigating the effect of MBSR on mental and physical health outcomes in various cancer patients, Ledesma and Kumano (2009) found an overall mean effect size of d = 0.48 (95% CI 0.38–0.59) for mental health and an overall mean effect size of d = 0.18 (95% CI 0.08–0.28) for physical health, suggesting that MBSR was more effective in treating psychosocial health issues relative to physical problems; however that could largely be due to the lack of studies that have included physical health outcomes as end-points, and the specific measures that have typically been chosen (most often biomarkers such as cytokine function and cortisol secretion rather than health status or functional measures).

A subsequent meta-analysis of 19 studies reported similar effect sizes on mood (d = 0.42) and distress (d = 0.48) (Musial et al., 2011). Two other 2012 and 2013 meta-analyses focused on breast cancer patients exclusively, reporting large effect sizes on stress (d = 0.71) and anxiety (d = 0.73) across nine studies with various designs (Zainal et al., 2013) and medium effects on depression and anxiety in three RCTs (Cramer et al., 2012). Piet et al. (2012) examined 22 randomized and non-randomized studies across cancer types and reported moderate effect sizes on anxiety and depression in non-randomized studies (0.6 and 0.42, respectively), and slightly smaller effects for RCTs (Piet et al., 2012).

Applying a biopsychosocial approach, several MBSR studies have attempted to clarify the physiological underpinnings that may drive the observed changes in psychosocial health following an MBSR intervention. Carlson et al. (2013, 2015) examined the effects of an 8-week MBSR intervention for breast (N = 59) and prostate (N = 10) cancer survivors and reported that participants experienced improved QoL and sleep quality and attenuated stress symptoms. Enhanced QoL was associated with attenuated levels of afternoon cortisol, although no significant effects were found for DHEAS or melatonin, suggesting that the MBSR intervention was effective in treating psychosocial issues, which may be modulated by alterations in HPA axis function (Carlson et al., 2004). In a longer-term follow-up of the same participants, Carlson et al. (2007) showed that following the MBSR intervention, patients experienced improvements in stress symptoms, as well as reduced salivary cortisol and Th1 pro-inflammatory cytokines over the course of 12 months, again suggesting that MBSR affects immune and endocrine function, which may underpin the observed improvements in QoL and symptoms of stress.

Similarly, in their study, examining the effects of a 6-week MBSR intervention, Lengacher et al. (2019) reported that immediately following MBSR classes, cortisol was attenuated at both the 1- and 6-week (first vs. last class) time points, while IL-6 was significantly reduced following MBSR class at week 6, also suggesting that MBSR modulates salivary cortisol and pro-inflammatory cytokine IL-6, in the short term. In another study, Witek-Janusek et al. (2008) found that patients undergoing MBSR interventions had attenuated salivary cortisol levels and improved scores of QoL and coping, and over time also improved immune function as shown by normalized natural killer cell activity (NKCA) and cytokine production (Witek-Janusek et al., 2008).

In addition, an RCT with 88 psychologically distressed breast cancer survivors, with a diagnosis of stage I–III cancer, showed that, relative to survivors in the control group, psychosocial interventions (both MBSR and supportive-expressive therapy) helped to maintain telomere length, which is important for protecting the ends of chromosomes and has been associated with breast cancer prognosis (Carlson et al., 2015). In another RCT comparing the effect of a 6-week MBI to usual care on telomere length and telomere activity in 142 breast cancer patients, Lengacher et al. (2014) found no difference in TL length pre-post intervention. However, they reported increase in telomere activity (TA) in the peripheral blood monocular cells. The findings have potential implications for understanding the role of MBSR in extending cell longevity at the cellular level. Taken together, these findings suggested that MBSR affects immune and endocrine function, as well as epigenetic mechanisms, and these biological factors may underpin the positive psychosocial effects experienced by participants.

Research to date suggests that MBIs are effective in helping to improve cancer patients' psychosocial health and that the immune and endocrine systems, and epigenetic factors, may play an important role in mediating or modulating MBI effects. Future research may benefit from also examining the effects of MBI on the gut microbiota, given its established role in immune and endocrine function and health (Dinan & Cryan, 2017; Ledesma & Kumano, 2009; Rahmani et al., 2014; Rahmani & Talepasand, 2015), and mounting evidence suggesting that cancer itself, and cancer treatments, can adversely affect the human gut microbiome (Bajic et al., 2018; Vivarelli et al., 2019).

Pain

Chronic pain, or pain that lasts longer than 3 months, is related to increased psychological distress and diminished quality of life. Unlike acute pain, chronic pain is often unresponsive to pharmacological intervention. While MBIs have demonstrated effectiveness in reducing pain intensity in acute pain research (Garland et al., 2017; Zeidan et al., 2010), current evidence of the effect of MBIs on chronic pain is variable. Given the complex, multidimensional nature of chronic pain, these findings are not unexpected. Indeed, there is promising evidence to suggest that mindfulness can help manage pain and pain-related symptoms in a range of medical conditions like musculoskeletal pain, cancer, chronic fatigue syndrome (CFS), and irritable bowel syndrome (IBS) (Garland et al., 2012; Zou et al., 2018).

Evidence on the effect of MBIs on musculoskeletal pain has been inconsistent in regard to reducing pain intensity but has shown effects on other pain-related symptoms in these populations (Grossman et al., 2004; Veehof et al., 2011). In individuals with chronic low back pain, MBSR has been shown to decrease short-term back pain and increase function, but improvements in function have not held up in long-term follow-ups (Chou et al., 2017). Among trials of mindfulness meditation for women with chronic pelvic pain, MBIs have reduced affective pain, sensory pain, anxiety, and depression, but had no effect on pain intensity or quality of life (Ball et al., 2017). Also, MBIs have also been shown to effectively improve quality of life, stress, pain, and symptom severity in individuals with fibromyalgia (Adler-Neal & Zeidan, 2017).

While research on MBIs for adolescents with chronic pain is lacking compared to the adult literature, there is still promising evidence of its effects on pain and pain-related symptoms. MBIs have been shown to improve mood and anxiety symptoms, coping with pain, functional disability, pain perception, and social and psychological impacts of chronic pain in adolescents (Lin et al., 2019). However, high-quality RCTs with larger sample sizes are needed to draw definitive conclusions about MBIs for younger populations with chronic pain.

Pain is a complex, multidimensional construct. Mechanisms by which MBIs may reduce pain are yet to be fully understood, considering the complex nature of both pain and mindfulness interventions. The most prevalent evidence-based viewpoint is that endogenous opioid pathways may mediate the analgesic effect of MBIs on pain intensity and unpleasantness through thalamic deactivation and orbitofrontal cortex activation (Sharon et al., 2016). Additionally, psychosocial mechanisms that may explain the effect of MBIs on pain include increased cognitive flexibility and emotion regulation using nonjudgmental and nonreactive focus (Adler-Neal & Zeidan, 2017). These psychosocial mechanisms could explain the current inconsistencies in the effect of MBIs on pain intensity (Anheyer et al., 2019; Ball et al., 2017; Chou et al., 2017; Khoo et al., 2019; McClintock et al., 2019), wherein the interventions are influencing other pain-related symptoms rather than the pain itself.

In sum, the current state of evidence suggests that MBIs are a promising intervention option for individuals living with chronic pain. Many important considerations should be made when choosing MBIs for pain management as well as conducting future research. For example, while mindfulness may not directly reduce the pain itself in the short term, MBIs could indirectly reduce pain by managing symptoms which are known to exacerbate pain, such as anxiety, depression, and fatigue over time. Additionally, MBIs may have a greater effect on the participant's pain modulation system when mindfulness is practiced regularly over time (Zeidan & Vago, 2016). Once mastered, these analgesic effects could potentially be sustained in the long term as MBIs essentially "rewire" the brain's pain appraisal system (Zeidan & Vago, 2016), and effects may be sustained even in non-meditative states (Grant, 2014). It is therefore concluded that mindfulness can indirectly reduce pain by targeting pain-related symptoms in the short term and there is promising evidence of the long-lasting effects of mindfulness on the pain modulation system if practiced over time.

Cardiovascular Conditions

Cardiovascular disease (CVD) refers to a group of disorders that affect the heart and blood vessels (including hypertension, stroke, heart attack and failure, peripheral vascular disease, rheumatic and congenital heart disease, and cardiomyopathies), and is a leading cause of death worldwide (World Health Organization [WHO], 2017). Research suggests that higher perceived stress, smoking, poor diet, lack of physical activity, and lack of medication adherence place individuals at higher risk for CVD (Janssen et al., 2013; Richardson et al., 2012). Importantly, programs targeting these risk factors are associated with reduced mortality among individuals with CVD (Janssen et al., 2013).

Loucks et al. (2015b) present a theoretical framework outlining mechanisms by which mindfulness may influence CVD risk. Preliminary evidence suggests that mindfulness may affect overall health through enhanced attention control, emotion regulation, and self-awareness, ultimately leading to improved self-regulation (Tang et al., 2015). In CVD specifically, mindfulness may improve adherence to health behaviors that ameliorate CVD risk via these proposed pathways. For example, Loucks et al. (2015b) propose that improving attention control by training the mind in moment-to-moment awareness can highlight short- and long-term effects of CVD risk behaviors, generating cognitive dissonance and increasing intrinsic motivation. Mindfulness practice may also help individuals notice cravings without acting on them (e.g., smoking cessation; Elwafi et al., 2013), increase self-efficacy and sense of control (Loucks et al., 2015a), promote self-compassion (Terry et al., 2013), and provide a social support network (when practiced in a group; Thurston & Kubzansky, 2009; Uchino, 2006), all of which are associated with healthier behaviors. Additionally, MBIs often address stress through normalization and enhancing self-awareness of responses to stress (e.g., as in MBSR; Kabat-Zinn, 2013), representing another plausible pathway via which mindfulness may reduce CVD risk (Goyal et al., 2014).

MBIs have been examined as potential interventions to reduce CVD risk and as interventions to improve the health and well-being of those with CVD. Systematic reviews and meta-analyses have reported on average 3–7 mmHg reductions in systolic and diastolic blood pressure (SBP/DBP) following MBIs that are structured (e.g., MBSR), centered on specific types of meditation (e.g., transcendental meditation), or combine meditation with physical activity (e.g., yoga, tai chi, and qigong), relative to a variety of control groups (e.g., active, attention, education, usual care) among prehypertensive individuals or individuals with stage I or II hypertension (Park & Han, 2017; Rainforth et al., 2007; Solano Lopez, 2018; Yang et al., 2017). In addition, there is some preliminary evidence that MBIs can influence other health behaviors and biomarkers that contribute to lower CVD risk, such as smoking cessation, increased physical activity, reduced atherosclerosis progression, and improved glucose regulation (Fulwiler et al., 2015; Levine et al., 2017).

Systematic reviews and meta-analyses have examined MBIs (most often MBSR or Mindfulness-Based Cognitive Therapy (MBCT)) for improving health and wellness of individuals with various types of CVD, including transient ischemic attack, stroke, coronary heart disease, heart failure, angina, and myocardial infarction. Overall, results of reviews support improvements in well-being, quality of life, depressive and anxious symptoms, distress, stress, and BP (Abbott et al., 2014; Lawrence et al., 2013; Scott-Sheldon et al., 2019a, b; Younge et al., 2015). Other physical or biological markers have been examined with mixed results, where medium-large effect-size improvements in exercise tolerance measures (e.g., VO2 max, 6-minute walk test), small effect-size improvements in heart rate (Younge et al., 2015), and no change in albuminuria, hemoglobin A1C, or stress hormones have been reported following MBIs (Abbott et al., 2014). Common methodological shortcomings in the literature include small sample sizes, lack of long-term follow-up data, non-reporting of adherence, and lack of blinded outcome assessment (Fulwiler et al., 2015; Levine et al., 2017; Solano Lopez, 2018).

Recently, methodologically rigorous trials have focused on mobile MBIs delivered though telephone or smartphone apps, with the aim of creating more accessible, cost-effective, and acceptable intervention options. Adams et al. (2018) randomized 64 prehypertensive adults to use a smartphone app for breathing awareness meditation for 5, 10, or 15 min twice daily over 6 months in a dose-response feasibility trial. They reported large, medium, and small effect-size reductions in SBP, DBP, and heart rate, respectively, across all groups, noting that a higher meditation dose resulted in greater reductions in SBP within the first month (Adams et al., 2018). They also noted that after a month, higher doses of meditation were associated with poorer adherence, suggesting that an initially higher dose of meditation for 1 month followed by a reduced dose may be ideal for this population (Adams et al., 2018).

Cox et al. (2019) randomized 80 cardiorespiratory failure patients post-discharge from intensive care to four sessions of mobile mindfulness, telephone mindfulness, or education. They reported that the mobile MBI performed similarly to therapist-led telephone MBI for reducing depressive, anxious, and posttraumatic symptoms (but not mindfulness skills or coping skills) relative to education (effect sizes not reported) (Cox et al., 2019). Although they noted that adherence and retention were slightly better in the telephone group, the authors suggested that smartphone-delivered MBIs may be an appropriate, feasible alternative (Cox et al., 2019). Both trials were pilot or feasibility trials meant to inform larger, appropriately powered trials which, if successful, will provide compelling evidence for the effectiveness of modern, mobile MBIs for well-being and risk factors in CVD.

In sum, MBIs consistently demonstrate psychological benefits and at least short-term improvements in BP, but their overall impact on CVD risk is unclear (Abbott et al., 2014; Scott-Sheldon et al., 2019a, b; Younge et al., 2015). In light of methodological limitations (particularly small sample sizes and lack of long-term follow-up data), the American Heart Association considers meditation to have possible (rather than definite) benefit for cardiovascular risk (Levine et al., 2017). Ultimately, however, they recommend that meditation can be considered as an adjunct to conventional CVD risk reduction strategies given the low-cost and low-risk inherent in MBIs (Levine et al., 2017).

HIV

Exposure to the human immunodeficiency virus (HIV) causes infection that targets the immune system, and results in an increased susceptibility to an array of other infections, cancers, and diseases (World Health Organization, 2019). One relevant marker of immune function often measured in people with HIV infection is CD4+ T-cell count, as when levels of CD4+ T cells reach a critical nadir, symptoms of AIDS may emerge. As of 2018, 37.9 million people were diagnosed and living with HIV, and 62% of adults and 54% of children who were diagnosed were also receiving antiretroviral therapy (ART; WHO, 2019). The development and widespread use of ART medications has turned this once highly fatal infection into a chronic disease that in most cases can be managed over the long term. Those diagnosed and receiving treatment have a life expectancy that is nearly as high as uninfected individuals. Increased life expectancy means that individuals living with HIV will experience prolonged symptoms associated with HIV.

People living with HIV/AIDS experience a wide range of symptoms, both associated with the chronic illness and treatment side effects from ART. Some of the symptoms include treatment side effects and inflammatory process (decreased CD4+ T-cell count), chronic pain and psychological symptoms such as high prevalence of depression and anxiety, increased stress and fatigue, and overall lower quality of life (Scott-Sheldon et al., 2019b; Wilson et al., 2016).

As with other chronic medical conditions, many of these disease and treatment-related symptoms respond well to mindfulness training. In an recent RCT of MBSR in people living with HIV not on ART, Hecht and colleagues (Hecht et al., 2018) examined whether MBSR (n = 84) would slow the naturally declining rate of CD4+ cell counts and improve psychological outcomes when compared to an active control of self-management skill (n = 88). Both interventions included one weekly class for 8 weeks as well as daily home practice, and participants were assessed at baseline and 3 and 12 months post-intervention. Those in MBSR improved from baseline to 3 months in depression, positive and negative affect, perceived stress, and mindfulness; however because the skills group was also beneficial, between-group differences were greater in the MBSR group only for positive affect. At the 12-month follow-up, improvements in depression and negative affect were still seen in the MBSR group and perceived stress for the control group. There were no between-group differences on any of the immune outcomes (CD4, c-reactive protein, IL-6, HIV-1 viral load, and d-dimer).

Bhochhibhoya et al.'s systematic review (Bhochhibhoya et al., 2018) of RCTs of mindfulness-based interventions included practices such as MBSR, MBCT, yoga, meditation, and qigong for stress management and disease progression for people living with HIV. It highlighted that the efficacy of MBIs within people living with HIV/AIDS is largely understudied; past literature has primarily focused on yoga and MBSR, and is inconsistent in finding improved stress-related outcomes. Bhochhibhoya et al.'s review included 13 MBI studies and found that mindfulness-based interventions were more effective in improving quality of life (n = 8) and reducing stress (n = 4) than control conditions. Comparatively, among nine studies that examined immune and virological status, only four studies reported increased CD4+ counts, but they were not statistically significant.

Although there are a wide range of studies examining the effects of mindfulness meditation for people living with HIV/AIDs, due to the complexity and wide range of symptoms associated with the chronic illness and the ART treatment, more rigorous study design is needed to explore mindfulness meditation as a symptom management tool. Future study designs should also compare different types of mindfulness-based and mind-body interventions in an attempt to achieve some specificity of intervention effect for specific side effects. Finally, relationships between quality of life variables and immune counts should be examined more in-depth as well.

IBD/IBS

Over the past decade, MBIs have gained significant traction among researchers in exploring psychological treatment modalities for people suffering from irritable bowel syndrome (IBS) and irritable bowel disease (IBD). IBS is a functional disorder characterized by disabling physical and psychological symptoms associated with altered bowel function, including abdominal pain, diarrhea, and constipation, while IBD represents a range of diseases that show long-lasting inflammation in the digestive tract which can cause more serious digestive problems and gross structural changes (Drossman, 2006). IBD is generally presented in two main types, Crohn's disease (CD) and ulcerative colitis (UC), which are both characterized by flare-ups, combined with periods of asymptomatic remission (Drossman, 2006).

Coping with the relapsing and unexpected symptomatic nature of the disease is a source of significant distress in IBS/IBD sufferers, who report high rates of anxiety and depression which persist even in times of remission (Bannaga & Selinger, 2015). While IBS and IBD present clear pathophysiological differences, both disorders cause chronic pain and a high level of psychiatric comorbidity involved in symptom presentation (Ballou & Keefer, 2017). Due to the level of psychological distress caused, both disorders are good candidates for mindfulness-based interventions as a possible treatment option (Kuo et al., 2015). Mindfulness-based interventions can buffer patients from catastrophizing and ruminating about flare-ups, as well as from focusing on abdominal pain (Yeh et al., 2017).

Several large-scale randomized controlled trials have investigated the application of MBIs for psychotherapeutic IBS/IBD treatment. The first was conducted by

Ljotsson et al. (2011a) and compared online MBI treatment to a waitlist control group among 85 participants. Intention-to-treat analysis showed improvement on outcome measures including IBS symptoms, quality of life, and anxiety related to IBS symptoms in the MBI group relative to control (Ljotsson et al., 2011a). This group went on to investigate their online mindfulness intervention compared to online stress management matched in time and format in a large RCT with 195 patients (Ljotsson et al., 2011b). At post-treatment and 6-month follow-up, the MBI group improved more than stress management on IBS symptom severity, IBS quality of life, visceral sensitivity, and the cognitive scale for functional bowel disorders. Both groups improved similarly on the perceived stress scale and hospital anxiety and depression scale subscales.

Two North American groups have also evaluated in-person traditional MBSR for IBS. In an RCT of MBSR vs. waitlist for 90 IBS suffers, Zernicke et al. (2012) found that MBSR group participants improved more than controls on symptom severity, with clinically meaningful decreases from constantly to occasionally present symptoms, which were maintained in the MBSR group 6 months later. In an active comparison trial, 75 women with MBSR were randomized to MBSR or a support group matched for time and other nonspecific factors (Gaylord et al., 2011). Women in MBSR, compared to the support group, showed greater reductions in IBS symptom severity post-training (26.4% vs. 6.2% reduction) and at 3-month follow-up (38.2% vs. 11.8%). Changes in quality of life, psychological distress, and visceral anxiety favoring MBSR emerged at the 3-month follow-up.

Most recently, a 2016 RCT showed further evidence for the efficacy of an 8-week MBSR protocol for 60 participants with IBD, showing significant improvements in anxiety ($p < 0.05$), depression ($p < 0.05$), quality of life ($p < 0.01$), and mindfulness ($p < 0.01$), when compared to a control group. The changes remained clinically meaningful during a 6-month follow-up after the intervention (Neilson et al., 2016). Summarizing this data, in 2014 a meta-analysis reviewed seven existing RCTs evaluating MBIs for IBS/IBD treatment, with pooled effects showing a decrease in symptom severity (0.59, 95% CI 0.33 to 0.86) and increase in quality of life (0.56, 95% CI 0.47 to 0.79) among all participants, suggesting that MBIs provide benefit for IBS/IBD sufferers (Aucoin et al., 2014).

Additional research has explored the hypothesized mechanisms of benefit of MBIs on IBS/IBD, suggesting improvements may be achieved through improving comorbid symptoms of psychological distress, such as decreased thought/pain reactivity and relatedly decreased visceral sensitivity in participants (Garland et al., 2012). Gaylord et al. found in a path analysis of their data that MBSR seemed to help by promoting nonreactivity to gut-focused anxiety and less catastrophic appraisals of the significance of abdominal sensations, as well as refocusing attention onto interoceptive data without the high levels of emotional reactivity often characteristic of the disorder (Gaylord et al., 2011).

Although additional RCTs are needed to add further support, these studies demonstrate feasibility and efficacy of applying MBIs toward treating not only the psychological symptoms of living with disease but also the symptom severity of the disorders themselves.

Methodological Issues and Recommendations

In terms of recommendations for research directions in this area, continued comparisons to gold-standard active interventions would constitute more difficult tests of the specificity of MBIs. As seen in some of the pain studies in particular, MBIs may not prove superior to other cognitive-behavioral approaches. Nonspecific factors such as group support, the therapeutic alliance, expectancy for improvement, psychoeducation, self-monitoring, and self-empowerment are likely also important drivers of change. Further research evaluating mechanisms of change, processes of change, and mediating factors would help improve understanding of what is happening within these complex multidimensional interventions.

Also, researchers are recommended to employ actual measures of the physiological stress systems that they intend to discuss with regard to their mechanisms of action. For example, previous reviews have noted that although researchers would reference the science of psychoneuroimmunology to provide a theoretical basis for their intervention, few studies actually measure all three aspects of the framework: the psychosocial, nervous/neuroendocrine, and immune systems. Furthermore, use of advanced statistical tools such as structural equation modeling and path analyses will help provide a richer understanding of the underlying mediators and moderators of these complex integrative interventions. Also, uniform standards are needed for dealing with missing data regarding outcomes of integrative interventions, which will help provide future reviewers with tools to synthesize evidence and compare interventions.

The populations studied need to be broadened, including people with late-stage or advanced disease who may benefit from integrative therapies, as well as people with a variety of cancer types and a wider breadth of ages (including children, adolescents and young adults, as well as older adults). We also need to purposively include people from disadvantaged backgrounds, with lower incomes, and include indigenous and ethnically diverse patient groups, as well as caregivers and family members.

Further adapting and tailoring MBIs to individual interventions, home-study programs, online adaptations, and investigating the efficacy of shorter groups would also be beneficial to reach larger groups of underserved patients in rural and remote locations. At the same time, treatment integrity is essential, including the training of professionals delivering the interventions. The principles of training facilitators who are grounded in the practices, practice themselves, and receive adequate training and supervision continue to be essential, and in fact are likely more important as various adaptations in form and delivery continue to evolve.

Finally, while strongly designed efficacy studies are needed to determine the causality of effects on outcomes of interest, once empirically supported treatments are identified, we need to work on the science of implementation and dissemination into healthcare systems and society at large. It is only with pragmatic real-world research of this type that we will move toward the goal of improving access and reducing barriers to the provision of these treatments for the many people who may benefit from them.

Conclusions

This paper reviewed the rationale for including MBIs in the treatment of a variety of medical conditions, and while many other conditions have been studied with smaller bodies of research, the most convincing evidence of the efficacy of such interventions come from studies of people living with cancer, pain, CVD, HIV/AIDS, and IBS/IBD. Overall, MBIs seem most effective in helping people cope with the psychological symptoms of living with chronic illness such as anxiety, depression, fear of the future, symptom catastrophizing, anger, and self-blame. They may also help with more disease-specific outcomes which can be exacerbated by autonomic nervous system arousal such as symptom severity, fatigue and sleep disorders, as well as blood pressure elevation. While the body of evidence is relatively convincing, many trials have been small without strong active comparators, so the specificity of MBIs compared to other mind-body or cognitive-behavioral therapies remains unknown. Suggestions for future research directions are provided.

References

Abbott, R. A., Whear, R., Rodgers, L. R., Bethel, A., Thompson Coon, J., Kuyken, W., … Dickens, C. (2014). Effectiveness of mindfulness-based stress reduction and mindfulness based cognitive therapy in vascular disease: A systematic review and meta-analysis of randomised controlled trials. *Journal of Psychosomatic Research, 76*(5), 341–351. https://doi.org/10.1016/j.jpsychores.2014.02.012

Adams, Z. W., Sieverdes, J. C., Brunner-Jackson, B., Mueller, M., Chandler, J., Diaz, V., … Treiber, F. A. (2018). Meditation smartphone application effects on prehypertensive adults' blood pressure: Dose-response feasibility trial. *Health Psychology, 37*(9), 850–860. https://doi.org/10.1037/hea0000584

Adler-Neal, A. L., & Zeidan, F. (2017). Mindfulness meditation for fibromyalgia: Mechanistic and clinical considerations. *Current Rheumatology Reports, 19*(9), 59. https://doi.org/10.1007/s11926-017-0686-0

Anheyer, D., Leach, M. J., Klose, P., Dobos, G., & Cramer, H. (2019). Mindfulness-based stress reduction for treating chronic headache: A systematic review and meta-analysis. *Cephalalgia: An International Journal of Headache, 39*(4), 544–555. https://doi.org/10.1177/0333102418781795

Antoni, M. H., & Dhabhar, F. S. (2019). The impact of psychosocial stress and stress management on immune responses in patients with cancer. *Cancer, 125*(9), 1417–1431. https://doi.org/10.1002/cncr.31943

Aucoin, M., Lalonde-Parsi, M. J., & Cooley, K. (2014). Mindfulness-based therapies in the treatment of functional gastrointestinal disorders: A meta-analysis. *Evidence-based Complementary and Alternative Medicine: Ecam, 2014*, 140724. https://doi.org/10.1155/2014/140724

Bajic, J. E., Johnston, I. N., Howarth, G. S., & Hutchinson, M. R. (2018). From the bottom-up: Chemotherapy and gut-brain Axis dysregulation. *Frontiers in Behavioral Neuroscience, 12*, 104.

Ball, E. F., Nur Shafina Muhammad Sharizan, E., Franklin, G., & Rogozinska, E. (2017). Does mindfulness meditation improve chronic pain? A systematic review. *Current Opinion in Obstetrics & Gynecology, 29*(6), 359–366. https://doi.org/10.1097/GCO.0000000000000417

Ballou, S., & Keefer, L. (2017). Psychological interventions for irritable bowel syndrome and inflammatory bowel diseases. *Clinical and Translational Gastroenterology, 8*(1), e214. https://doi.org/10.1038/ctg.2016.69

Bannaga, A. S., & Selinger, C. P. (2015). Inflammatory bowel disease and anxiety: Links, risks, and challenges faced. *Clinical and Experimental Gastroenterology, 8*, 111–117. https://doi.org/10.2147/CEG.S57982

Bhochhibhoya, A., Stone, B., & Li, X. (2018). Mindfulness-based intervention among people living with HIV/AIDS: A systematic review. *Complementary Therapies in Clinical Practice, 33*, 12.

Bower, J. E., Garet, D., & Sternlieb, B. (2011). Yoga for persistent fatigue in breast cancer survivors: Results of a pilot study. *Evidence-based Complementary and Alternative Medicine: Ecam, 2011*, 623168. https://doi.org/10.1155/2011/623168

Carlson, L. E., Speca, M., Patel, K. D., & Goodey, E. (2003). Mindfulness-based stress reduction in relation to quality of life, mood, symptoms of stress, and immune parameters in breast and prostate cancer outpatients. *Psychosomatic Medicine, 65*(4), 571–581.

Carlson, L. E., Speca, M., Patel, K. D., & Goodey, E. (2004). Mindfulness-based stress reduction in relation to quality of life, mood, symptoms of stress and levels of cortisol, dehydroepiandrosterone sulfate (DHEAS) and melatonin in breast and prostate cancer outpatients. *Psychoneuroendocrinology, 29*(4), 448–474.

Carlson, L. E., Speca, M., Patel, K. D., & Faris, P. (2007). One year pre-post intervention follow-up of psychological, immune, endocrine and blood pressure outcomes of mindfulness-based stress reduction (MBSR) in breast and prostate cancer outpatients. *Brain, Behavior, and Immunity, 21*(8), 1038–1049. https://doi.org/10.1016/j.bbi.2007.04.002

Carlson, L. E., Doll, R., Stephen, J., Faris, P., Tamagawa, R., Drysdale, E., & Speca, M. (2013). Randomized controlled trial of mindfulness-based cancer recovery versus supportive expressive group therapy for distressed survivors of breast cancer. *Journal of Clinical Oncology, 31*(25), 3119–3126.

Carlson, L. E., Beattie, T. L., Giese-Davis, J., Faris, P., Tamagawa, R., Fick, L. J., Degelman, E. S., & Speca, M. (2015). Mindfulness-based cancer recovery and supportive-expressive therapy maintain telomere length relative to controls in distressed breast cancer survivors. *Cancer, 121*(3), 476–484.

Chou, R., Deyo, R., Friedly, J., Skelly, A., Hashimoto, R., Weimer, M., ... Brodt, E. D. (2017). Nonpharmacologic therapies for low back pain: A systematic review for an american college of physicians clinical practice guideline. *Annals of Internal Medicine, 166*(7), 493–505. https://doi.org/10.7326/M16-2459

Cole, S. W., Nagaraja, A. S., Lutgendorf, S. K., Green, P. A., & Sood, A. K. (2015). Sympathetic nervous system regulation of the tumour microenvironment. *Nature Reviews. Cancer, 15*(9), 563–572. https://doi.org/10.1038/nrc3978

Cox, C. E., Hough, C. L., Jones, D. M., Ungar, A., Reagan, W., Key, M. D., ... Porter, L. S. (2019). Effects of mindfulness training programmes delivered by a self-directed mobile app and by telephone compared with an education programme for survivors of critical illness: A pilot randomised clinical trial. *Thorax, 74*(1), 33–42. https://doi.org/10.1136/thoraxjnl-2017-211264

Cramer, H., Haller, H., Lauche, R., & Dobos, G. (2012). Mindfulness-based stress reduction for low back pain. A systematic review. *BMC Complementary & Alternative Medicine, 12*, 162.

Dinan, T. G., & Cryan, J. F. (2017). The microbiome-gut-brain Axis in health and disease. *Gastroenterology Clinics of North America, 46*(1), 77–89.

Drossman, D. A. (2006). The functional gastrointestinal disorders and the rome III process. *Gastroenterology, 130*(5), 1377–1390. doi:S0016-5085(06)00503-8 [pii].

Elwafi, H. M., Witkiewitz, K., Mallik, S., Thornhill, T. A., 4th, & Brewer, J. A. (2013). Mindfulness training for smoking cessation: Moderation of the relationship between craving and cigarette use. *Drug and Alcohol Dependence, 130*(1–3), 222–229. https://doi.org/10.1016/j.drugalcdep.2012.11.015

Fulwiler, C., Brewer, J. A., Sinnott, S., & Loucks, E. B. (2015). Mindfulness-based interventions for weight loss and CVD risk management. *Current Cardiovascular Risk Reports, 9*(10). https://doi.org/10.1007/s1217-1. Epub 2015 Aug 29. doi:46 [pii].

Garland, E. L., Gaylord, S. A., Palsson, O., Faurot, K., Douglas Mann, J., & Whitehead, W. E. (2012). Therapeutic mechanisms of a mindfulness-based treatment for IBS: Effects on visceral sensitivity, catastrophizing, and affective processing of pain sensations. *Journal of Behavioral Medicine, 35*(6), 591–602. https://doi.org/10.1007/s10865-011-9391-z

Garland, E. L., Baker, A. K., Larsen, P., Riquino, M. R., Priddy, S. E., Thomas, E., ... Nakamura, Y. (2017). Randomized controlled trial of brief mindfulness training and hypnotic suggestion for acute pain relief in the hospital setting. *Journal of General Internal Medicine, 32*(10), 1106–1113. https://doi.org/10.1007/s11606-017-4116-9

Gaylord, S. A., Palsson, O. S., Garland, E. L., et al. (2011). Mindfulness training reduces the severity of irritable bowel syndrome in women: Results of a randomized controlled trial. *American Journal of Gastroenterology, 106*(9), 1678–1688.

Goyal, M., Singh, S., Sibinga, E. M., Gould, N. F., Rowland-Seymour, A., Sharma, R., ... Haythornthwaite, J. A. (2014). Meditation programs for psychological stress and well-being: A systematic review and meta-analysis. *JAMA Internal Medicine, 174*(3), 357–368. https://doi.org/10.1001/jamainternmed.2013.13018

Grant, J. A. (2014). Meditative analgesia: The current state of the field. *Annals of the New York Academy of Sciences, 1307*, 55–63. https://doi.org/10.1111/nyas.12282

Grossman, P., Niemann, L., Schmidt, S., & Walach, H. (2004). Mindfulness-based stress reduction and health benefits. A meta-analysis. *Journal of Psychosomatic Research, 57*(1), 35–43. https://doi.org/10.1016/S0022-3999(03)00573-7

Hecht, F. M., Moskowitz, J. T., Moran, P., Epel, E. S., Bacchetti, P., Acree, M., ... Levy, J. A. (2018). A randomized, controlled trial of mindfulness-based stress reduction in HIV infection. *Brain, Behavior, and Immunity, 73*, 331–339.

Janssen, V., De Gucht, V., Dusseldorp, E., & Maes, S. (2013). Lifestyle modification programmes for patients with coronary heart disease: A systematic review and meta-analysis of randomized controlled trials. *European Journal of Preventive Cardiology, 20*(4), 620–640. https://doi.org/10.1177/2047487312462824

Kabat-Zinn, J. (1990). *Full catastrophe living: Using the wisdom of your body and mind to face stress, pain and illness*. Delacourt.

Kabat-Zinn, J. (2013). *Full catastrophe living: Using the wisdom of your body and mind to face stress, pain, and illness*. Bantam.

Kenne Sarenmalm, E., Mårtensson, L. B., Andersson, B. A., Karlsson, P., & Bergh, I. (2017). Mindfulness and its efficacy for psychological and biological responses in women with breast cancer. *Cancer Medicine, 6*(5), 1108–1122.

Khoo, E. L., Small, R., Cheng, W., Hatchard, T., Glynn, B., Rice, D. B., ... Poulin, P. A. (2019). Comparative evaluation of group-based mindfulness-based stress reduction and cognitive behavioural therapy for the treatment and management of chronic pain: A systematic review and network meta-analysis. *Evidence-Based Mental Health, 22*(1), 26–35. https://doi.org/10.1136/ebmental-2018-300062

Kuo, B., Bhasin, M., Jacquart, J., Scult, M. A., Slipp, L., Riklin, E. I., ... Denninger, J. W. (2015). Genomic and clinical effects associated with a relaxation response mind-body intervention in patients with irritable bowel syndrome and inflammatory bowel disease. *PLoS One, 10*(4), e0123861. https://doi.org/10.1371/journal.pone.0123861

Lawrence, M., Booth, J., Mercer, S., & Crawford, E. (2013). A systematic review of the benefits of mindfulness-based interventions following transient ischemic attack and stroke. *International Journal of Stroke, 8*(6), 465–474. https://doi.org/10.1111/ijs.12135

Ledesma, D., & Kumano, H. (2009). Mindfulness-based stress reduction and cancer: A meta-analysis. *Psycho-Oncology: Journal of the Psychological, Social and Behavioral Dimensions of Cancer, 18*(6), 571–579.

Lengacher, C. A., Reich, R. R., Kip, K. E., Barta, M., Ramesar, S., Paterson, C. L., et al. (2014). Influence of mindfulness-based stress reduction on telomerase activity in women with breast cancer. *Biological Research in Nursing, 16*(4), 438–447.

Lengacher, C. A., Reich, R. R., Paterson, C. L., Ramesar, S., Park, J. Y., Alinat, C., ... Kip, K. E. (2016). Examination of broad symptom improvement resulting from mindfulness-based stress reduction in breast cancer survivors: A randomized controlled trial. *Journal of Clinical Oncology, 34*(24), 2827–2834. https://doi.org/10.1200/JCO.2015.65.7874

Lengacher, C. A., Reich, R. R., Paterson, C. L., Shelton, M., Shivers, S., Ramesar, S., ... Post-White, J. (2019). A large randomized trial: Effects of mindfulness-based stress reduction (MBSR) for breast cancer (BC) survivors on salivary cortisol and IL-6. *Biological Research for Nursing, 21*(1), 39–49.

Levine, G. N., Lange, R. A., Bairey-Merz, C. N., Davidson, R. J., Jamerson, K., Mehta, P. K., ... and Council on Hypertension. (2017). Meditation and cardiovascular risk reduction: A scientific statement from the american heart association. *Journal of the American Heart Association, 6*(10), e002218 [pii]. https://doi.org/10.1161/JAHA.117.002218

Lin, J., Chadi, N., & Shrier, L. (2019). Mindfulness-based interventions for adolescent health. *Current Opinion in Pediatrics, 31*(4), 469–475. https://doi.org/10.1097/MOP.0000000000000760

Ljotsson, B., Hedman, E., Lindfors, P., Hursti, T., Lindefors, N., Andersson, G., & Ruck, C. (2011a). Long-term follow-up of internet-delivered exposure and mindfulness based treatment for irritable bowel syndrome. *Behaviour Research and Therapy, 49*(1), 58–61. https://doi.org/10.1016/j.brat.2010.10.006

Ljotsson, B., Hedman, E. A., et al. (2011b). Internet-delivered exposure-based treatment vs. stress management for irritable bowel syndrome: A randomized trial. *American Journal of Gastroenterology, 106*(8), 1481–1491.

Loucks, E. B., Britton, W. B., Howe, C. J., Eaton, C. B., & Buka, S. L. (2015a). Positive associations of dispositional mindfulness with cardiovascular health: The New England family study. *International Journal of Behavioral Medicine, 22*(4), 540–550. https://doi.org/10.1007/s12529-014-9448-9

Loucks, E. B., Schuman-Olivier, Z., Britton, W. B., Fresco, D. M., Desbordes, G., Brewer, J. A., & Fulwiler, C. (2015b). Mindfulness and cardiovascular disease risk: State of the evidence, plausible mechanisms, and theoretical framework. *Current Cardiology Reports, 17*(12), 11–17. https://doi.org/10.1007/s11886-015-0668-7

McClintock, A. S., McCarrick, S. M., Garland, E. L., Zeidan, F., & Zgierska, A. E. (2019). Brief mindfulness-based interventions for acute and chronic pain: A systematic review. *Journal of Alternative and Complementary Medicine (New York, N.Y.), 25*(3), 265–278. https://doi.org/10.1089/acm.2018.0351

Melhem-Bertrandt, A., Chavez-Macgregor, M., Lei, X., Brown, E. N., Lee, R. T., Meric-Bernstam, F., ... Gonzalez-Angulo, A. M. (2011). Beta-blocker use is associated with improved relapse-free survival in patients with triple-negative breast cancer. *Journal of Clinical Oncology, 29*(19), 2645–2652. https://doi.org/10.1200/JCO.2010.33.4441

Musial, F., Bussing, A., Heusser, P., Choi, K. E., & Ostermann, T. (2011). Mindfulness-based stress reduction for integrative cancer care: A summary of evidence. *Forschende Komplementarmedizin (2006), 18*(4), 192–202. https://doi.org/10.1159/000330714

Neilson, K., Ftanou, M., Monshat, K., Salzberg, M., Bell, S., Kamm, M. A., ... Castle, D. (2016). A controlled study of a group mindfulness intervention for individuals living with inflammatory bowel disease. *Inflammatory Bowel Diseases, 22*(3), 694–701. https://doi.org/10.1097/MIB.0000000000000629

Park, S. H., & Han, K. S. (2017). Blood pressure response to meditation and yoga: A systematic review and meta-analysis. *Journal of Alternative and Complementary Medicine (New York, N.Y.), 23*(9), 685–695. https://doi.org/10.1089/acm.2016.0234

Piet, J., Wurtzen, H., & Zachariae, R. (2012). The effect of mindfulness-based therapy on symptoms of anxiety and depression in adult cancer patients and survivors: A systematic review and meta-analysis. *Journal of Consulting and Clinical Psychology, 80*(6), 1007–1020.

Rahmani, S., & Talepasand, S. (2015). The effect of group mindfulness-based stress reduction program and conscious yoga on the fatigue severity and global and specific life quality in women with breast cancer. *Medical Journal of the Islamic Republic of Iran, 29*, 175.

Rahmani, S., Talepasand, S., & Ghanbary-Motlagh, A. (2014). Comparison of effectiveness of the metacognition treatment and the mindfulness-based stress reduction treatment on global and specific life quality of women with breast cancer. *Iranian Journal of Cancer Prevention, 7*(4), 184.

Rainforth, M. V., Schneider, R. H., Nidich, S. I., Gaylord-King, C., Salerno, J. W., & Anderson, J. W. (2007). Stress reduction programs in patients with elevated blood pressure: A systematic review and meta-analysis. *Current Hypertension Reports, 9*(6), 520–528.

Reich, R. R., Lengacher, C. A., Kip, K. E., Shivers, S. C., Schell, M. J., Shelton, M. M., ... Paterson, C. L. (2014). Baseline immune biomarkers as predictors of MBSR (BC) treatment success in off-treatment breast cancer patients. *Biological Research for Nursing, 16*(4), 429–437.

Richardson, S., Shaffer, J. A., Falzon, L., Krupka, D., Davidson, K. W., & Edmondson, D. (2012). Meta-analysis of perceived stress and its association with incident coronary heart disease. *The American Journal of Cardiology, 110*(12), 1711–1716. https://doi.org/10.1016/j.amjcard.2012.08.004

Rouleau, C. R., Garland, S. N., & Carlson, L. E. (2015). The impact of mindfulness-based interventions on symptom burden, positive psychological outcomes, and biomarkers in cancer patients. *Cancer Management and Research, 7*, 121–131. https://doi.org/10.2147/CMAR.S64165

Scott-Sheldon, L. A. J., Gathright, E. C., Donahue, M. L., Balletto, B., Feulner, M. M., DeCosta, J., ... Salmoirago-Blotcher, E. (2019a). Mindfulness-based interventions for adults with cardiovascular disease: A systematic review and meta-analysis. *Annals of Behavioral Medicine: A Publication of the Society of Behavioral Medicine.* kaz020 [pii].

Scott-Sheldon, L. A., Balletto, B. L., Donahue, M. L., Feulner, M. M., Cruess, D. G., Salmoirago-Blotcher, E., ... Carey, M. P. (2019b). Mindfulness-based interventions for adults living with HIV/AIDS: A systematic review and Meta-analysis. *AIDS and Behavior, 23*(1), 60–75.

Shapiro, S. L., & Carlson, L. E. (2017). *The art and science of mindfulness: Integrating mindfulness into psychology and the helping professions* (2nd ed.). American Psychological Association Books.

Sharon, H., Maron-Katz, A., Ben Simon, E., Flusser, Y., Hendler, T., Tarrasch, R., & Brill, S. (2016). Mindfulness meditation modulates pain through endogenous opioids. *The American Journal of Medicine, 129*(7), 755–758. https://doi.org/10.1016/j.amjmed.2016.03.002

Sloan, E. K., Priceman, S. J., Cox, B. F., Yu, S., Pimentel, M. A., Tangkanangnukul, V., ... Cole, S. W. (2010). The sympathetic nervous system induces a metastatic switch in primary breast cancer. *Cancer Research, 70*(18), 7042–7052. https://doi.org/10.1158/0008-5472.CAN-10-0522

Solano Lopez, A. L. (2018). Effectiveness of the mindfulness-based stress reduction program on blood pressure: A systematic review of literature. *Worldviews on Evidence-Based Nursing, 15*(5), 344–352. https://doi.org/10.1111/wvn.12319

Tang, Y. Y., Holzel, B. K., & Posner, M. I. (2015). The neuroscience of mindfulness meditation. *Nature Reviews. Neuroscience, 16*(4), 213–225. https://doi.org/10.1038/nrn3916

Terry, M. L., Leary, M. R., Mehta, S., & Henderson, K. (2013). Self-compassionate reactions to health threats. *Personality & Social Psychology Bulletin, 39*(7), 911–926. https://doi.org/10.1177/0146167213488213

Thurston, R. C., & Kubzansky, L. D. (2009). Women, loneliness, and incident coronary heart disease. *Psychosomatic Medicine, 71*(8), 836–842. https://doi.org/10.1097/PSY.0b013e3181b40efc

Uchino, B. N. (2006). Social support and health: A review of physiological processes potentially underlying links to disease outcomes. *Journal of Behavioral Medicine, 29*(4), 377–387. https://doi.org/10.1007/s10865-006-9056-5

Veehof, M. M., Oskam, M. J., Schreurs, K. M., & Bohlmeijer, E. T. (2011). Acceptance-based interventions for the treatment of chronic pain: A systematic review and meta-analysis. *Pain, 152*(3), 533–542. https://doi.org/10.1016/j.pain.2010.11.002

Vivarelli, S., Salemi, R., Candido, S., Falzone, L., Santagati, M., Stefani, S., Torino, F., Banna, G. L., Tonini, G., & Libra, M. (2019). Gut microbiota and Cancer: From pathogenesis to therapy. *Cancers (Basel), 11*(1), 38.

Wilson, N. L., Azuer, A., Vance, D. E., Richman, J. S., Moneyham, L. D., Raper, J. L., ... Kempf, M. C. (2016). Identifying symptom patterns in people living with HIV disease. *Journal of the Association of Nurses in AIDS Care, 27*(2), 121–132.

Witek-Janusek, L., Albuquerque, K., Chroniak, K. R., Chroniak, C., Durazo-Arvizu, R., & Mathews, H. L. (2008). Effect of mindfulness based stress reduction on immune function, quality of life and coping in women newly diagnosed with early stage breast cancer. *Brain, Behavior, & Immunity, 22*(6), 969–981.

World Health Organization. (2017). *Cardiovascular diseases (CVDs)*. Retrieved from https://www.who.int/news-room/fact-sheets/detail/cardiovascular-diseases-(cvds)

World Health Organization. (2019). *Human immunodeficiency virus (HIV)*. Retrieved from https://www.who.int/gho/hiv/en/

Yang, H., Wu, X., & Wang, M. (2017). The effect of three different meditation exercises on hypertension: A network meta-analysis. *Evidence-based Complementary and Alternative Medicine: Ecam, 2017*, 9784271. https://doi.org/10.1155/2017/9784271

Yeh, A. M., Wren, A., & Golianu, B. (2017). Mind-body interventions for pediatric inflammatory bowel disease. *Children (Basel, Switzerland), 4*(4), E22 [pii]. https://doi.org/10.3390/children4040022

Younge, J. O., Gotink, R. A., Baena, C. P., Roos-Hesselink, J. W., & Hunink, M. G. (2015). Mind-body practices for patients with cardiac disease: A systematic review and meta-analysis. *European Journal of Preventive Cardiology, 22*(11), 1385–1398. https://doi.org/10.1177/2047487314549927

Zainal, N. Z., Booth, S., & Huppert, F. A. (2013). The efficacy of mindfulness-based stress reduction on mental health of breast cancer patients: A meta-analysis. *Psycho-Oncology, 22*(7), 1457–1465. https://doi.org/10.1002/pon.3171

Zeidan, F., & Vago, D. R. (2016). Mindfulness meditation-based pain relief: A mechanistic account. *Annals of the New York Academy of Sciences, 1373*(1), 114–127. https://doi.org/10.1111/nyas.13153

Zeidan, F., Gordon, N. S., Merchant, J., & Goolkasian, P. (2010). The effects of brief mindfulness meditation training on experimentally induced pain. *The Journal of Pain, 11*(3), 199–209. https://doi.org/10.1016/j.jpain.2009.07.015

Zernicke, K. A., Campbell, T. S., Blustein, P. K., Fung, T. S., Johnson, J. A., Bacon, S. L., & Carlson, L. E. (2012). Mindfulness-based stress reduction for the treatment of irritable bowel syndrome symptoms: A randomized wait-list controlled trial. *International Journal of Behavioral Medicine 2013, 20*(3), 385–396.

Zou, L., Yeung, A., Quan, X., Boyden, S. D., & Wang, H. (2018). A systematic review and meta-analysis of mindfulness-based (baduanjin) exercise for alleviating musculoskeletal pain and improving sleep quality in people with chronic diseases. *International Journal of Environmental Research and Public Health, 15*(2), E206 [pii]. https://doi.org/10.3390/ijerph15020206

Chapter 8
Mindfulness-Based Interventions for Traumatic Stress

Daniel Szoke, Asha Putnam, and Holly Hazlett-Stevens

Introduction

Growing evidence supports the claim that mindfulness, a purposeful and present-moment awareness, reduces symptoms of posttraumatic stress disorder (PTSD) faced by survivors of trauma (Gallegos et al., 2017; Hilton et al., 2017; Hopwood & Schutte, 2017). Intentional attention to present moment experience with an attitude of patience, compassion, and nonjudgment may specifically address symptoms of avoidance, hyperarousal, intrusions, and negative moods and thoughts commonly experienced by those suffering from PTSD. Mindfulness-based interventions have been tested both as stand-alone (e.g., Kelly & Garland, 2016) and adjunctive treatments (e.g., King et al., 2016) for survivors of trauma, with promising results. Further, a specific attitudinal foundation of mindfulness, self-compassion, has been linked to lower PTSD symptoms and has been found to prospectively, negatively predict levels of PTSD symptoms in trauma-exposed populations (Hiraoka et al., 2015; Maheux & Price, 2016; Seligowski et al., 2014; Thompson & Waltz, 2008).

Two prominent theoretical camps have emerged in the use of mindfulness-based interventions for survivors of trauma. Follette, Palm, and Pearson (2006) argued that mindfulness may enhance exposure-based treatment for survivors of trauma, and this theory has received empirical support in the form of mindfulness-based exposure therapy with samples of combat veterans reporting clinically significant reductions in PTSD symptoms (King et al., 2016). A second theoretical approach has pointed specifically to the fact that mindfulness practices lead to a natural exposure to both internal and external trauma reminders, leading to the same inhibitory learning experienced in exposure therapy (Brown et al., 2007; Holzel et al., 2011). This raises the possibility that mindfulness-based interventions can benefit trauma survivors in isolation without added imaginal or in vivo exposure.

D. Szoke · A. Putnam · H. Hazlett-Stevens (✉)
University of Nevada, Reno, NV, USA
e-mail: hhazlett@unr.edu

© Springer Nature Switzerland AG 2021   177
H. Hazlett-Stevens (ed.), *Biopsychosocial Factors of Stress, and Mindfulness for Stress Reduction*, https://doi.org/10.1007/978-3-030-81245-4_8

Mindfulness-based interventions also have received empirical support by producing reductions in PTSD symptoms, even when used as a stand-alone treatment (Kelly & Garland, 2016), and have received support from meta-analytic studies (Gallegos et al., 2017; Hilton et al., 2017; Hopwood & Schutte, 2017).

The following chapter reviews theories of PTSD treatment, mindfulness, and self-compassion, and empirical support for each theoretical perspective. First, the symptoms of PTSD are reviewed as they are reported in the *Diagnostic and Statistical Manual of Mental Disorders – 5* (DSM-5; American Psychological Association, 2013). Next, theoretical models of PTSD including Mowrer's (1947) two-factor theory, Foa and Kozak's (1986) emotional processing model, and Craske et al.'s (2014) inhibitory learning model are discussed. This information will lay the groundwork for the following section, which details theoretical models for the ways in which mindfulness may aid survivors of trauma. As a central focal point, Shapiro et al.'s (2006) intention, attention, and attitude (IAA) model is reviewed, as well as possible mechanisms of mindfulness including enhanced cognitive, emotional, and behavioral flexibility, enhanced self-regulation, and natural exposure to feared stimuli. Next, empirical findings are reviewed to support the claim that mindfulness-based interventions ameliorate symptoms of PTSD, and do so consistently, leading to significantly lower levels of PTSD than treatment as usual. Then, the relationship between PTSD symptoms and self-compassion, an attitudinal foundation of mindfulness, is reviewed, along with evidence to support the negative relationship between the two. Finally, the relationship between PTSD and inflammation is described, along with linkages to mindfulness-based interventions as a way to reduce both inflammation in the body and symptoms of PTSD.

Traumatic Stress

Traumatic stress and its related disorders are unique in the field of psychopathology due to the clarity of their etiology. Traumatic stress follows traumatic events. According to the Diagnostic and Statistical Manual of Mental Disorders (fifth ed.), traumatic events are those that involve "actual or threatened death, serious injury, or sexual violence" (American Psychiatric Association, 2013, p. 271). Stress that develops following events such as these qualifies as traumatic stress. Traumatic stress can result from directly experiencing or witnessing a traumatic event, from hearing about the extreme details of a traumatic event through work (e.g., police officers, first responders, victim advocates), or from a family member or close friend who experienced the event. Symptoms fall into four main categories: intrusion, avoidance, negative alterations in thoughts and mood, and changes in levels of arousal. If symptoms are present for less than a month following the traumatic event, a person may qualify for a diagnosis of acute stress disorder; if symptoms persist for longer than 1 month, they may qualify for a diagnosis of posttraumatic stress disorder (American Psychiatric Association, 2013).

Symptoms of intrusion include disturbances related to the traumatic event that a person has experienced. This aspect of traumatic stress involves the re-experiencing of the traumatic event through thoughts, dreams, psychological distress, physiological arousal, and flashbacks. Specific symptoms include intrusive and involuntary memories and/or dreams about the traumatic event, psychological distress and/or marked physiological reactions to internal and external reminders of the traumatic event, and dissociative reactions that give a survivor the feeling that the traumatic event is happening again (American Psychiatric Association, 2013). While some survivors may experience all of the symptoms described, others will experience only a subset.

Symptoms of avoidance include evasion of trauma reminders. This includes both avoidance of external stimuli that evoke memories, thoughts, and feelings related to the traumatic event and avoidance of internal stimuli, such as the memories, thoughts, and feelings themselves (American Psychiatric Association, 2013). For example, a person who has experienced a physical assault in a parking garage may choose street parking instead of risking entry into the adjacent garage which could in turn elicit memories related to their assault. Additionally, this person may listen to loud music or say a silent prayer at the first sign of a memory related to their trauma arising.

Negative alterations in cognitions and mood that are related to the traumatic event make up the third symptom group of trauma-related stress. In brief, this set of symptoms relates to negative emotional states such as fear, anger, and guilt as well as negative thoughts about oneself, the world, other people, and the traumatic event itself that occur following trauma. For example, following a traumatic event, a person may feel frequent guilt and shame related to the traumatic event, believe that the world is an extremely dangerous place, and feel detached from others. The following symptoms are included in this group: failure to remember important aspects of the traumatic event; distorted, negative beliefs about oneself, the world, and others; distorted cognitions about the cause and/or consequences of the traumatic event; negative emotional states that are persistent; diminished interest in significant activities; feelings of estrangement from others; and an inability to experience positive emotions that are persistent (American Psychiatric Association, 2013).

The final group of symptoms relates to changes in arousal and reactivity that is associated with the traumatic event. Traumatic stress can lead to irritable outbursts, hypervigilance, and sleep difficulties that are related to a higher level of physiological arousal. Additionally, survivors may experience an exaggerated startle response and difficulty concentrating and may engage in self-destructive behavior (American Psychiatric Association, 2013). This state of heightened arousal related to the traumatic event is a visceral experience for survivors of trauma and can cause major disturbances in interpersonal relationships as well as functioning at work or school.

Theoretical Models of PTSD Mowrer's two-factor theory provided an early theoretical framework for the acquisition and maintenance of fear in response to traumatic events (Mowrer, 1947). First, a person acquires a fear response through classical conditioning. A neutral stimulus (seeing a traffic light turn green) is paired with an unconditioned stimulus (a life-threatening car accident in the middle of an intersection) which elicits an unconditioned response (fear for one's life.) In the example provided, seeing a traffic light turn green becomes a conditioned stimulus capable of eliciting fear, which is now a conditioned response. Classical conditioning explains why this example trauma survivor may feel fearful when driving through an intersection.

Mowrer's second factor involves the maintenance of the fear response through operant conditioning (Mowrer, 1947). With enough repeated presentations of the conditioned stimulus in the absence of the unconditioned stimulus, the pairing between the conditioned stimulus and conditioned response will go extinct. Following the previous motor vehicle accident example, driving through many intersections after the light turns green without experiencing any life-threatening car accidents should lead to the extinction of fear in response to traffic lights turning green. However, for many survivors of trauma, avoiding the conditioned stimulus all together provides relief from negative emotions, which in turn negatively reinforces the avoidance behavior. As noted previously, avoidance symptoms make up one of the four primary categories of symptoms for trauma-related disorders. After experiencing a motor vehicle accident, a person may simply refuse to drive on roads with traffic lights, or perhaps refuse to drive under any circumstances. Choosing not to drive allows this person to escape the discomfort elicited by reminders of their trauma, but also robs them of their chance to experience extinction of the fear response.

Foa and Kozak's (1986) emotional processing model was built on Mowrer's theory. The emotional processing model rests on the assumption that fear structures exist within the brain that link stimuli, responses, and meaning (Rauch & Foa, 2006). The person who was involved in the motor vehicle accident may have a fear network that includes seeing a traffic light turn green (stimulus), activation of the sympathetic nervous system (physiological response) and driving away from the sound (behavioral response), and thoughts such as "I am in danger of dying" (meaning). Foa and Kozak (1986) posited that the structure itself must be modified through activation of the fear network by means of exposure to trauma reminders and the presentation of new information that is not compatible with the current fear structure. Clients will need to experience activation of their fear networks, for example, by driving through many intersections just after the light turns green, in order to integrate new information that objectively safe trauma reminders are not signs of danger.

However, newer studies have questioned the emotional processing model due to the frequency of relapse after treatment. The emotional processing model implies that those who successfully complete treatment should have a healthy, properly functioning fear network. This model therefore fails to explain why some clients experience relapse. Craske et al. (2014) suggested an alternative explanation for the success of exposure-based treatment: the inhibitory learning model. This model suggests that the original conditioned stimulus-unconditioned stimulus (CS-US) pairing, for example, driving through an intersection and colliding with another vehicle, cannot be erased through exposure. Rather, a new conditioned stimulus-no unconditioned stimulus (CS-noUS) pairing is created through repeated presentations of the conditioned stimulus without the unconditioned stimulus, for example, driving through an intersection without a collision. Each pairing gives a separate instruction: the learned pairing of CS-US activates fear, and the newly learned CS-noUS pairing inhibits fear (Craske et al., 2014). With enough repeated presentations of traffic lights turning green without car accidents, the inhibitory learning will be strengthened, and the fear response will not be elicited.

Mindfulness and Trauma

Mindfulness-based interventions have been successfully implemented for those who suffer from anxiety and mood disorders across many clinical trials (see Chap. 9). However, early concerns that mindfulness practice could worsen trauma symptoms led to a delay in their implementation with survivors of trauma (Lustyk et al., 2009; Germer, 2005). In previous editions of MBSR treatment manuals, Santorelli and Kabat-Zinn (2009) cautioned that those with a current diagnosis of PTSD may not be ready to participate in the program. A large qualitative study interviewed current meditators about adverse experiences that have been encountered over the course of practice. After conducting a large qualitative study, authors reported that it "was not uncommon" for participants with a trauma history to report re-experiencing traumatic memories (Lindahl et al., 2017, p. 19).

One author from this qualitative study, mindfulness researcher Willoughby Britton, wrote in more detail about one such story of a survivor of trauma who experienced paralyzing flashbacks when attending a 10-day meditation retreat. Britton continued to describe several stories she had encountered from survivors who experienced dissociation during mindfulness practice as well as the shame that often accompanies these survivors' stories and a sense that they somehow "failed" mindfulness practice (Treleaven, 2008, p. xi). David Treleaven, a psychotherapist and author of the book *Trauma-Sensitive Mindfulness*, offers several anecdotes from individuals who experienced an aggravation and intensification of trauma symptoms when engaging with mindfulness practice. Many of these stories involve participants who attended intensive, multiday silent retreats, and those who received mindfulness training as part of other classes, such as a high school student who participated in weekly mindfulness lessons with his classmates and reported re-experiencing symptoms and hyperarousal in response to the practices (Treleaven, 2008). However, as is noted later in this chapter, re-experiencing of traumatic memories during meditation may actually be ameliorative for survivors of trauma with proper guidance from a psychotherapist. Treleaven pointed to Peter Levine's work, who also recognized the potential benefits of mindfulness practice for survivors of trauma but noted that it could be harmful if participants are not adequately prepared to encounter re-experiencing symptoms during meditation (Levine, 2010). Importantly, no iatrogenic effects were found in the five quantitative studies that measured the potential adverse effects of conducting a mindfulness-based treatment with survivors of trauma (Bormann et al., 2013; Kearney et al., 2013; Mitchell et al., 2014; Niles et al., 2012; Polusny et al., 2015; see Hilton et al., 2017 for a complete review).

In contrast, mindfulness may provide benefits to either supplement concurrent trauma treatment or possibly treat traumatic stress directly (Gallegos et al., 2017; Hilton et al., 2017; Hopwood & Schutte, 2017). Specifically, mindfulness practice allows for greater cognitive, emotional, and behavioral flexibility, self-regulation and self-management, and exposure which all uniquely benefit survivors of trauma. Before seeing how these benefits arise and work to counter symptoms of PTSD, it is first important to provide a conceptual model of mindfulness. What follows is an exploration of different components that come together to describe mindful awareness, and synthesis of these ideas with what is already known about the treatment of survivors of trauma.

Intention, Attention, and Attitude In order to understand the role that mindfulness plays in the amelioration of mental illness broadly, Shapiro et al. (2006) introduced a model that breaks mindfulness down into three smaller, interconnected components. The model describes three axioms of mindfulness: intention, attention, and attitude (IAA). Intention is the purpose that a practitioner ascribes to the practice. For example, a meditator may practice with the intention of learning more about their habitual patterns of responding to stress. A previous study that investigated intentions of long-term meditators found a common progression of intentions starting with self-regulation, transitioning to self-exploration, and ending with the intention of self-liberation (Shapiro, 1992). The second axiom, attention, refers to the process of observing present moment experience. Attending to moment-to-moment experiences includes both external phenomena taken in through sensory data and internal events such as thoughts, emotions, and physical sensations. The final axiom, attitude, is the quality with which a person attends to the present moment. Attitudinal foundations of mindfulness include patience, compassion, and nonjudgment. An open and accepting quality of awareness is the difference between mindfulness and other states that involve intentionally attending to the present moment with judgment, such as hypervigilance or anxious overanalysis.

Taken together, these three axioms describe the essential features of mindfulness, a purposeful, present moment awareness that is nonjudgmental (Kabat-Zinn, 2013). Shapiro et al. (2006) coined the term "reperceiving" to describe the shift in perspective that occurs when these three mechanisms come together. This shift allows the practitioner to experience greater clarity of mind, and to recognize that thoughts themselves are distinct from the mind in which they were formed (Shapiro et al., 2006). This "de-fusion" between the observer and the observed allows space for disidentification, (e.g., "I am not my thoughts," "I am not my fear").

Cognitive, Emotional, and Behavioral Flexibility
Shapiro et al. (2006) suggested that increases in cognitive, emotional, and behavioral flexibility may partially explain why mindfulness is beneficial for those

suffering from mental illness. Here, the term flexibility refers to the ability to see multiple options for responding, the willingness to respond adaptively, and the self-efficacy to believe that it is possible to adapt (Martin & Rubin, 1995). A cognitively flexible person can see that there are several ways to respond to being cut off in traffic, selects the most adaptive and situationally appropriate option, and believes they have the power to follow through on their decision to assume the person has somewhere important to be rather than responding habitually by cursing their fellow driver's performance.

Mindfulness is positively related to cognitive flexibility (Moore & Malinowski, 2009). The skill of reperceiving provides a chance to deidentify from habitual patterns of reactivity while also seeing a broader range of options for responses. Intentionally attending to thought patterns with an attitude of nonjudgment allows a person to notice habitual, maladaptive ways of responding to stimuli. Those high in trait mindfulness can be flexible and chose to see the situation in a new light in order to adjust prior beliefs. For example, a combat veteran may experience symptoms of intrusion, such as reoccurring, unwanted memories about taking the life of an enemy combatant. Perhaps this person has a tendency to feel frustrated with themselves for having these intrusive memories and thinks things like, "I am a broken and damaged individual." With mindfulness, this veteran has an opportunity to observe the habitual thought and reperceive the intrusive memories with a broader range of response options. Perhaps they say to themselves, "These memories are difficult to face but they will not last forever." In this example, this person was able to see several options of responding to the intrusive memory and select a more adaptive response thanks to their intention to attend to the memories with an attitude of openness and acceptance. Empirical findings support this conclusion; researchers found cognitive flexibility to be a negative predictor of PTSD severity in survivors of interpersonal violence (Palm & Follette, 2011).

As seen in the example above, greater flexibility across the domains of thoughts, behaviors, and emotions can aid survivors in their attempts to cope with symptoms, including intrusive thoughts. Flexibility is also likely to have a positive impact on the group of symptoms known as negative alterations in mood and cognitions. As a survivor becomes more aware of their negative thought patterns and selects more adaptive responses, they should encounter changes in their beliefs about themselves, the world, and others. However, no studies that have applied mindfulness-based interventions to trauma populations have investigated this specific question. More research is needed to determine whether cognitive flexibility is the mechanism of change for improvement of the negative alterations in cognitions that occur following trauma.

Self-Regulation
Self-regulation refers to the processes involved in maintaining stability in functioning while also adapting to a constantly changing world (Shapiro & Schwartz, 2000). This model of self-regulation identified a chain of events essential to achieve and maintain health:

Intention → Attention → Connection → Regulation → Order → Health

Shapiro and Schwartz (2000) upheld that attention is a necessary precursor but emphasized the importance of the intention behind the attention being afforded. The authors offer mindfulness – a specific way of paying attention – as an avenue for achieving the correct intentions and attention that inevitably lead to health. For example, someone who aims to lose weight can choose to attend to their caloric intake and the interoceptive experience of hunger. Without an attitude that is in accordance with mindfulness, this person may inhabit judgments such as "My body will never be satisfied with such a small portion," or "I hate this body." Ultimately, these judgments are more likely to lead to disconnection between the mind and body, possibly leading to greater dysregulation and eventually disorder. On the other hand, an attitude of warmth, nonjudgment, and loving-kindness may lead to greater self-acceptance and self-compassion, and in turn connection, regulation, order, and health. The attitude behind the attention being offered functions as a major determining factor as to which outcomes are available: regulation vs. dysregulation, order vs. disorder, and ultimately health vs. an absence of health.

This model applies to both physical and mental health. Shapiro et al. (2006) elaborated on ways that mindful self-regulation can benefit those suffering from mental illness. Cultivating mindfulness allows for a person to attend to data that may have previously been too uncomfortable to examine. This includes habitual, maladaptive patterns of thoughts, behaviors, and emotions that may be maintaining dysregulation in the system. For example, a survivor of trauma who is introduced to mindfulness may begin attending to thoughts and emotions that arise when reminded of their traumatic experience. While sensing into an experience of fear, a survivor may notice a thought such as "I am broken and weak for feeling this way," which is representative of the negative cognitions commonly experienced by survivors. Mindfulness allows for reperceiving, the freedom to demarcate oneself from one's thoughts and perceive the emotion in a new way. Now perhaps the emotion can function as a signal for a need that is not being met or a sign that it is time to implement a coping skill, thus putting into motion greater self-regulation.

Exposure

Exposure is the best empirically supported treatment for a mix of anxiety and trauma-related disorders (Craske et al., 2014). In the case of trauma, survivors have learned to fear objectively safe stimuli through classical conditioning which then can generalize to other similar, objectively safe stimuli. This fear learning is maintained by avoidance of the feared stimuli which removes the chance for exposure and the inhibitory learning that follows. Exposure is the process of facing one's fears; experiencing repeated inhibitory learning (CS-noUS pairings) results in the inhibition of the fear response.

Follette, Palm, and Pearson (2006) note that the avoidance behaviors that prevent successful exposure could be resolved with mindful awareness. As noted previously, the DSM-5 identifies two types of avoidance symptoms for survivors experiencing traumatic stress: avoidance of external reminders and avoidance of internal reminders. Mindful awareness can reduce avoidance of both types of reminders. As survivors learn to attend intentionally to the present moment without judgment, they will have the opportunity to experience internal reminders (i.e., thoughts, memories, and emotions) as they arise. This willingness to be with whatever arises allows a survivor to be exposed to internal reminders (CS) in the absence of real danger (noUS) which allows for inhibitory learning to take place. Survivors realize that memories and feelings that serve as reminders of their traumatic experience are not inherently dangerous. The same principles apply to external reminders. With mindful awareness, clients can become aware of what situations, people, and places they are avoiding, and attempt to approach these situations. Further, when a client approaches a feared external stimulus, they may cope by using covert avoidance tactics (e.g., silently praying, distracting with other thoughts) which serve to maintain their original fear of the external stimulus. Mindfulness provides the opportunity to be with whatever arises in the present moment, likely optimizing planned exposures as part of exposure-based treatment for PTSD. As avoidance symptoms decrease and survivors approach additional feared stimuli with mindful awareness, they will have increased opportunities for inhibitory learning in natural settings, that is, greater number of naturally occurring "exposures" in day-to-day living.

Empirical Findings

Experimental studies concerning the implementation of mindfulness-based interventions with survivors of trauma broadly fall into two theoretical camps. First, along the lines of Follette, Palm, and Pearson (2006), mindfulness has been added as an adjunct to exposure-based protocols as a way to enhance the effectiveness of gold-standard treatments for trauma survivors. A clear example of this is King et al.'s (2016) mindfulness-based exposure therapy (MBET), which significantly reduced symptoms of PTSD in populations of combat veterans. The second group of studies attempts to test the theory that mindfulness practice will naturally lead to increased exposure to previously avoided thoughts, feelings, and even environments, as suggested by Brown, Ryan, and Cresswell's (2007) theoretical work and enhanced by the psychophysiological work of Hölzel et al. (2011). An example can be found in the work of Kelly and Garland (2016), who crafted a modified version of MBSR for survivors of trauma, which they refer to as Trauma-Informed Mindfulness-Based Stress Reduction (TI-MBSR). When compared to a waitlist control, TI-MBSR resulted in significantly greater reductions in symptoms of PTSD in a group of survivors of intimate partner violence. The following section details findings to support each theoretical position and then explores future directions for research in this area.

Mindfulness as an Adjunct Treatment There are four main ways in which mindfulness has been evaluated as an adjunct to treatment as usual conditions, in order to test whether mindfulness interventions enhance treatment for survivors of trauma. First, as mentioned previously, MBET adds mindfulness practices to intervention components from prolonged exposure therapy, a gold-standard treatment for treating symptoms of trauma, with promising results (King et al., 2016). Next, the practices of mantram repetition and yoga have been added to treatment as usual, also showing promising results (Bormann et al., 2013; van der Kolk et al., 2014). Finally, researchers have explored the effects of enrolling participants in both treatment as usual and MBSR simultaneously, without significant results (Kearney et al., 2013).

Mindfulness-based exposure therapy (King et al., 2016) began after a research group at the Ann Arbor VA tested the effectiveness of MBCT with a group of combat veterans (King et al., 2013). Using a matched control support group focused on present difficulties and coping skills, researchers found that MBCT resulted in roughly the same statistically significant reduction in PTSD symptoms as the control group. Following the successful trial of MBCT, King et al. (2016) designed and tested a protocol that merged elements of MBCT with elements of prolonged exposure therapy for PTSD. In line with MBCT, MBET is a group program that covers topics including mindfulness and self-compassion, while heavily emphasizing mindfulness practices, including mindfulness of breathing and mindfulness of emotions practices. In accordance with prolonged exposure therapy, MBET offers psychoeducation about PTSD and implements in vivo exposures, in which survivors of trauma collaborate to identify objectively safe situations and places that they have been avoiding and purposefully approach such situations in order to gain the benefits of inhibitory learning (i.e., CS-noUS pairings). The intervention is broken up into four modules as follows: psychoeducation about PTSD and relaxation, in vivo exposure and mindfulness of the breath and body, in vivo exposure and mindfulness of emotions, and training in self-compassion.

Results from King et al.'s (2016) trial compared MBET with the same matched control as in their 2013 study, Present-Centered Group Therapy, with each group consisting of combat veterans diagnosed with PTSD (MBET, N = 14; PCGT, N = 9). Both the group who underwent MBET and the group who participated in Present-Centered Group Therapy showed statistically significant reductions in trauma symptoms, as measured by Clinically Administered PTSD Scale (CAPS). Additionally, only those in the MBET group showed statistically significant increases in connectivity between the default mode network, which is a network of interconnected brain regions associated with both mind-wandering and self-reference (Kiviniemi et al., 2003), and the regions of the brain that are involved in executive control. King and colleagues explained that this increase in connectivity points to an enhanced ability to volitionally shift attention, a common element of formal mindfulness practices. This psychophysiological finding implicates the ability to choose the focus of the spotlight of one's attention as a key to understanding how it is that mindfulness may enhance treatment for PTSD. Shapiro et al. (2006) emphasized the importance of attention in the IAA model discussed previously in this chapter. The ability to choose where attention is focused allows for greater opportunities for reperceiving, self-regulation, and present moment awareness.

Mantram repetition, or the meditation practice of repeating mantras, has been tested as an adjunctive treatment with a group of veterans undergoing treatment as usual, using a randomized control design (experimental group, N = 66; control group, N = 70; Bormann et al., 2013). Veterans who were randomized to the added mantram repetition condition showed statistically significant improvements in PTSD symptoms when compared to the control group. Those in the control group and those in the experimental group had identical dropout rates, 7%, which could indicate that this meditation adjunct is an acceptable treatment for combat veterans with PTSD.

Yoga has been explored as an asynchronous add-on to treatment as usual for female survivors with treatment resistant PTSD (van der Kolk et al., 2014). This study randomized 64 female survivors of various traumas to either a 10-week trauma-informed yoga class or an active control condition, a women's health education program. In order to be included in the study, survivors had to report at least 3 years of previous therapy focused on symptoms of PTSD, yet still qualify for diagnosis. Those in the yoga condition showed significantly greater reductions in PTSD symptoms at the midpoint of treatment (yoga, d = 1.07; health education, d = 0.66). However, improvements maintained only for those in the yoga condition by the end of the treatment period. This study indicates that mindfulness-based interventions could be beneficial specifically in treatment-resistant groups; however, more studies are needed, especially studies including male survivors of trauma, to support this conclusion.

One study explored the utility of adding MBSR to a treatment as usual condition. Kearney et al. (2013) examined a sample of 47 veterans diagnosed with PTSD, randomizing half to receive treatment as usual, while the other half were assigned to receive both treatment as usual and MBSR. The results showed no significant differences between the two conditions in the reduction of PTSD symptoms, failing to support the hypothesis that adding MBSR to treatment as usual would further decrease symptoms of PTSD. Those who participated in MBSR were more likely to report increases in health-related quality of life, and decreases in symptoms of depression (Kearney et al., 2013).

In sum, adding mindfulness practices to treatment as usual often yields statistically significant reductions in symptoms of PTSD when compared to treatment as usual on its own (Bormann et al., 2013; King et al., 2016). Psychophysiological evidence points to an increased ability to volitionally shift attention as a possible mechanism to explain the added power of mindfulness practices to treatment as usual. Nascent support also exists for the use of trauma-informed yoga interventions for women with treatment-resistant PTSD (van der Kolk, 2014). The limited research available does not support adding MBSR to treatment as usual for those diagnosed with PTSD (Kearney et al., 2013).

Mindfulness as a Stand-Alone Treatment Mindfulness may work well as a stand-alone treatment because the practice itself leads to exposure to internal experiences, such as thoughts and emotions (Brown et al., 2007; Holzel et al., 2011), consistent with Shapiro et al.'s (2006) IAA model. Intentionally attending to objectively safe internal and external trauma reminders with an attitude of openness, curiosity, and nonjudgment provides a converse approach to the way that survivors usually respond to such stimuli (i.e., avoidance). Other authors have also noted that the grounding component of mindfulness, connecting back to the present moment by purposefully attending to sensory data in the present moment, may further ameliorate symptoms of PTSD such as hyperarousal (Kelly & Garland, 2016; Lang et al., 2012).

While results from programs such as MBET show the virtue of adding mindfulness as an adjunctive treatment, Lang et al. (2012) made a case for the removal of formal exposure exercises from the treatment of PTSD. Authors describe that exposure-based protocols, while effective, suffer from high dropout rates and clinicians have reported that their own discomfort leads them away from implementing exposure exercises with clients (Becker et al., 2004; Schottenbauer et al., 2008). The MBET developers specifically mentioned to participants that imaginal exposure, an intervention in which survivors are asked to repeat their trauma memory with as many details as possible, would not be a part of treatment in hopes of recruiting more participants (King et al., 2016). The IAA model and the theoretical work of Brown, Ryan, and Cresswell (2007) suggest that mindfulness practice on its own could lead to increased cognitive, emotional, and behavioral flexibility, improved self-regulation, and increased opportunities for natural exposure to external and internal stimuli with an attitude of kindness, openness, and curiosity.

While relatively few studies have explored the effectiveness of mindfulness practices as an adjunct to treatment as usual, many more studies have tested the effectiveness of mindfulness-based interventions as a stand-alone treatment for PTSD. Two recent, largely overlapping, meta-analyses were conducted to examine the effects mindfulness for PTSD (Gallegos et al., 2017; Hopwood & Schutte, 2017).

Hopwood and Schutte (2017) conducted a meta-analysis that included studies that compared mindfulness-based interventions to waitlist, active, and placebo control conditions. Their analysis ultimately included 18 studies which reported the results from a total of 21 samples. Almost all studies included required that participants received a diagnosis of PTSD prior to enrolling in the study, and used a variety of mindfulness-based interventions, including MBSR, yoga, and mindfulness interventions that were specifically tailored to trauma. Interventions ranged from as short as 2 hours to as long as 27 hours. Most of the experiments reported in this meta-analysis conducted follow-up assessments 1 week after treatment, and some conducted follow-up assessments as long as 1 year after the intervention took place. The primary focus of this meta-analysis was to compare the effectiveness of mindfulness-based interventions to controls; however, researchers also analyzed the impact of mindfulness-based interventions on survivors' self-reported levels of mindful awareness and explored other possibly influential variables such as treatment length to further elucidate possible mechanisms and moderators.

The overall results of Hopwood and Schutte's (2017) meta-analysis supported the conclusion that mindfulness-based interventions yield a statistically significant reduction in symptoms of PTSD. The overall mean weighted effect size was Hedges' $g = -0.44$, which shows that mindfulness-based interventions had a significant, moderate, negative effect on symptoms of PTSD when compared to control conditions. This effect size remained in a similar range regardless of whether symptoms of PTSD were assessed by a clinician, with measures such as the Clinician-Administered PTSD Scale ($g = -0.43$), or a self-report measure, such as the PTSD Checklist ($g = -0.38$). Further, no statistically significant difference in effect size was found between studies that compared mindfulness-based interventions to wait-list controls, placebo controls, nor active treatment controls. The only significant moderator was treatment length, with interventions of longer duration showing greater reduction in PTSD symptoms. Authors describe these meta-analytic findings as evidence that PTSD should be added to the list of clinical disorders that mindfulness-based interventions treat.

Further exploratory analyses conducted by Hopwood and Schutte (2017) looked at the 12 studies that measured mindfulness for increases that resulted from treatment. The overall mean effect size for mindfulness was Hedge's $g = 0.52$, which indicates that mindfulness-based interventions increased mindfulness with a moderate significant effect size, compared to control conditions. The sample of 12 studies was underpowered to properly test whether greater increases in self-reported mindfulness would be correlated with greater decreases in PTSD symptoms. Future studies involving mindfulness-based interventions for survivors of trauma should include self-reported measures of mindfulness as a manipulation check, and to further investigate self-reported mindfulness as a mediator between practicing meditation and PTSD symptom reduction.

A second meta-analysis was published in the same year, conducted by Gallegos et al. (2017). In this largely overlapping analysis, 19 studies were included that compared mindfulness-based interventions to active and waitlist control groups. This meta-analysis resulted in a similar effect size, Hedge's $g = -0.39$, indicating that meditation and yoga interventions outperformed their comparison active and non-active controls in the amelioration of symptoms of PTSD. Authors separated interventions by type as follows in an exploratory analysis: mindfulness-based (e.g., MBSR), other meditation (e.g., transcendental meditation), and yoga. While the yoga interventions yielded a larger effect size than the other two categories of interventions (yoga, Hedge's $g = -0.71$; mindfulness-based, Hedge's $g = -0.34$; other meditation, Hedge's $g = -0.38$), the combined effect size of the yoga interventions was only marginally significant ($p = 0.055$).

Of the 19 studies included in Gallegos et al.'s (2017) meta-analysis, eight compared MBSR to control conditions for the treatment of PTSD. The overall mean weighted effect size from these studies was Hedge's $g = -0.33$. MBSR applies different mindfulness practices throughout the course of the 8-week intervention. Colgan, Christopher, Michael, and Wahbeh (2016) tested whether practices presented in MBSR result in different outcomes for survivors of trauma than matched, non-mindfulness practices (e.g., sitting quietly). Researchers collected a sample of 102 combat veterans diagnosed with PTSD. Random assignment was used to compare four practices, two from MBSR, body scan and breathing meditation, and two matched, non-mindfulness interventions, slow breathing and sitting quietly. Only those in the body scan and breathing meditation groups showed significant decreases in PTSD symptoms and significant increases in self-reported mindfulness. A follow-up qualitative analysis reported that veterans in the body scan group were most likely to report enhanced present moment awareness, reduced anger, and reduced hyperarousal. Those in the mindful breathing condition were the most likely of any group to report increased coping skills and increased nonjudgmental acceptance. Both groups had about an equal number of participants reporting increased non-reactivity as a benefit of their assigned practice (Colgan et al., 2017).

While MBSR has consistently shown benefits as a stand-alone treatment for PTSD, a modified version of MBSR for survivors of trauma, Kelly and Garland's (2016) Trauma-Informed Mindfulness-Based Stress Reduction (TI-MBSR), yielded a large effect size of Hedge's $g = -0.92$. While the effect size calculated in meta-analysis is more robust and is of higher validity, the effect size of TI-MBSR may point to the importance of tailoring mindfulness-based interventions specifically to trauma.

Kelly and Garland (2016) reported that their intervention preserved all content from the original MBSR curriculum, and adds psychoeducation about psychological, neurophysiological, and relational effects of trauma. Authors aimed to reduce self-blame and increase self-efficacy of survivors by adding this educational component about surviving trauma. The only other piece added to TI-MBSR is specific coping strategies to help survivors better regulate their physiological arousal. It is not clear whether these regulation techniques conflict with traditional mindfulness teachings, such as allowing what is here to be here, and treating every experience, even an unpleasant experience, as a welcome guest. Additionally, TI-MBSR was created for survivors of intimate partner violence, childhood sexual abuse, and/or childhood physical abuse. Added psychoeducation included about parenting, attachment style, and the trauma triangle (victim, victimizer, and bystander) may or may not be relevant to survivors of other forms of trauma such as motor vehicle accidents or natural disasters.

In a sample of 39 survivors of intimate partner violence, childhood sexual abuse, and/or childhood physical abuse, researchers randomly assigned half to receive TI-MBSR and the other half was assigned to a waitlist control condition. Participants in both groups completed measures of PTSD at the beginning and end of the 8-week intervention. While both groups reported significant decreases in PTSD over the course of 8 weeks, the TI-MBSR group showed a significantly greater reduction compared to the waitlist group. All participants qualified for a diagnosis of PTSD upon enrolling in the study; 8 weeks later, only 20% of those in the TI-MBSR group still qualified for the diagnosis, compared to 80% in the waitlist condition (Kelly & Garland, 2016).

In conclusion, large meta-analyses resulted in small to medium effect sizes for mindfulness-based interventions when compared to active, placebo, and waitlist control conditions in the reduction of PTSD symptoms. These studies provide support for the theory that mindfulness-based interventions cause significant reductions in symptoms of PTSD for survivors of trauma. However, many moderators explored in these meta-analyses (e.g., gender, type of trauma, etc.) were nonsignificant. This is either due to type II error, the evidence for which is the relatively small number of RCTs in this area, or because mindfulness-based interventions produce similar effects across diverse groups. More RCTs are needed, and a future, adequately powered, meta-analysis would help draw conclusions on this topic. Further, researchers designing such RCTs should consider collecting a diverse sample of trauma survivors. Across many of the studies reviewed in this chapter, survivors of a single type of trauma are examined in isolation (i.e., combat veterans or survivors of IPV). Combat veteran studies in these reviews are predominately male, while studies concerning the treatment of survivors of IPV are predominately female. This is not a sampling error, but rather due to base rates for both types of trauma. An ideal RCT would include diverse traumas and diverse genders to test the mindfulness-based intervention, unless authors have justification for the separate examination of different categories of traumatic events. The investigations listed in Table 8.1 allow the reader to compare results across different types of trauma.

Table 8.1 Suggested readings by type of population served

Type of trauma	Suggested reading
Military	Bormann et al. (2013)
Military	Bremner et al. (2017)
Military	Cole et al. (2015)
Military	Davis et al. (2019)
Military	Harding et al. (2018)
Military	Heffner et al. (2016)
Military	Held et al. (2017)
Military	Kearney et al. (2013)
Military	King et al. (2013)
Military	Nakamura et al. (2011)
Military	Niles et al. (2012)
Military	Polusny et al. (2015)
Military	Possemato et al. (2016)
Military	Rice et al. (2018)
Military	Wahbeh et al. (2016)
Sexual/domestic violence	Centeno (2013)
Sexual/domestic violence	Gallegos et al. (2020)
Sexual/domestic violence	Kelly and Garland (2016)
Sexual/domestic violence	Müller-Engelmann et al. (2019)
Sexual/domestic violence	Valdez et al. (2016)
Sexual/domestic violence	Kimbrough et al. (2010)
Mixed	Bränström et al. (2012)
Mixed	Earley et al. (2014)
Mixed	Goldsmith et al. (2014)
Mixed	Kim et al. (2013)
Mixed	Mitchell et al. (2014)
Mixed	Müller-Engelmann (2017)

Best Practices in Mindfulness Interventions for Trauma

In reviewing studies related to the treatment of trauma with mindfulness as a stand-alone treatment, and in recognition of Lindahl et al.'s (2017) finding that survivors of trauma often report re-experiencing symptoms during meditation, it is important to recognize emerging best practices in this area. Several have emerged, from both Kelly and Garland's (2016) TI-MBSR and suggestions made by Treleaven (2008) in his work, *Trauma-Sensitive Mindfulness.*

Kelly and Garland (2016) added psychoeducation about common reactions and effects of trauma to traditional MBSR education. Survivors are informed about the re-experiencing, avoidance, negative alterations in mood and cognition, and arousal symptoms before beginning mindfulness practice. Once symptoms are normalized for survivors, authors suggested that they may experience less self-blame when encountering symptoms such as flashbacks or intrusive thoughts during practice and may benefit from an enhanced self-efficacy when educated about the potential to ameliorate symptoms by resisting avoidance symptoms. TI-MBSR also provides participants with strategies to regulate physiological arousal. This may assist survivors if they encounter significant hyperarousal or other dysregulation during mindfulness practice.

Treleaven (2008) builds on the concept of the regulation of physiological arousal during meditation practice for survivors of trauma. He describes that when a survivor experiences hyperarousal during meditation, such as after encountering an intrusive memory of their traumatic experience, their sympathetic nervous system becomes active, leading to traumatic sensations and disorganized cognition. Treleaven calls on those who teach mindfulness to monitor participants for signs of hyperarousal. For example, noticing hyperventilation, excessive sweating, and ridged muscle tone during meditation and checking in which any survivors exhibiting such symptoms in a one-on-one setting after class. He encourages teachers to use a "phase-oriented approach," in which stabilization and safety need to be established before a survivor can continue to subsequent phases of processing trauma memories and reintegrating with family, culture, and normal daily life (p. 103). Another helpful tip comes in the suggestion to participants that they notice their own arousal, gauge approaching hyperarousal, and "apply the brakes," meaning slowing down their mindfulness practice by opening their eyes during practice, taking a break, engaging in self-soothing behaviors, focusing on external objects instead of internal, and considering shorter practices (p.106–107).

Self-Compassion

Survivors of trauma commonly experience negative changes in their beliefs about themselves (American Psychiatric Association, 2013). Negative beliefs can change as immediate reactions to trauma; survivors often believe that they are helpless, and that they are to blame for what has happened to them. Long-term reactions to trauma include grief, shame, and feelings of fragility (Center for Substance Abuse Treatment, 2014). These common experiences of self-blame and shame often warrant special attention in trauma therapies (e.g., cognitive processing therapy; Resick & Schnicke, 1992).

Shapiro et al.'s (2006) IAA model specifies several components to the attitude of mindfulness. Self-compassion can be thought of as the attitudinal foundation of one's relationship with oneself. Self-compassion is a complementary Buddhist practice recently added to the western cannon of Buddhist psychological research (Neff, 2003). Neff conceptualized self-compassion as having three pairs of traits, with one half of each pair representing a self-compassionate trait, and the other half of the pair representing its opposite. The first pair includes self-kindness and self-judgment. Self-compassion requires a level of kindness toward all parts of the self, including those parts that often feel inadequate and experiences that are painful. Self-kindness is the opposite of the all too common habitual response of self-judgment, or a tendency to find fault and provide criticism toward oneself. The second pair presented by Neff (2003) is common humanity and isolation. Common humanity is the recognition that no person is alone in their experience of suffering. If one is unable to recognize suffering is universal, a feeling of isolation in one's own pain is evoked. The final pair is mindfulness and overidentification. The non-judgmental and welcoming awareness of mindfulness stands opposite to a tendency to become fused to thoughts and emotions, which Neff (2003) refers to as being overidentified with experience. In sum, the self-compassionate person treats all parts of oneself with kindness, recognizes one is not alone in the experience of suffering, and neither ignores nor overindulges in emotions and thoughts.

Thompson and Waltz (2008) conducted a study to test the theory that self-compassion can benefit survivors of trauma. Data from 210 undergraduate students was collected, 100 of whom endorsed one or more traumatic event falling under Criterion A for PTSD. Participants self-reported on their levels of posttraumatic stress symptoms and level of self-compassion. The measure used for posttraumatic stress symptoms unfortunately drew from DSM-IV criteria, which did not include the negative alterations in mood and cognition criterion (added in DSM-5). Working with the other three symptom sets (i.e., re-experiencing, avoidance, and hyper-arousal), only avoidance symptoms were significantly related to self-compassion. Seligowski, Miron, and Orcutt (2014) conducted a similar study with undergraduate students, although with a larger sample of 453 students, and found significant relationships between self-compassion and re-experiencing, avoidance, and hyper-arousal symptoms. Again, this study did not include the added DSM-5 criterion of negative alterations in mood and cognition. These significant relationships support the claim that symptoms of PTSD have a negative relationship with self-compassion – but these studies cannot support the theory that self-compassion negatively predicts the negative alterations in mood and cognition experienced by survivors of trauma.

Maheux and Price (2016) conducted a study involving two samples comparing the predictive validity of self-compassion to both DSM-IV and DSM-5 criteria for

PTSD. The first sample, composed of 74 trauma-exposed individuals, received measures for self-compassion and DSM-IV PTSD symptoms. Similar to the results from Thomson and Waltz (2008), self-compassion was only significantly associated with avoidance symptoms. The second sample, comprising 152 trauma survivors, completed measures of self-compassion and DSM-5 PTSD symptoms. In this sample, self-compassion was significantly, negatively related to all four symptom clusters. Authors described that the discrepancy could be due to the different sampling methods. On the whole, the study provides further support that self-compassion is negatively related to symptoms of PTSD, including the DSM-5 addition of negative alterations in mood and cognition.

Building on correlational studies, a prospective study was conducted with 115 Iraq and Afghanistan combat veterans who reported being exposed to at least one traumatic event (Hiraoka et al., 2015). Self-compassion and level of combat exposure were assessed at baseline; PTSD symptoms were assessed at baseline, and again at 12-month follow-up. Self-compassion negatively predicted symptoms of PTSD at both baseline (β: -0.59, $p < 0.05$) and 12-month follow-up (β: -0.24, $p < 0.05$), above and beyond the predictive ability of level of trauma exposure. This provides evidence that self-compassion is negatively related to PTSD symptoms, but also that self-compassion prospectively predicts levels of PTSD symptoms.

Growing evidence supports the negative relationship between self-compassion and PTSD symptoms (Hiraoka et al., 2015; Maheux & Price, 2016; Seligowski et al., 2014; Thompson & Waltz, 2008). This attitude toward oneself may deserve extra attention in therapy with those suffering from PTSD. For a review of possible techniques to integrate into therapy with trauma survivors, see Germer and Neff (2015).

Mindfulness, Trauma, and Inflammation

Chronic inflammation has been termed the "common soil," or mutual underlying factor, of many diseases including type 2 diabetes, cardiovascular disease, and cancer (Scrivo et al., 2011). Recent findings suggested that stress-triggered psychological disorders, such as posttraumatic stress disorder (PTSD), may both maintain and be maintained by chronic inflammation (Michopoulos et al., 2017). Those with PTSD experience heightened and more frequent stress responding in the body. This chronic stress responding leads to dysregulation of the physiological systems involved in the stress response, including the hypothalamic-pituitary-adrenal (HPA) axis and the sympathetic-adrenal-medullary (SAM) system (Cohen et al., 2007). The dysregulation in these physiological systems leads to overall increased activity in the sympathetic nervous system, and decreased parasympathetic nervous system responding. This pattern of responding feeds back into the area of the brain that regulates the fear response (e.g., insula, amygdala, hippocampus), which maintains PTSD symptoms. PTSD has specifically been associated with increased risk for a wide array of inflammatory disorders (Boscarino, 2004).

Kadziolka, Di Pierdomenico, and Miller (2016) measured participants' self-reported levels of mindfulness before having them vividly describe a personal example of a stressful event. While undergoing the recall of a stressful event, participants' heart rate variability (HRV) was measured using an electrocardiogram (ECG), and their skin conductance response (SCR) was measured using galvanic skin response finger electrodes. Participants with higher self-rated dispositional mindfulness showed lower sympathetic nervous system activation, as measured by their SCR, and greater parasympathetic nervous system responding, as measured by their HRV and their return to a neutral state. In sum, self-reported mindfulness is negatively associated with the pattern of responding that is believed to both maintain and be maintained by PTSD.

MBSR has been found to reduce inflammation (Rosenkranz et al., 2013). MBSR has been compared to an active control intervention, the Health Enhancement Program (HEP), designed for the purposes of testing the unique effects of MBSR. While both MBSR and HEP participants experienced similar levels of stress-evoked cortisol responses following participation in their respective interventions, those in the MBSR group had significantly smaller inflammatory responses compared to those in the HEP group (Rosenkranz et al., 2013). The hidden cost of PTSD may be the inflammation-related illnesses acquired by survivors of trauma that simultaneously cause harm to the body and likely maintain PTSD symptoms. Mindfulness-based interventions offer a psychotherapeutic option for reducing both symptoms of PTSD and chronic inflammation in the body.

Summary and Future Directions

In sum, there is now significant evidence to support the use of mindfulness-based interventions with survivors of trauma suffering from PTSD (Gallegos et al., 2017; Hilton et al., 2017; Hopwood & Schutte, 2017). Survivors of trauma likely benefit from increased emotional, behavioral, and cognitive flexibility, enhanced self-regulation, and natural exposure through mindfulness practice (Shapiro et al., 2006). Additionally, self-compassion has been found to significantly negatively predict symptoms of PTSD in both cross-sectional and longitudinal studies (Hiraoka et al., 2015; Maheux & Price, 2016; Seligowski et al., 2014; Thompson & Waltz, 2008). These empirical findings support theoretical positions that mindfulness and self-compassion improve symptoms of PTSD experienced by survivors of trauma. Additionally, recent research concerning the bidirectional relationship between PTSD and chronic inflammation reveal a new avenue for psychophysiological studies connecting mindful awareness, PTSD symptoms, and inflammation.

More work is needed to determine significant variables that may moderate the effectiveness of mindfulness-based interventions. The only significant moderator found in meta-analytic reviews of mindfulness-based interventions for survivors of trauma included length of intervention, with longer interventions performing better than shorter interventions (Hopwood & Schutte, 2017). Future studies should aim to include large samples of participants across genders, culturally diverse samples, and participants who have survived different types of Criterion A traumatic events. These more diverse samples would allow for the identification of individual difference variables that moderate the effectiveness of mindfulness-based interventions for survivors of trauma.

Additionally, dismantling study designs, such as the one conducted by Colgan, Christopher, Michael, and Wahbeh (2016), could investigate which features of multicomponent programs such as MBSR are most effective in the reduction of PTSD symptoms. This nomothetical information can inform idiographic treatment recommendations for survivors of trauma. For example, for a survivor who reports experiencing the most distress when exhibiting hyperarousal symptoms, the body scan practice is likely to be the most helpful (Colgan et al., 2017). Further research, including qualitative studies about the experiences of trauma survivors in mindfulness-based treatments, can help inform and better tailor future interventions.

Furthermore, a clear negative association exists between self-compassion and symptoms of PTSD, even in prospective studies (Hiraoka et al., 2015). Questions remain about the role that self-compassion may already be playing in mindfulness-based intervention with survivors of trauma. In future studies that involve a mindfulness-based intervention for PTSD, self-compassion could be as closely monitored as each participant's levels of mindful awareness and PTSD symptoms. Intensive longitudinal methods (see Bolger & Laurenceau, 2013) also may shed light on the between-subject and within-subject relationships between PTSD, mindfulness practice, and self-compassion.

Finally, while PTSD has been linked to inflammation in the body in a bidirectional relationship, and mindfulness has been associated with both reductions in PTSD and inflammation in separate studies, no study examines the relationship among mindfulness, inflammation, and PTSD symptoms (Michopoulos et al., 2017; Rosenkranz et al., 2013). More research examining the interplay between these three variables is needed in order to address the mental and physical health of survivors of trauma. Findings in this area could be the key to unlocking the psychophysiological mechanisms behind mindfulness-based interventions for survivors of trauma.

References

American Psychiatric Association. (2013). *Diagnostic and statistical manual of mental disorders* (5th ed.).

Becker, C. B., Zayfert, C., & Anderson, E. (2004). A survey of psychologists' attitudes towards and utilization of exposure therapy for PTSD. *Behaviour Research and Therapy, 42*, 277–292.

Bolger, N., & Laurenceau, J. P. (2013). *Intensive longitudinal methods: An introduction to diary and experience sampling research.* Guilford Press.

Bormann, J. E., Thorp, S. R., Wetherell, J. L., Golshan, S., & Lang, A. J. (2013). Meditation-based mantram intervention for veterans with posttraumatic stress disorder: A randomized trial. *Psychological Trauma Theory Research Practice and Policy, 5*(3), 259–267.

Boscarino, J. A. (2004). Posttraumatic stress disorder and physical illness: Results from clinical and epidemiologic studies. *Annals of the New York Academy of Sciences, 1032*(1), 141–153.

Bränström, R., Kvillemo, P., & Moskowitz, J. T. (2012). A randomized study of the effects of mindfulness training on psychological well-being and symptoms of stress in patients treated for cancer at 6-month follow-up. *International Journal of Behavioral Medicine, 19*(4), 535–542.

Bremner, J. D., Mishra, S., Campanella, C., Shah, M., Kasher, N., Evans, S., ... Vaccarino, V. (2017). A pilot study of the effects of mindfulness-based stress reduction on post-traumatic stress disorder symptoms and brain response to traumatic reminders of combat in Operation Enduring Freedom/Operation Iraqi Freedom combat veterans with post-traumatic stress disorder. *Frontiers in Psychiatry, 8*, 157.

Brown, K. W., Ryan, R. M., & Creswell, J. D. (2007). Mindfulness: Theoretical foundations and evidence for its salutary effects. *Psychological Inquiry, 18*(4), 211–237.

Centeno, E. (2013). *Mindfulness meditation and its effects on survivors of intimate partner violence* (Doctoral dissertation, Saybrook University).

Center for Substance Abuse Treatment (US). Trauma-Informed Care in Behavioral Health Services. Rockville (MD): Substance Abuse and Mental Health Services Administration (US); 2014. (Treatment Improvement Protocol (TIP) Series, No. 57.) Chapter 3, Understanding the Impact of Trauma. Available from: https://www.ncbi.nlm.nih.gov/books/NBK207191/

Cohen, S., Janicki-Deverts, D., & Miller, G. E. (2007). Psychological stress and disease. *JAMA : The Journal of the American Medical Association, 298*(14), 1685–1687.

Cole, M. A., Muir, J. J., Gans, J. J., Shin, L. M., D'Esposito, M., Harel, B. T., & Schembri, A. (2015). Simultaneous treatment of neurocognitive and psychiatric symptoms in veterans with post-traumatic stress disorder and history of mild traumatic brain injury: A pilot study of mindfulness-based stress reduction. *Military Medicine, 180*(9), 956–963.

Colgan, D. D., Christopher, M., Michael, P., & Wahbeh, H. (2016). The body scan and mindful breathing among veterans with PTSD: Type of intervention moderates the relationship between changes in mindfulness and post-treatment depression. *Mindfulness, 7*(2), 372–383.

Colgan, D. D., Wahbeh, H., Pleet, M., Besler, K., & Christopher, M. (2017). A qualitative study of mindfulness among veterans with posttraumatic stress disorder: Practices differentially affect symptoms, aspects of well-being, and potential mechanisms of action. *Journal of Evidence-Based Complementary & Alternative Medicine, 22*(3), 482–493.

Craske, M. G., Treanor, M., Conway, C. C., Zbozinek, T., & Vervliet, B. (2014). Maximizing exposure therapy: An inhibitory learning approach. *Behaviour Research and Therapy, 58*, 10–23.

Davis, L. L., Whetsell, C., Hamner, M. B., Carmody, J., Rothbaum, B. O., Allen, R. S., & Bremner, J. D. (2019). A multisite randomized controlled trial of mindfulness-based stress reduction in the treatment of posttraumatic stress disorder. *Psychiatric Research and Clinical Practice, 1*(2), 39–48.

Earley, M. D., Chesney, M. A., Frye, J., Greene, P. A., Berman, B., & Kimbrough, E. (2014). Mindfulness intervention for child abuse survivors: A 2.5-year follow-up. *Journal of Clinical Psychology, 70*(10), 933–941.

Foa, E. B., & Kozak, M. J. (1986). Emotional processing of fear: Exposure to corrective information. *Psychological Bulletin, 99*(1), 20.

Follette, V., Palm, K. M., & Pearson, A. N. (2006). Mindfulness and trauma: Implications for treatment. *Journal of Rational-Emotive and Cognitive-Behavior Therapy, 24*(1), 45–61.

Gallegos, A. M., Crean, H. F., Pigeon, W. R., & Heffner, K. L. (2017). Meditation and yoga for posttraumatic stress disorder: A meta-analytic review of randomized controlled trials. *Clinical Psychology Review, 58*, 115–124.

Gallegos, A. M., Heffner, K. L., Cerulli, C., Luck, P., McGuinness, S., & Pigeon, W. R. (2020). Effects of mindfulness training on posttraumatic stress symptoms from a community-based pilot clinical trial among survivors of intimate partner violence. *Psychological Trauma: Theory, Research, Practice, and Policy.* Advance online publication.

Germer, C. (2005). Teaching mindfulness in therapy. In *Mindfulness and psychotherapy* (2nd ed., pp. 113–129). https://doi.org/10.1007/978-1-4614-3033-9.

Germer, C. K., & Neff, K. D. (2015). Cultivating self-compassion in trauma survivors. In V. M. Follette, J. Briere, D. Rozelle, J. W. Hopper, & D. I. Rome (Eds.), *Mindfulness-oriented interventions for trauma: Integrating contemplative practices* (pp. 43–58). The Guilford Press.

Goldsmith, R. E., Gerhart, J. I., Chesney, S. A., Burns, J. W., Kleinman, B., & Hood, M. M. (2014). Mindfulness-based stress reduction for posttraumatic stress symptoms: Building acceptance and decreasing shame. *Journal of Evidence-Based Complementary & Alternative Medicine, 19*(4), 227–234.

Harding, K., Simpson, T., & Kearney, D. J. (2018). Reduced symptoms of post-traumatic stress disorder and irritable bowel syndrome following mindfulness-based stress reduction among veterans. *The Journal of Alternative and Complementary Medicine, 24*(12), 1159–1165.

Heffner, K. L., Crean, H. F., & Kemp, J. E. (2016). Meditation programs for veterans with posttraumatic stress disorder: Aggregate findings from a multi-site evaluation. *Psychological Trauma Theory Research Practice and Policy, 8*(3), 365.

Held, P., Owens, G. P., Monroe, J. R., & Chard, K. M. (2017). Increased mindfulness skills as predictors of reduced trauma-related guilt in treatment-seeking veterans. *Journal of Traumatic Stress, 30*(4), 425–431.

Hilton, L., Maher, A. R., Colaiaco, B., Apaydin, E., Sorbero, M. E., Booth, M., … Hempel, S. (2017). Meditation for posttraumatic stress: Systematic review and meta-analysis. *Psychological Trauma Theory Research Practice and Policy, 9*(4), 453–460.

Hiraoka, R., Meyer, E. C., Kimbrel, N. A., DeBeer, B. B., Gulliver, S. B., & Morissette, S. B. (2015). Self-compassion as a prospective predictor of PTSD symptom severity among trauma-exposed US Iraq and Afghanistan war veterans. *Journal of Traumatic Stress, 28*(2), 127–133.

Hölzel, B. K., Lazar, S. W., Gard, T., Schuman-Olivier, Z., Vago, D. R., & Ott, U. (2011). How does mindfulness meditation work? Proposing mechanisms of action from a conceptual and neural perspective. *Perspectives on Psychological Science, 6*(6), 537–559.

Hopwood, T. L., & Schutte, N. S. (2017). A meta-analytic investigation of the impact of mindfulness-based interventions on post traumatic stress. *Clinical Psychology Review, 57*, 12–20.

Kabat-Zinn, J. (2013). *Full catastrophe living: Using the wisdom of your body and mind to face stress, pain, and illness.* New York, NY: Bantam Books.

Kadziolka, M. J., Di Pierdomenico, E. A., & Miller, C. J. (2016). Trait-like mindfulness promotes healthy self-regulation of stress. *Mindfulness, 7*(1), 236–245.

Kearney, D. J., Mcdermott, K., Malte, C., Martinez, M., & Simpson, T. L. (2013). Effects of participation in a mindfulness program for veterans with posttraumatic stress disorder: A randomized controlled pilot study. *Journal of Clinical Psychology, 69*(1), 14–27. https://doi.org/10.1002/jclp.21911.

Kelly, A., & Garland, E. L. (2016). Trauma-informed mindfulness-based stress reduction for female survivors of interpersonal violence: Results from a stage I RCT. *Journal of Clinical Psychology, 72*(4), 311–328.

Kim, S. H., Schneider, S. M., Bevans, M., Kravitz, L., Mermier, C., Qualls, C., & Burge, M. R. (2013). PTSD symptom reduction with mindfulness-based stretching and deep breathing exercise: Randomized controlled clinical trial of efficacy. *The Journal of Clinical Endocrinology & Metabolism, 98*(7), 2984–2992.

Kimbrough, E., Magyari, T., Langenberg, P., Chesney, M., & Berman, B. (2010). Mindfulness intervention for child abuse survivors. *Journal of Clinical Psychology, 66*(1), 17–33.

King, A. P., Erickson, T. M., Giardino, N. D., Favorite, T., Rauch, S. A. M., Robinson, E., … Liberzon, I. (2013). A Pilot Study of Group Mindfulness-Based Cognitive Therapy (MBCT) for Combat Veterans with Posttraumatic Stress Disorder (PTSD). *Depression and Anxiety, 30*(7), 638–645.

King, A. P., Block, S. R., Sripada, R. K., Rauch, S., Giardino, N., Favorite, T., … Liberzon, I. (2016). Altered default mode network (DMN) resting state functional connectivity following a mindfulness-based exposure therapy for posttraumatic stress disorder (PTSD) in combat veterans of Afghanistan and Iraq. *Depression and Anxiety, 33*(4), 289–299.

Kiviniemi, V., Kantola, J., Jauhiainen, J., Hyvärinen, A., & Tervonen, O. (2003). Independent component analysis of nondeterministic fMRI signal sources. *Neuro Image, 19*, 253–260.

Lang, A. J., Strauss, J. L., Bomyea, J., Bormann, J. E., Hickman, S. D., Good, R. C., & Essex, M. (2012). The theoretical and empirical basis for meditation as an intervention for PTSD. *Behavior Modification, 36*(6), 759–786.

Levine, P. A. (2010). *In an unspoken voice: How the body releases trauma and restores goodness.* North Atlantic Books.

Lindahl, J. R., Fisher, N. E., Cooper, D. J., Rosen, R. K., & Britton, W. B. (2017). The varieties of contemplative experience: A mixed-methods study of meditation-related challenges in Western Buddhists. *PLoS One, 12*(5), e0176239.

Lustyk, M. K., Chawla, N., Nolan, R., & Marlatt, G. A. (2009). Mindfulness meditation research: Issues of participant screening, safety procedures, and researcher training. *Advances in Mind-Body Medicine, 24*(1), 20–30.

Maheux, A., & Price, M. (2016). The indirect effect of social support on post-trauma psychopathology via self-compassion. *Personality and Individual Differences, 88*, 102–107.

Martin, M. M., & Rubin, R. B. (1995). A new measure of cognitive flexibility. *Psychological Reports, 76*(2), 623–626.

Michopoulos, V., Powers, A., Gillespie, C. F., Ressler, K. J., & Jovanovic, T. (2017). Inflammation in fear-and anxiety-based disorders: PTSD, GAD, and beyond. *Neuropsychopharmacology, 42*(1), 254–270.

Mitchell, K. S., Dick, A. M., DiMartino, D. M., Smith, B. N., Niles, B., Koenen, K. C., & Street, A. (2014). A pilot study of a randomized controlled trial of yoga as an intervention for PTSD symptoms in women. *Journal of Traumatic Stress, 27*(2), 121–128.

Moore, A., & Malinowski, P. (2009). Meditation, mindfulness and cognitive flexibility. *Consciousness and Cognition, 18*(1), 176–186.

Mowrer, O. H. (1947). On the dual nature of learning: A re-interpretation of "conditioning" and "problem-solving.". *Harvard Educational Review, 17*, 102–148.

Müller-Engelmann, M., Wünsch, S., Volk, M., & Steil, R. (2017). Mindfulness-based stress reduction (MBSR) as a standalone intervention for posttraumatic stress disorder after mixed traumatic events: A mixed-methods feasibility study. *Frontiers in Psychology, 8*, 1407.

Müller-Engelmann, M., Schreiber, C., Kümmerle, S., Heidenreich, T., Stangier, U., & Steil, R. (2019). A trauma-adapted mindfulness and loving-kindness intervention for patients with PTSD after interpersonal violence: A multiple-baseline study. *Mindfulness, 10*(6), 1105–1123.

Nakamura, Y., Lipschitz, D. L., Landward, R., Kuhn, R., & West, G. (2011). Two sessions of sleep-focused mind–body bridging improve self-reported symptoms of sleep and PTSD in veterans: A pilot randomized controlled trial. *Journal of Psychosomatic Research, 70*(4), 335–345.

Neff, K. D. (2003). The development and validation of a scale to measure self-compassion. *Self and Identity, 2*(3), 223–250.

Niles, B. L., Klunk-Gillis, J., Ryngala, D. J., Silberbogen, A. K., Paysnick, A., & Wolf, E. J. (2012). Comparing mindfulness and psychoeducation treatments for combat-related PTSD using a telehealth approach. *Psychological Trauma Theory Research Practice and Policy, 4*(5).

Palm, K. M., & Follette, V. M. (2011). The roles of cognitive flexibility and experiential avoidance in explaining psychological distress in survivors of interpersonal victimization. *Journal of Psychopathology and Behavioral Assessment, 33*(1), 79–86.

Polusny, M. A., Erbes, C. R., Thuras, P., Moran, A., Lamberty, G. J., Collins, R. C., … Lim, K. O. (2015). Mindfulness-based stress reduction for posttraumatic stress disorder among veterans a randomized clinical trial. *JAMA : The Journal of the American Medical Association, 314*(5), 456–465.

Possemato, K., Bergen-Cico, D., Treatman, S., Allen, C., Wade, M., & Pigeon, W. (2016). A randomized clinical trial of primary care brief mindfulness training for veterans with PTSD. *Journal of Clinical Psychology, 72*(3), 179–193.

Rauch, S., & Foa, E. (2006). Emotional processing theory (EPT) and exposure therapy for PTSD. *Journal of Contemporary Psychotherapy, 36*(2), 61.

Resick, P. A., & Schnicke, M. K. (1992). Cognitive processing therapy for sexual assault victims. *Journal of Consulting and Clinical Psychology, 60*(5), 748.

Rice, V. J., Liu, B., & Schroeder, P. J. (2018). Impact of in-person and virtual world mindfulness training on symptoms of post-traumatic stress disorder and attention deficit and hyperactivity disorder. *Military Medicine, 183*, 413–420.

Rosenkranz, M. A., Lutz, A., Perlman, D. M., Bachhuber, D. R., Schuyler, B. S., MacCoon, D. G., & Davidson, R. J. (2013). Reduced stress and inflammatory responsiveness in experienced meditators compared to a matched healthy control group. *Psychoneuroendocrinology, 68*, 117–125.

Santorelli, S. F., & Kabat-Zinn, J. (2009). Mindfulness-based stress reduction (MBSR) professional education and training: MBSR curriculum and supporting materials. *University of Massachusetts Medical School, Center for Mindfulness in Medicine, Health Care, and Society.*

Schottenbauer, M. A., Glass, C. R., Arnkoff, D. B., Tendick, V., & Gray, S. H. (2008). Nonresponse and dropout rates in outcome studies on PTSD: Review and methodological considerations. *Psychiatry: Interpersonal and Biological Processes, 71*(2), 134–168.

Scrivo, R., Vasile, M., Bartosiewicz, I., & Valesini, G. (2011). Inflammation as "common soil" of the multifactorial diseases. *Autoimmunity Reviews, 10*(7), 369–374.

Seligowski, A. V., Miron, L. R., & Orcutt, H. K. (2014). Relations among self-compassion, PTSD symptoms, and psychological health in a trauma-exposed sample. *Mindfulness, 6*(5), 1033–1041.

Shapiro, D. H. (1992). A preliminary study of long-term meditators: Goals, effects, religious orientation, cognitions. *Journal of Transpersonal Psychology, 24*(1), 23–39.

Shapiro, S. L., & Schwartz, G. E. (2000). The role of intention in self-regulation: Toward intentional systemic mindfulness. In *Handbook of self-regulation* (pp. 253–273). Academic Press.

Shapiro, S. L., Carlson, L. E., Astin, J. A., & Freedman, B. (2006). Mechanisms of mindfulness. *Journal of Clinical Psychology, 62*(3), 373–386.

Thompson, B. L., & Waltz, J. (2008). Self-compassion and PTSD symptom severity. *Journal of Traumatic Stress, 21*(6), 556–558.

Treleaven, D. A. (2008). *Trauma-sensitive mindfulness: Practices for safe and transformative healing.* New York, NY: WW Norton & Company.

Valdez, C. E., Sherrill, A. M., & Lilly, M. (2016). Present moment contact and nonjudgment: Pilot data on dismantling mindful awareness in trauma-related symptomatology. *Journal of Psychopathology and Behavioral Assessment, 38*(4), 572–581.

van der Kolk, B. A., Stone, L., West, J., Rhodes, A., Emerson, D., Suvak, M., & Spinazzola, J. (2014). Yoga as an adjunctive treatment for posttraumatic stress disorder: A randomized controlled trial. *Journal of Clinical Psychiatry, 75*(6), 3559–e565.

Wahbeh, H., Goodrich, E., Goy, E., & Oken, B. S. (2016). Mechanistic pathways of mindfulness meditation in combat veterans with posttraumatic stress disorder. *Journal of Clinical Psychology, 72*(4), 365–383.

Chapter 9
Mindfulness-Based Interventions for Clinical Anxiety and Depression

Holly Hazlett-Stevens

Introduction

Mindfulness training first appeared in Western medicine when Jon Kabat-Zinn developed mindfulness-based stress reduction (MBSR) at the University of Massachusetts Medical School in 1979. Kabat-Zinn originally designed the MBSR curriculum to teach patients mindfulness as a means of working more effectively with the stress of daily life, chronic pain, and illness. MBSR is a public health course, delivered in large patient groups over approximately 28 total hours of instruction, in which patients attend weekly class sessions for 8 weeks and a full-day retreat following the sixth weekly session. With mindfulness practice, patients become increasingly aware of their habitual personal patterns of stress appraisal and reactivity, eventually allowing them to respond to early cues of stress reactivity with the more adaptive and intentional "mindfulness-mediated stress response" (Kabat-Zinn, 2013). Subsequent research trials demonstrated clinical benefits of MBSR for a wide variety of patient groups, and MBSR now appears in the Substance Abuse and Mental Health Services Administration (SAMHSA) National Registry of Evidence-based Programs and Practices (NREPP). MBSR investigations often included self-report measures of subjective distress, including symptoms of stress, anxiety, and depression. A meta-analysis of 26 MBSR randomized controlled trials including 1456 participants (de Vibe et al., 2012) yielded Hedges' g effect sizes of 0.53 for anxiety, 0.54 for depression, and 0.56 for stress or distress outcome measures. Similarly, Hofmann et al. (2010) conducted a meta-analysis of 20 MBSR investigations that included anxiety and/or depression outcome measures and reported Hedges' g effect sizes of 0.55 for anxiety and 0.49 for depression. Such studies suggested that MBSR not

H. Hazlett-Stevens (✉)
University of Nevada, Reno, NV, USA
e-mail: hhazlett@unr.edu

© Springer Nature Switzerland AG 2021   201
H. Hazlett-Stevens (ed.), *Biopsychosocial Factors of Stress, and Mindfulness for Stress Reduction*, https://doi.org/10.1007/978-3-030-81245-4_9

only improved patients' general health and well-being, but also may alleviate symptoms of anxiety and depression.

Researchers in the field of Western clinical psychology began to take note of such findings, leading to clinical trials examining effectiveness and potential change mechanisms of MBSR for individuals diagnosed with clinical anxiety disorders. In addition, Zindel Segal, Mark Williams, and John Teasdale adapted the original MBSR protocol specifically for patients recovered from recurrent major depressive episodes to develop a maintenance form of therapy that prevents future relapse. Although their mindfulness-based cognitive therapy (MBCT; Segal et al., 2013) kept much of the curriculum structure and mindfulness meditation practice schedule of MBSR, MBCT incorporates psychoeducation specific to the nature of depression. MBCT didactic components include discussion topics such as how ruminative thought patterns can lead to depression, how negative interpretive biases can impact emotion, and how mood can influence thoughts and behavior. A systematic program of psychotherapy outcome research established the effectiveness of MBCT for clinical depression as well as for some anxiety disorders. The Hofmann et al. (2010) meta-analysis also included ten MBCT investigations yielding Hedges' g effect sizes of 0.85 for depression measures and 0.79 for anxiety measures. Across all 39 mindfulness-based intervention studies included in this meta-analysis, effect sizes were especially large for patient groups diagnosed with anxiety and/or mood disorders (Hedges' g of 0.97 for anxiety symptoms and 0.95 for mood symptoms) and remained at follow-up. Thus, MBSR and MBCT appear especially promising in the treatment of clinical anxiety and depression.

Mindfulness training soon became a viable treatment option for clinical anxiety and depressive disorders. In many ways, Western psychology views of anxiety and depression are compatible with emerging theoretical mechanisms of change identified in the mindfulness research literature. Guided by such developing theories, a growing body of clinical research has examined the effects of MBSR and MBCT among individuals diagnosed with clinical anxiety and depressive disorders. This chapter briefly reviews leading theoretical models of stress, anxiety, and depression with an emphasis on how increased mindfulness may effect clinical change for individuals suffering from anxiety disorders and clinical depression. Key theoretical concepts in the mindfulness literature are presented in the context of anxiety and depression symptom improvement as well. Results from outcome investigations comparing mindfulness-based interventions to no treatment or to comparison treatments for anxiety disorders and for clinical depression will be summarized next. An extensive review of all relevant theories and an exhaustive list of each clinical trial conducted for anxiety and depression are both beyond the scope of this chapter. However, this discussion explores the leading theoretical reasons mindfulness training appears promising in the treatment of clinical anxiety and depression as well as the available empirical evidence supporting this clinical practice.

Theoretical Perspectives

Several related theories from different research areas elucidate how mindfulness training might reduce anxiety and depressive disorder symptomology. Perspectives from the stress research literature emphasize stress reactivity and how mindfulness may reduce allostatic load while improving whole-system self-regulation. Theories from the field of clinical psychology point to more adaptive cognitive-behavioral responding, improved negative and positive emotion regulation, and cultivating a more decentered and compassionate relationship to thoughts, feelings, and other private internal experience.

Stress Reactivity Versus Responding

MBSR is grounded in principles of mind-body medicine and a holistic view of health, in which physical, mental, and emotional phenomena are considered intricately interconnected and not truly separate from each other. Jon Kabat-Zinn (1990, 2013) originally proposed that MBSR teaches patients how to respond to stressors less automatically and with greater intentionality and awareness, thereby enabling patients to respond to stressful situations – and to any signs of internal stress reactivity – skillfully and adaptively. Kabat-Zinn considered *stress reactivity* a physiological and emotional full-system occurrence: an individual perceives external and/or internal events as stressful, triggering "fight-or-flight" autonomic nervous system activity. Attempts to inhibit initial stress reactions and acute hyperarousal lead to further dysregulation of the entire mind-body system and can result in chronic problems such as hypertension, sleep problems, chronic pain, and anxiety symptoms. Over time, such automatic habit patterns eventually can lead to maladaptive coping behaviors such as overworking, overeating, or substance dependence, and ultimately result in further breakdown in the form of exhaustion, depression, and/or various stress-related chronic medical conditions. From this viewpoint, symptoms such as anxiety, panic, and depression are possible manifestations along the trajectory of this stress-reaction cycle.

In contrast, Kabat-Zinn (1990, 2013) argued, individuals can learn to bring the nonjudgmental awareness of mindfulness to such situations instead of getting caught in one's original reactive habit patterns. Rather than trying to inhibit arousal and engaging in self-destructive attempts to cope, individuals instead can allow themselves to feel any sensations and emotions as they recognize thoughts as mere mental phenomena. As they allow themselves to feel any uncomfortable sensations of arousal, while present, such feelings eventually pass and individuals no longer feel the need to attempt to inhibit arousal. This change in perspective enables individuals to *respond* more adaptively to the situation with intentionality and awareness. Instead of automatically engaging old maladaptive coping strategies, they recover more quickly and become able to try new and more adaptive coping strategies. Kabat-Zinn (2013) coined the term "mindfulness-mediated stress response" to describe this alternative of *responding*, rather than *reacting*, to stress that becomes possible with mindfulness training. Kabat-Zinn updated his framework over the years to incorporate McEwen's (1998, 2005) more modern conceptualization of chronic stress reactivity as "allostatic load." As discussed in greater detail in previous chapters of this book (see Chaps. 1, 2, and 4), *allostatic load* reflects the biological "wear and tear" of an organism's ongoing adaptation to changing circumstances (i.e., *allostasis*). From this perspective, allostatic "overload" can develop over time as stress reactivity patterns lead to a breakdown across various systems of functioning, whereas learning to implement mindfulness-mediated responding promotes optimal allostasis and reduces allostatic load (Kabat-Zinn, 2013).

Cognitive-Behavioral Model of Anxiety and Depression

Kabat-Zinn's (1990, 2013) views of the stress-reaction cycle and the possibility of a mindfulness-mediated stress response are compatible with cognitive-behavioral models of anxiety and depressive disorders from Western clinical psychology. Cognitive-behavioral theories of anxiety and depression build upon general models of stress reactivity by identifying individual difference variables that make certain individuals more prone to developing anxiety and mood disorders as they encounter internal and external stressors. For example, Barlow's (2002) triple vulnerability model of emotional disorders proposed that certain individuals are more prone to developing anxiety and mood disorders when a general biological vulnerability, characterized by a genetically based temperament of neuroticism and low extraversion, is coupled with a general psychological vulnerability. Such general psychological vulnerability stems from early childhood experiences of unpredictable and/or stressful environments or parenting, which inhibit the development of adaptive coping strategies and instead create a tendency to perceive life events as unpredictable and uncontrollable. Vulnerable individuals, therefore, are more likely to perceive events as threatening, are biologically predisposed toward elevated sympathetic nervous system arousal in reaction to stress, and are at increased risk of falling into dysregulated reactive habit patterns with poorer coping resources, increasing their risk for symptoms of generalized anxiety and/or depression. Furthermore, a third disorder-specific psychological vulnerability may develop when an individual learns to focus this distress in a particular way that leads to a specific clinical manifestation. As examples, an individual who learned to believe that certain thoughts, images, and impulses are threatening may develop obsessive-compulsive disorder, whereas another individual who learned to fear certain interoceptive physical sensations of arousal may develop panic disorder.

One large research investigation of this model (Brown & Naragon-Gainey, 2013) found that the biologically based temperament of neuroticism had significant direct effects across anxiety disorders and depression. They further found that a specific psychological vulnerability called *thought-action fusion* (e.g., a tendency to believe that thinking about a disturbing event will increase the likelihood it will occur and that thinking about a disturbing action is morally equivalent to carrying out that action; Shafran, Thordarson, & Rachman, 1996) was significantly and specifically related to obsessive-compulsive disorder. Given the importance of general vulnerabilities across the emotional disorders, the field of clinical psychology increasingly has emphasized transdiagnostic cognitive-behavioral therapies that target these broader underlying general vulnerabilities (e.g., Barlow et al., 2011). From a cognitive-behavioral perspective, mindfulness training that increases awareness of anxiety- and/or depression-maintaining cognitive-behavioral reactive habit patterns could augment patients' self-monitoring efforts, thereby creating new opportunities for more adaptive cognitive appraisals and behavioral responding.

Self-Regulation Theories

Kabat-Zinn (1990, 2013) viewed MBSR as a means of restoring self-regulation of the whole mind-body system. Kabat-Zinn applied the Self-Regulation Theory of Gary Schwartz (1984, 1990) which viewed health and disease from a systems perspective, in which complex mind-body living systems maintain health through their natural capacity to self-regulate via interconnected feedback loops within and across component systems. Health is compromised when this self-regulating system becomes imbalanced due to a disconnection between systems, potentially leading to various forms of disorder and disease. Kabat-Zinn suggested MBSR teaches patients how to re-establish connections between systems with increased conscious attention, allowing for greater attending and skillful responding to relevant feedback messages from the body and mind needed for effective self-regulation. As discussed in the previous chapter (Chap. 8), Shapiro and Schwartz (2000a) later argued that mindfulness improves whole system self-regulation by increasing the *intention* to pay mindful attention to such messages. Their Intentional Systemic Mindfulness model (Shapiro & Schwartz, 2000b) highlighted the role of intentionality cultivated by mindfulness training and claimed that an individual must first form an intention to pay attention for the purpose of restoring connection, eventually leading to greater order and ease across the whole system.

Emotion Regulation Theories

A number of Western psychological theories have emphasized the regulation of emotion-related systems in particular for optimal mental health and psychological well-being. Thus, emotion dysregulation and/or disruption in effective emotion regulation processes has been implicated across multiple mental disorder diagnoses, and psychologists have incorporated mindfulness training as a means of improving emotion regulation across such clinical conditions. For example, dialectical behavior therapy (DBT; Linehan, 1993) teaches mindfulness as its core skill. In DBT, mindfulness provides the foundation and first steps toward emotion regulation skills and other skills (distress tolerance, interpersonal skills, etc.) to promote adaptive emotional and behavioral responding to daily life events among patients struggling with chronic suicidality and self-harming behaviors (see Linehan, 2015 and Hazlett-Stevens & Fruzzetti, 2021 for further discussion of the role of mindfulness in DBT). In another example specific to anxiety and mood disorders, Mennin et al. (2015) conceptualized generalized anxiety disorder (GAD) from an emotion regulation theoretical framework. They proposed that the hallmark GAD symptoms of chronic anticipatory anxiety and worry stem from a breakdown in the normative functioning of motivational and regulatory emotional response system mechanisms, as well as from a reduced ability to learn from situational contexts resulting in a lack of flexible behavioral response options. Their individual psychotherapy protocol, emotion regulation therapy (ERT), therefore aims to promote development of these emotion regulation capacities through an integration of cognitive-behavioral, experiential, and mindfulness-oriented therapy practices (Mennin et al., 2015). A randomized controlled trial (RCT) of ERT supported its effectiveness in reducing GAD severity and related measures of anxiety and depression (Mennin et al., 2018), and an open trial including ethnically diverse young adults diagnosed with any anxiety or mood disorder reported significant clinical gains across a range of anxiety and depression measures (Renna et al., 2018). These researchers since have extended their emotion regulation theoretical framework transdiagnostically beyond a specific diagnosis such as GAD, arguing that this psychotherapy approach applies across the "distress disorder" diagnoses of GAD, major depressive disorder, persistent depressive disorder, and posttraumatic stress disorder (Renna et al., 2020).

Goldin and colleagues proposed that MBSR ameliorated social anxiety disorder (SAD) symptoms by enhancing emotion regulation. In an open trial (Goldin & Gross, 2010), individuals diagnosed with SAD reported clinical improvements following MBSR. Importantly, these participants also completed functional magnetic resonance imaging (fMRI) assessments pre- and post-MBSR, during which they were instructed to regulate emotion in reaction to negative self-beliefs using breath-focused and distraction-focused attention tasks. Participants reported decreased negative emotion while exhibiting reduced amygdala activation and increased activity in brain regions important for deployment of attention during only the breath-focused attention task following MBSR. In a subsequent RCT comparing MBSR to an aerobic exercise stress reduction program for SAD, Goldin et al. (2013) further found that MBSR led to decreased negative emotion upon exposure to negative self-beliefs during a receptive awareness attention regulation task when compared to aerobic exercise. Neuroimaging assessment results showed increased activation in attention-related parietal cortex regions for the MBSR group compared to the aerobic exercise group, highlighting the potential role of improved attention regulation systems underlying observed emotion regulation improvements in the treatment of social anxiety. These investigations are presented with additional detail in the SAD outcome research section appearing later in this chapter. See Chap. 10 for further discussion of underlying neurobiological mechanisms of mindfulness meditation training.

In addition to enhanced regulation of negative emotions, mindfulness also may enhance positive emotion regulation by promoting positive appraisals of adverse events as well as the savoring of naturally positive aspects of experience. Mindfulness-to-Meaning Theory (Garland et al., 2015) posits that mindfulness practice allows for a broader and more flexible meta-cognitive perspective that promotes reappraising adverse events in positive ways, such as promoting personal growth, enriching meaning in life, and increasing appreciation for life. Garland et al. noted that most modern mindfulness theories were limited in their focus on disengagement from negative mental states and maladaptive habitual behavior, yet in its original historical context, people practiced mindfulness to cultivate positive mental states to promote well-being and ethical virtues. From this perspective, mindfulness not only interrupts negative cognitive appraisals and reactive habit patterns but also enhances positive emotion regulation processes to promote meaning in life, characterized by a sense of meaningful purpose when activities are congruent with deeply held values. Thus, mindfulness promotes positive reappraisals, which in turn generate meaning, creating a self-reinforcing cycle of positive emotion and cognition, and this upward spiral increases resilience and happiness over time.

Initial empirical support for this theory came from post hoc analysis of longitudinal data from a RCT of MBSR for adults with SAD (Garland et al., 2017). Increased attentional control immediately following MBSR predicted increased cognitive non-reactivity (also referred to as *decentering*; see below) 3 months later, which in turn predicted increased broadened awareness of intero- and exteroceptive experience an additional 3 months later. This broadened awareness 6 months post-MBSR then predicted increased reappraisal measured 9 months after MBSR, which then predicted greater positive affect at 12 months post-MBSR. Mediational analyses found that increased cognitive non-reactivity indeed mediated the effect of MBSR on broadened awareness, which in turn mediated enhanced reappraisal efficacy. Importantly, the increased cognitive non-reactivity and broadened awareness found among MBSR participants was not found among participants randomized to cognitive-behavioral group therapy (CBGT) instead. A replication study that delivered MBSR to students provided further empirical support for the Mindfulness-to-Meaning Theory (Hanley et al., 2021). MBSR promoted well-being over a period of 6 years by increasing the trajectory of positive reappraisal. As predicted, MBSR increased cognitive non-reactivity, which in turn broadened awareness, which then promoted positive reappraisal, which ultimately increased well-being. Taken together, these findings support a positive emotion regulation model of mindfulness, in which mindfulness training initiates a cascade of adaptive processes that build over time to promote continued quality of life for prolonged periods of time.

Changing One's Relationship to Thoughts

Leading cognitive theories of anxiety and depression implicate anxio- and depressogenic thoughts as a key factor maintaining anxiety and depressive disorder symptomology. Cognitive therapy techniques therefore were developed to restructure dysfunctional thoughts and core beliefs underlying anxiety and depressive disorders (e.g., Beck et al., 1979, 1985). Based on the fundamental observation that "a thought is just a thought" instead of fact, cognitive therapists teach clients to identify, examine, investigate, and question their fundamental assumptions, beliefs, and automatic thoughts believed to contribute to anxious and depressive symptoms. Beck et al. (1985) famously described a cognitive therapy process of "decentering," in which the cognitive therapy client challenges the basic assumption that the client is the focal point of others' attention and considers that the client might not be the center of the social world. This concept of *decentering* later expanded to capture a possible decentered relationship to one's own thoughts and feelings as well; that is, we are capable of observing our own thoughts as mere mental phenomena without believing their content, personally identifying with them, reacting to them, or otherwise getting "caught," "hooked," or adversely impacted by them (Safran & Segal, 1990). Cognitive therapists increasingly questioned whether this very premise of cognitive therapy – commonly labeled "decentering," the related concept of "cognitive distancing," or simply put, "a thought is just a thought, not a fact" – may indeed be largely responsible for cognitive therapy effectiveness (Safran & Segal; Ingram & Hollon, 1986).

Another psychotherapy approach for clinical anxiety and depression focuses on decentering without cognitive restructuring techniques. Acceptance therapy strategies aim to change the client's relationship to thoughts rather than to change thought content itself. For example, acceptance and commitment therapy (ACT; Hayes et al., 1999) emphasizes the acceptance of thoughts and feelings over trying to change them and later adopted mindfulness principles to enhance acceptance further (Hayes et al., 2012). Similarly, Roemer, Orsillo, and Salters-Pedneault (2008) integrated components of cognitive-behavioral therapy (CBT) with MBSR, MBCT, DBT, and ACT procedures to develop an acceptance-based behavior therapy protocol specifically for generalized anxiety and worry. These effective acceptance-based therapies teach mindfulness to help clients relate to thoughts as just thoughts and allow thoughts to pass through the mind as ephemeral phenomena rather than becoming emotionally entangled with them or believing them as truth.

Cognitive therapy researchers Segal, Williams, and Teasdale adapted the mindfulness meditation training protocol of MBSR for the explicit purpose of fostering a more decentered relationship to thoughts and feelings to prevent relapse of clinical depression. Their 8-week group therapy protocol, mindfulness-based cognitive therapy (MBCT; 2013), largely contains the mindfulness meditation practice schedule and curriculum of MBSR, but the stress-related didactic material of MBSR was replaced with psychoeducation and further practices specific to depression. MBCT teaches clients to relate to experience from a decentered perspective with a welcoming and friendly attitude toward all direct experience, including a willingness to allow themselves to feel the effects of any thoughts and feelings within the body. From a theoretical perspective, as previously depressed patients learn this new "mode of mind" characterized by "being," they become able to shift out of the "driven-doing" mode characterized by rumination and motivation to try to fix or change uncomfortable experience. Segal and colleagues therefore identified the core skill taught in MBCT as the ability to recognize and disengage from the "driven-doing" mode responsible for rumination and self-perpetuating destructive patterns that could lead to depression relapse, and instead engage "being" mode with the nonjudgmental and present-centered awareness of mindfulness.

This MBCT view of mindfulness is consistent with leading theoretical models of mindfulness, such as Shapiro and colleagues' (2006) intention-attention-attitude (IAA) model. According to the IAA model, three core elements of mindfulness continuously inform each other over the course of moment-to-moment practice: 1) intention, in which one reflects on why one is practicing mindfulness and an awareness of the underlying personal values motivating practice; 2) attention, paying close attention to moment-to-moment internal and external experience; and 3) attitude, including the seven attitudinal foundations of mindfulness described by Kabat-Zinn (1990) – nonjudging, patience, "beginner's mind" or openness, trust, non-striving, acceptance, and nonattachment or letting go – and additional qualities of curiosity, gentleness, non-reactivity, and loving-kindness. Over the course of repeated intentional mindfulness practice, a shift in perspective termed *reperceiving* develops, in which the individual becomes able to witness the contents of one's consciousness and to disidentify from them (Shapiro & Carlson, 2009). Shapiro and Carlson likened the process of reperceiving associated with mindfulness meditation practice to the concept of decentering found in Western psychology and psychotherapy literatures. See Chap. 6 for further review of leading psychological theories of mindfulness.

Effectiveness of Mindfulness-Based Interventions

Mindfulness-based interventions may be clinically indicated for a variety of psychological disorders, either alone or combined with other psychotherapy. For example, the original MBSR protocol has been adapted for eating disorders (mindfulness-based eating awareness therapy, MB-EAT; Kristeller et al., 2006), substance use (mindfulness-based relapse prevention, MBRP; Witkeiwitz et al., 2005), and insomnia (mindfulness-based therapy for insomnia, MBT-I; Ong, 2017). In addition, some interventions specifically target symptoms of PTSD, such as mindfulness-based exposure therapy (MBET) developed for veterans (King et al., 2016) and trauma-informed mindfulness-based stress reduction (TI-MBSR) developed for survivors of childhood sexual and physical abuse and/or intimate partner violence in adulthood (Kelly & Garland, 2016). See Chap. 8 for further review of mindfulness treatment approaches for survivors of trauma.

Mindfulness-based interventions reduced symptoms of anxiety and depression across individuals with various medical and psychiatric conditions (Hofmann et al., 2010), and a growing body of research trials conducted specifically with individuals diagnosed with anxiety disorders have established the effectiveness of mindfulness-based interventions for clinical anxiety. One meta-analysis of studies including either mindfulness-based interventions or acceptance-based psychotherapy for anxiety disorders supported both intervention approaches (Vøllestad et al., 2012). Another meta-analysis investigation comparing mindfulness-based interventions to control interventions for anxiety- and stress-related disorders found that mindfulness-based interventions were superior to control interventions and equivalent to CBT on internalizing symptom and distress measure outcomes but not for fear symptoms (de Abreu Costa et al., 2019). Since the development of MBCT, many research trials support this protocol for the prevention of major depressive relapse, the reduction of residual depressive symptoms, and possibly improved chronic depression. Mindfulness-based interventions not only reduce anxiety and depression symptoms but also may increase dispositional mindfulness, personal growth, and life satisfaction and improve quality of life (Hazlett-Stevens, 2018a).

This next section reviews outcome research of both MBSR and MBCT conducted among adults diagnosed with anxiety and/or depressive disorders. Less is known about the effectiveness of mindfulness-based interventions for children, although MBSR reduced symptoms of anxiety, depression, somatic distress, and sleep disturbance significantly more than psychiatric treatment as usual among adolescent psychiatric outpatients (Biegel et al., 2009). A specialized mindfulness training protocol that reduced adolescent attention-deficit/hyperactivity disorder (ADHD) symptoms also improved comorbid anxiety symptoms (Haydicky et al., 2012), and MBCT has been adapted for anxious children with promising results (Semple et al., 2010; Semple & Lee, 2011).

MBSR and MBCT were selected for this review because these two protocols were most often studied among adults diagnosed with anxiety disorders or clinical depression. However, another meditation-based stress reduction program (MBSM) reduced panic disorder and generalized anxiety disorder symptoms more than anxiety education (Lee et al., 2007). Three separate mindfulness-based interventions specifically tailored for social anxiety improved social anxiety symptoms as expected: 1) mindfulness and task concentration training (Bögels et al., 2006), 2) mindfulness and acceptance-based group therapy (MAGT; Kocovski et al., 2009; Kocovski et al., 2013), and 3) mindfulness-based intervention for social anxiety disorder (MBI-SAD; Koszycki et al., 2016). Both MBSR and MBCT require instructors to develop their own mindfulness meditation practice and to have a daily formal mindfulness practice in their own lives before teaching mindfulness meditation to others. Both protocols follow the eight weekly sessions and one full-day retreat schedule. MBSR was originally designed as large public health course, and although different treatment settings and research trial group sizes may vary, MBSR class sizes can exceed 30 participants. In contrast, MBCT was designed as group therapy limited to approximately 8–15 clients per group, and MBCT also requires that instructors have previous mental health professional training.

GAD, SAD, and panic disorders were the specific anxiety disorders most often studied, with the possible exception of PTSD (see Chap. 8). Some adapted mindfulness-based protocols are showing promise for obsessive-compulsive disorder (Fairfax, 2018), and although little research has examined mindfulness training for specific phobias, brief mindfulness training might facilitate the unlearning of conditioned fear responses (Björkstrand et al., 2019). The following sections therefore review the available outcome research conducted with panic disorder, GAD, and/or SAD individuals as well as investigations testing the effectiveness of mindfulness training for clinical depression.

Mixed Anxiety Disorder Samples

Kabat-Zinn, Massion, Kristeller, and Peterson (1992) conducted an early uncontrolled study at the original UMass Stress Reduction Clinic among a subset of enrolled MBSR participants screened positive for GAD and/or panic disorder. Significant reductions in both generalized and acute anxiety symptoms following MBSR were found. Importantly, these same participants reported continued anxiety symptom improvement 3 years later (Miller et al., 1995).

In a subsequent RCT including a mixed clinical sample with diagnoses of SAD, GAD, and/or panic disorder, Vøllestad, Sivertsen, and Nielsen (2011) found that participants randomized to MBSR reported significant transdiagnostic anxiety, depression, and insomnia symptom reduction when compared to a waitlist control group. A randomly selected subset of 16 MBSR participants were interviewed about their experience of the intervention afterward (Schanche et al., 2020). Qualitative analysis revealed five main themes. First, MBSR provided something useful to do when anxiety appears, including concrete focus on bodily sensations and practical tasks and letting go of negative thinking. Second, MBSR led to feeling more at ease. Third, MBSR helped participants do things anxiety had prevented before, such as going places or doing activities previously avoided and engaging with greater agency and sensitivity in interpersonal relationships. Fourth, MBSR increased their ability to meet what is there in the present, including greater awareness in everyday activities, pausing and noticing in times of difficulty, relating to distress more constructively, and cultivating a more friendly relationship with oneself. The fifth theme revealed that most participants appreciated these positive changes but also acknowledged that MBSR did not fully solve their problems, a theme the authors labeled "Better – But not there yet."

Another RCT comparing a ten-session modified MBSR protocol to a ten-session group-administered CBT (Arch et al., 2013) included veteran patients with a principal diagnosis of panic disorder (with or without agoraphobia), GAD, SAD, obsessive-compulsive disorder, or civilian PTSD. Significant reductions in clinician-rated diagnostic severity of the principal anxiety disorder were demonstrated post-intervention for both the modified MBSR and group-administered CBT groups, and these gains were maintained through 3-month follow up. These improvements, as well as treatment credibility and therapist adherence and competency, were equivalent between the two groups. Group-administered CBT yielded greater reductions in anxious arousal at follow-up compared to modified MBSR, and modified MBSR yielded greater reductions in worry and comorbid emotional disorders compared to group-administered CBT.

Four studies have examined the effects of MBCT among mixed anxiety disorder samples. In an open trial conducted with patients diagnosed with GAD and/or panic disorder (Yook et al., 2008), MBCT was associated with significant improvements in insomnia, anxiety, and depression symptoms and with reduced worry and rumination. Although not properly randomized, Y.W. Kim et al. (2009) assigned psychiatric outpatients who were receiving adjuvant pharmacotherapy and diagnosed with either GAD or panic disorder to MBCT or to an anxiety disorder education program. The MBCT group demonstrated significant improvement compared to the education group across all clinical measures of anxiety and depression as well as obsessive-compulsive and phobic symptom rating scales. No improvements or group differences were found with other psychiatric symptom ratings of somatization, interpersonal sensitivity, paranoid ideation, or psychoticism. More recently, Strege, Swain, Bochicchio, Valdespino, and Richey (2018) conducted an open trial of MBCT with individuals diagnosed with an anxiety disorder, predominantly GAD and/or SAD. Reductions in social anxiety symptoms and worry and increased positive affect were found post-intervention compared to pre-intervention, and improved positive affect ratings predicted social anxiety symptom improvement. Finally, Ninomiya et al. (2020) randomized psychiatric outpatients diagnosed with either panic disorder/agoraphobia or SAD to MBCT or to a waitlist control group. MBCT led to significant improvements on state and trait anxiety, self-reported mindfulness, and psychological distress measures compared to the control group.

Panic Disorder

Most of the studies described in the previous section included individuals diagnosed with panic disorder. However, one investigation examined the effectiveness of MBCT as an adjunct to pharmacotherapy for psychiatric outpatients diagnosed with panic disorder specifically (Kim et al., 2010). This open trial documented improvements across multiple clinical anxiety and panic measures following MBCT when compared to baseline.

Generalized Anxiety Disorder

In one large RCT (Hoge et al., 2013), participants diagnosed as GAD following structured clinical interviews received either MBSR (n = 48) or an active stress management control intervention (n = 41). The MBSR group demonstrated greater improvement on clinical severity and clinical improvement clinician ratings and on self-reported anxiety measures than the active stress management control group post-intervention, even though both groups showed comparable reductions in Hamilton Anxiety Scale symptom ratings. During a laboratory Trier Social Stress Test (TSST), the MBSR group reported greater reductions in subjective anxiety and distress and increased positive self-statements in response to this challenge task when compared to the active stress management control group. Forty-two of the participants who received MBSR also completed a laboratory homophone task before and after the intervention to measure any changes in negative interpretation bias (Hoge et al., 2020). Results indicated that these participants exhibited significant reductions in negative interpretation bias and self-reported anxiety as well as improved dispositional mindfulness. However, no evidence of an indirect relationship between improved mindfulness and reduced anxiety via changes in negative interpretation bias was found, failing to support this hypothesized relationship. In another subsample of 38 participants from the original outcome study, Hoge et al. (2015) found that self-reported changes in decentering and in mindfulness significantly mediated the effect of MBSR on self-reported anxiety. Increased decentering, but not mindfulness, yielded a significant indirect effect when both of these were included in the same model. Furthermore, the direct effect of MBSR on decreased anxiety no longer remained statistically significant, suggesting that decentering fully mediated this observed intervention outcome effect. MBSR also appeared to reduce worry through increases in the awareness and the non-reactivity subscales of the self-report mindfulness measure.

Importantly, GAD symptom reduction following MBSR has been documented outside the controlled research context when delivered as originally designed to large and diagnostically heterogeneous patient groups in a general hospital setting (Hazlett-Stevens, 2020a). When provided in mixed diagnosis groups within a smaller community mental health setting, MBSR was associated with significant GAD symptom improvement even among the most severe cases of GAD and was not limited to patients reporting only mild or moderate GAD symptoms (Hazlett-Stevens, 2018b).

MBCT also appears promising for individuals diagnosed as GAD. In an open trial, Craigie, Rees, and Marsh (2008) delivered MBCT to adults with a primary GAD diagnosis. Results documented reductions across most symptom-related outcome measures and improved quality of life, and all gains either maintained or improved further by 3-month follow-up. Although statistically significant reductions on the measure of pathological worry were found, clinical significance analysis revealed that these improvements were only modest. In another uncontrolled open trial conducted with participants diagnosed as GAD, Evans et al. (2008) reported significant reductions on all anxiety and depression measures at the end of the MBCT intervention. Clinical significance analyses revealed that 45% of the sample fully recovered to fall within the nonclinical range on clinical anxiety measures, and 60% of those participants who also exhibited clinical levels of depression beforehand recovered to drop within the nonclinical range on the depression measure post-intervention. Finally, results from a large RCT (Wong et al., 2016) found that GAD participants randomized to MBCT reported significant reductions in anxiety and worry that maintained or further improved over the following 11 months. These effects were significant when compared to a treatment as usual control group but not when compared to an active control CBT psychoeducation group, and the CBT education group actually outperformed MBCT on additional measures of physical and depressive symptoms.

Social Anxiety Disorder

One open trial delivered MBSR to young adult college students diagnosed with SAD (Hjeltnes et al. (2017)). Compared to baseline, participants demonstrated significant reductions in SAD symptoms and global psychological distress coupled with increased mindfulness, self-compassion, and self-esteem. Clinical significance analyses revealed that of those participants in the clinical range at pretreatment, two-thirds reported either clinically significant change or reliable improvement on SAD symptoms after completing MBSR. Subsequent qualitative research exploring how participants experienced MBSR-related change (Hjeltnes et al., 2019) revealed an increased use of present-centered awareness to sense the body, see negative thoughts for what they are, allow fear and shyness, transform feelings of inferiority into kindness and self-acceptance, and engage with other people in new ways.

Goldin, Ramel, and Gross (2009) examined changes in brain activity with fMRI technology in an uncontrolled trial of MBSR for adult participants diagnosed as SAD. Following MBSR, participants reported increased self-esteem and decreased social anxiety, depression, rumination, and state anxiety when compared to baseline. Participants also exhibited increased positive and decreased negative self-endorsement when performing an information-processing laboratory task designed to measure self-endorsement of social traits. Functional MRI results demonstrated increased activity in an attention regulation related brain network and reduced activity in areas associated with conceptual-linguistic self-related thought following MBSR when compared to baseline. In an additional laboratory task, participants were presented with negative self-beliefs and instructed to regulate emotion using a breath-focused attention task and a distraction-focused attention task (Goldin & Gross, 2010). Following MBSR, participants exhibited decreased negative emotion, reduced amygdala activation, and increased activity in regions important for deployment of attention during the breath-focused attention task only. Taken together, results from this investigation suggested that MBSR might improve symptoms of SAD by supporting more adaptive self-views and self-referential cognitive processes, enhancing attention and emotion regulation, and reducing emotional reactivity.

A later trial randomly assigned SAD participants to MBSR or to an active control aerobic exercise stress reduction program (Jazaieri et al., 2012). Both groups reported significant and equivalent reductions in social anxiety and depression and increased well-being when compared to a healthy comparison group, immediately after the intervention and at 3-month follow-up. Despite these equivalent symptom outcomes and equivalent increases in positive self-views exhibited during the self-endorsement information-processing laboratory task, MBSR led to decreased negative self-views and increased neural responses in posterior cingulate cortex areas as compared to the aerobic exercise group (Goldin et al., 2012). MBSR also appeared to reduce negative emotional reactivity and improve emotion regulation when compared to the aerobic exercise group, as MBSR led to decreased negative emotion when faced with negative self-beliefs during a receptive awareness attention regulation laboratory task conducted during fMRI assessments (Goldin et al., 2013). Furthermore, Goldin et al. found associated neural responses characterized by increased activation in attention-related parietal cortex regions for the MBSR group compared to the aerobic exercise group, highlighting the potential role of enhanced attention regulation in response to negative self-beliefs in SAD recovery.

A subsequent RCT compared MBSR to cognitive-behavioral group therapy (CBGT) – often considered the gold-standard psychotherapy for SAD – and to a waitlist control group (Goldin et al., 2016). Both MBSR and CBGT produced significant and similar immediate improvements in social anxiety symptoms, related clinical measures such as rumination, cognitive reappraisal, and cognitive distortion when compared to the waitlist group, and these treatment gains maintained for both intervention groups 1 year later. MBSR therefore may be comparable to CBGT in effectiveness for SAD. However, these two interventions differentially impacted a couple of specific psychological processes measured, and a previous comparison of MBSR to CBGT demonstrated some additional potential benefits of CBGT (Koszycki et al., 2007).

Another RCT compared the MBCT protocol to CBGT among young adults (aged 18–25) diagnosed with SAD (Piet et al., 2010). Both interventions yielded statistically significant and equivalent improvements following treatment across outcome measures, although effect sizes tended to be larger for the CBGT intervention. This investigation employed a crossover design in which participants received both interventions in either order. Combining these treatments did not appear to yield much additional benefit, and both groups continued to improve posttreatment until 6-month follow-up and were still improved at 12-month follow-up.

Depression

Most research examining the effectiveness of mindfulness-based interventions for clinical depression delivered MBCT. This is not surprising, given that the MBCT protocol specifically aims to prevent future relapse among individuals recovered from recurrent major depressive episodes. Early RCTs demonstrated that MBCT without continued antidepressant medication significantly reduced the risk of subsequent depressive relapse among patients with three or more previous major depressive episodes when compared to treatment as usual (Teasdale et al., 2000; Ma & Teasdale, 2004). Initial RCTs conducted by independent researchers replicated these results (Godfrin & van Heeringen, 2010), although one yielded ambiguous results at 12-month follow-up (Bondolfi et al., 2010). Also of note, most of the patients in the Godfrin and van Heeringen study (approximately 75%) were still taking antidepressant medication, demonstrating that MBCT also effectively prevented relapse while patients continued medication. Meta-analysis including six such studies found that MBCT significantly reduced the risk of depressive relapse by 34% overall, and risk reduced by 43% among patients with three or more past episodes (Piet & Hougaard, 2011). In addition to reducing residual depressive symptoms, MBCT also reduced emotional reactivity exhibited during a laboratory Trier Social Stress Test (TSST), and this reduced emotional reactivity to social stress partially mediated the observed improvements in depressive symptoms (Britton et al., 2012).

Researchers also compared MBCT to antidepressant medication treatment directly. Kuyken et al. (2008) recruited recurrent depression patients already treated with antidepressant medication and randomized them to receive either MBCT as they discontinued medication or to continued maintenance antidepressant medication. Relapse rates in the MBCT condition over the next 15 months was 47% compared to 60% in the maintenance antidepressant medication condition, even though 75% of MBCT patients had discontinued medication completely. MBCT also reduced residual depressive symptoms and psychiatric comorbidities and increased quality of life more effectively than medication. Segal et al. (2010) also recruited depressed patients in remission following antidepressant medication treatment. Patients then were randomized to receive MBCT while discontinuing medication, to discontinue their medication unknowingly and switched to a placebo pill, or to continue maintenance antidepressant medication. Among a subgroup of patients who demonstrated unstable remission during acute treatment, both maintenance antidepressant medication and MBCT with discontinued medication significantly and equivalently reduced the risk of relapse over the following 18 months compared to the placebo condition. No treatment group differences were found among a subgroup of patients considered stable remitters. Taken together, these studies suggest that MBCT may be just as effective as medication in the prevention of major depressive relapse.

MBCT research continued to grow over the subsequent decade. UK's National Institute for Health and Clinical Excellence (NICE) endorsed MBCT as an evidence-based therapy. A number of systematic reviews and meta-analyses concluded that MBCT significantly reduced the risk of relapse and depressive symptoms during remission, improved current depressive symptoms, and may even benefit patients with chronic treatment-resistant depression (see Clarke et al., 2015; Kuyken et al. 2016; Perestelo-Perez et al., 2017; MacKenzie et al., 2018). In addition, MBCT appears promising in the treatment of mixed anxiety and depressive disorder samples (Strauss et al., 2014) and in lowering high levels of psychological distress during pregnancy (MacKinnon et al., 2021).

MBCT has been shown to increase day-to-day mindfulness, self-compassion, and meta-cognitive awareness compared to antidepressant medication, and these changes significantly predicted lower depression levels as long as 13 months later

(Kuyken et al., 2010; Bieling et al., 2012). Furthermore, levels of cognitive reactivity in response to a sad mood induction in the laboratory did not predict depression relapse among MBCT participants – whereas cognitive reactivity to the sad mood induction did predict subsequent depression among antidepressant medication participants – and this prophylactic MBCT effect was clearly linked to levels of self-compassion (Kuyken et al. 2010). Thus, the increased self-compassion associated with MBCT appears to reduce the depressogenic impact of cognitive reactivity. Another mindfulness-based protocol that specifically targets self-compassion, mindful self-compassion (MSC), significantly increased self-compassion as well as compassion for others, mindfulness, and life satisfaction while decreasing symptoms of depression, anxiety, and stress (Neff & Germer, 2013).

Although the vast majority of effectiveness research for clinical depression studied MBCT, one non-randomized controlled trial found that MBSR reduced ruminative thinking relative to a waitlist matched control group while also reducing depression and anxiety symptoms and dysfunctional beliefs among individuals with a history of mood disorders (Ramel et al., 2004). Therefore, patients with a history of clinical depression who do not have access to MBCT might still receive some benefit from the more generalized MBSR protocol.

Summary and Conclusions

Mindfulness-based interventions appear particularly helpful for individuals suffering from clinical anxiety and depressive disorders, with outcome research trials consistently yielding large effect sizes. Several interrelated and complimentary theoretical approaches explain how mindfulness training might alleviate anxiety and depression symptoms, ranging from reduced stress reactivity and allostatic load to improved mind-body self-regulation, more adaptive cognitive-behavioral responding, enhanced negative and positive emotion regulation, and cultivation of a decentered and compassionate relationship with experience. Research support for MBSR and MBCT in the treatment of GAD, panic disorder, and SAD continues to grow, and some specialized protocols for SAD, panic disorder, and OCD appear promising as well. The MBCT protocol originally was designed to target clinical depression symptoms, and several RCTs have established its effectiveness to prevent relapse and decrease depression while also identifying additional benefits beyond those associated with antidepressant medication. Future research is needed to examine the effectiveness of mindfulness-based interventions among individuals from diverse backgrounds and to address the acceptability and culturally sensitive delivery of mindfulness training when provided to minority groups (Hazlett-Stevens, 2020b). Other future directions may include alternative delivery formats, such as self-directed bibliotherapy and online delivery of mindfulness training (e.g., Hazlett-Stevens & Oren, 2017).

References

Arch, J. J., Ayers, C. R., Baker, A., Almklov, E., Dean, D. J., & Craske, M. G. (2013). Randomized clinical trial of adapted mindfulness-based stress reduction versus group cognitive behavioral therapy for heterogeneous anxiety disorders. *Behaviour Research and Therapy, 51*, 185–196.

Barlow, D. H. (2002). *Anxiety and its disorders: The nature and treatment of anxiety and panic* (2nd ed.). New York: Guilford Press.

Barlow, D. H., Farchione, T. J., Fairholme, C. P., Ellard, K. K., Boisseau, C. L., Allen, L. B., & Ehrenreich-May, J. (2011). *The unified protocol for transdiagnostic treatment of emotional disorders: Therapist guide*. New York: Oxford University Press.

Beck, A. T., Rush, A. J., Shaw, B. F., & Emery, G. (1979). *Cognitive therapy of depression*. New York: Guilford.

Beck, A., Emery, G., & Greenberg, R. (1985). *Anxiety disorders and phobias: A cognitive perspective*. New York: Basic Books.

Biegel, G. M., Brown, K. W., Shapiro, S. L., & Schubert, C. M. (2009). Mindfulness-based stress reduction for the treatment of adolescent psychiatric outpatients: A randomized clinical trial. *Journal of Consulting and Clinical Psychology, 77*, 855–866.

Bieling, P. J., Hawley, L. L., Bloch, R. T., Corcoran, K. M., Levitan, R. D., Young, L. T., MacQueen, G. M., & Segal, Z. V. (2012). Treatment-specific changes in decentering following mindfulness-based cognitive therapy versus antidepressant medication or placebo for prevention of depressive relapse. *Journal of Consulting and Clinical Psychology, 80*, 365–372.

Björkstrand, J., Schiller, D., Li, J., Davidson, P., Rosén, J., Mårtensson, J., & Kirk, U. (2019). The effect of mindfulness training on extinction retention. *Scientific Reports, 9*, 19896.

Bögels, S. M., Sijbers, G. F. V. M., & Voncken, M. (2006). Mindfulness and task concentration training for social phobia: A pilot study. *Journal of Cognitive Psychotherapy, 20*, 33–44.

Bondolfi, G., Jermann, F., der Linden, M. V., Gex-Fabry, M., Bizzini, L., Rouget, B. W., Myers-Arrazola, L., Gonzalez, C., Segal, Z., Aubry, J. M., & Bertschy, G. (2010). Depression relapse prophylaxis with mindfulness-based cognitive therapy: Replication and extension in the Swiss health care system. *Journal of Affective Disorders, 122*, 224–231.

Britton, W. B., Shahar, B., Szepsenwol, O., & Jacobs, W. J. (2012). Mindfulness-based cognitive therapy improves emotional reactivity to social stress: Results from a randomized controlled trial. *Behavior Therapy, 43*, 365–380.

Brown, T. A., & Naragon-Gainey, K. (2013). Evaluation of the unique and specific contributions of dimensions of the triple vulnerability model to the prediction of DSM-IV anxiety and mood disorder constructs. *Behavior Therapy, 44*, 277–292.

Clarke, K., Mayo-Wilson, E., Kenny, J., & Pilling, S. (2015). Can non-pharmacological interventions prevent relapse in adults who have recovered from depression? A systematic review and meta-analysis of randomised controlled trials. *Clinical Psychology Review, 39*, 58–70.

Craigie, M. A., Rees, C. S., & Marsh, A. (2008). Mindfulness-based cognitive therapy for generalized anxiety disorder: A preliminary evaluation. *Behavioural and Cognitive Psychotherapy, 36*, 553–568.

de Abreu Costa, M., D'Alò de Oliveira, G. S., Tatton-Ramos, T., Manfro, G. G., & Salum, G. A. (2019). Anxiety and stress-related disorders and mindfulness-based interventions: A systematic review and multilevel meta-analysis and meta-regression of multiple outcomes. *Mindfulness, 10*, 996–1005.

de Vibe, M., Bjørndal, A., Tipton, E., Hammerstrøm, K., & Kowalski, K. (2012). Mindfulness-based stress reduction (MBSR) for improving health, quality of life, and social functioning in adults. *Campbell Systematic Reviews, 3*.

Evans, S., Ferrando, S., Findler, M., Stowell, C., Smart, C., & Haglin, D. (2008). Mindfulness-based cognitive therapy for generalized anxiety disorder. *Journal of Anxiety Disorders, 22*, 716–721.

Fairfax, H. (2018). Mindfulness and obsessive compulsive disorder: Implications for psychological intervention. *Journal of Mental Health and Clinical Psychology, 2*, 55–63.

Garland, E. L., Farb, N. A., Goldin, P., & Fredrickson, B. L. (2015). Mindfulness broadens awareness and builds eudaimonic meaning: A process model of mindful positive emotion regulation.
Psychological Inquiry, 26, 293–314.
Garland, E. L., Hanley, A. W., Goldin, P. R., & Gross, J. J. (2017). Testing the mindfulness-to-
meaning theory: Evidence for mindful positive emotion regulation from a reanalysis of longitudinal data. *PLoS One, 12*, Article e0187727.
Godfrin, K. A., & van Heeringen, C. (2010). The effects of mindfulness-based cognitive therapy
on recurrence of depressive episodes, mental health and quality of life: A randomized con-
trolled study. *Behaviour Research and Therapy, 48*, 738–746.
Goldin, P. R., & Gross, J. J. (2010). Effects of mindfulness-based stress reduction (MBSR) on
emotion regulation in social anxiety disorder. *Emotion, 10*, 83–91.
Goldin, P., Ramel, W., & Gross, J. (2009). Mindfulness meditation training and self-referential
processing in social anxiety disorder: Behavioral and neural effects. *Journal of Cognitive
Psychotherapy, 23*, 242–257.
Goldin, P., Ziv, M., Jazaieri, H., & Gross, J. J. (2012). Randomized controlled trial of mindfulness-
based stress reduction versus aerobic exercise: Effects on the self-referential brain network in
social anxiety disorder. *Frontiers in Human Neuroscience, 6*, 295.
Goldin, P., Ziv, M., Jazaieri, H., Hahn, K., & Gross, J. J. (2013). MBSR vs aerobic exercise in
social anxiety: fMRI of emotion regulation of negative self-beliefs. *SCAN, 8*, 65–72.
Goldin, P. R., Morrison, A., Jazaieri, H., Brozovich, F., Heimberg, R., & Gross, J. J. (2016). Group
CBT versus MBSR for social anxiety disorder: A randomized controlled trial. *Journal of
Consulting and Clinical Psychology, 84*, 427–437.
Hanley, A. W., de Vibe, M., Solhaug, I., Farb, N., Goldin, P. R., Gross, J. J., & Garland, E. L. (2021).
Modeling the mindfulness-to-meaning theory's mindful reappraisal hypothesis: Replication
with longitudinal data from a randomized controlled study. *Stress and Health, 2021*, 1–12.
Haydicky, J., Wiener, J., Badali, P., Milligan, K., & Ducharme, J. M. (2012). Evaluation of a
mindfulness-based intervention for adolescents with learning disabilities and co-occurring
ADHD and anxiety. *Mindfulness, 3*, 151–164.
Hayes, S. C., Strosahl, K., & Wilson, K. G. (1999). *Acceptance and commitment therapy: An
experimental approach to behavior change*. New York: Guilford.
Hayes, S. C., Strosahl, K. D., & Wilson, K. G. (2012). *Acceptance and commitment therapy: The
process and practice of mindful change* (2nd ed.). New York: Guilford.
Hazlett-Stevens, H. (2018a). Mindfulness-based stress reduction in a mental health outpatient
setting: Benefits beyond symptom reduction. *Journal of Spirituality in Mental Health, 20*,
275–292.
Hazlett-Stevens, H. (2018b). Mindfulness-based stress reduction for generalized anxiety disorder:
Does pre-treatment symptom severity relate to clinical outcomes? *Journal of Depression and
Anxiety Forecast, 1, 1007*, 1–4.
Hazlett-Stevens, H. (2020a). Generalized anxiety disorder symptom improvement following
mindfulness-based stress reduction in a general hospital setting. *Journal of Medical Psychology,
22*, 21–29.
Hazlett-Stevens, H. (2020b). Cultural considerations when treating anxiety disorders with
mindfulness-based interventions. In L. Benuto, F. Gonzalez, & J. Singer (Eds.), *Handbook
of cultural factors in behavioral health* (pp. 277–292). Cham, Switzerland: Springer Nature.
Hazlett-Stevens, H., & Fruzzetti, A. E. (2021). Regulation of physiological arousal and emotion. In A. Wenzel (Ed.), *Handbook of cognitive behavioral therapy, volume 1: Overview and
approaches*. Washington, D.C.: American Psychological Association.
Hazlett-Stevens, H., & Oren, Y. (2017). Effectiveness of mindfulness-based stress reduction bibliotherapy: A preliminary randomized controlled trial. *Journal of Clinical Psychology, 73*,
626–637.
Hjeltnes, A., Molde, H., Schanche, E., Vøllestad, J., Lillebostad Svendsen, J., Moltu, C., & Binder,
P. E. (2017). An open trial of mindfulness-based stress reduction for young adults with social
anxiety disorder. *Scandinavian Journal of Psychology, 58*, 80–90.

Hjeltnes, A., Moltu, C., Schanche, E., Jansen, Y., & Binder, P.-E. (2019). Facing social fears: How do improved participants experience change in mindfulness-based stress reduction for social anxiety disorder? *Counselling and Psychotherapy Research, 19*, 35–44.

Hofmann, S. G., Sawyer, A. T., Witt, A. A., & Oh, D. (2010). The effect of mindfulness-based therapy on anxiety and depression: A meta-analytic review. *Journal of Consulting and Clinical Psychology, 78*, 169–183.

Hoge, E. A., Bui, E., Marques, L., Metcalf, C. A., Morris, L. K., Robinaugh, D. J., Worthington, J. J., Pollack, M. H., & Simon, N. M. (2013). Randomized controlled trial of mindfulness meditation for generalized anxiety disorder: Effects on anxiety and stress reactivity. *Journal of Clinical Psychiatry, 74*, 786–792.

Hoge, E. A., Bui, E., Goetter, E., Robinaugh, D. J., Ojserkis, R. A., Fresco, D. M., & Simon, N. M. (2015). Change in decentering mediates improvement in anxiety in mindfulness-based stress reduction for generalized anxiety disorder. *Cognitive Therapy and Research, 39*, 228–235.

Hoge, E. A., Reese, H. E., Oliva, I. A., Gabriel, C. D., Guidos, B. M., Bui, E., Simon, N. M., & Dutton, M. A. (2020). Investigating the role of interpretation bias in mindfulness-based treatment of adults with generalized anxiety disorder. *Frontiers in Psychology, 11*, 82.

Ingram, R. E., & Hollon, S. D. (1986). Cognitive therapy for depression from an information processing perspective. In R. E. Ingram (Ed.), *Personality, psychopathology, and psychotherapy series. Information processing approaches to clinical psychology* (pp. 259–281). San Diego, CA: Academic Press.

Jazaieri, H., Goldin, P. R., Werner, K., Ziv, M., & Gross, J. J. (2012). A randomized trial of MBSR versus aerobic exercise for social anxiety disorder. *Journal of Clinical Psychology, 68*, 715–731.

Kabat-Zinn, J. (1990). *Full catastrophe living*. New York: Bantam Books.

Kabat-Zinn, J. (2013). *Full catastrophe living, revised and* (updated ed.). New York: Bantam Books.

Kabat-Zinn, J., Massion, A. O., Kristeller, J., & Peterson, L. G. (1992). Effectiveness of a meditation-based stress reduction program in the treatment of anxiety disorders. *American Journal of Psychiatry, 149*, 936–943.

Kelly, A., & Garland, E. L. (2016). Trauma-informed mindfulness-based stress reduction for female survivors of interpersonal violence: Results from a stage I RCT. *Journal of Clinical Psychology, 72*, 311–328.

Kim, Y. W., Lee, S. H., Choi, T. K., Suh, S. Y., Kim, B., Kim, C. M., Cho, S. J., Kim, M. J., Yook, K., Ryu, M., Song, S. K., & Yook, K. H. (2009). Effectiveness of mindfulness-based cognitive therapy as an adjuvant to pharmacotherapy in patients with panic disorder or generalized anxiety disorder. *Depression and Anxiety, 26*, 601–606.

Kim, B., Lee, S. H., Kim, Y. W., Choi, T. K., Yook, K., Suh, S. Y., Cho, S. J., & Yook, K. H. (2010). Effectiveness of a mindfulness-based cognitive therapy program as an adjunct to pharmacotherapy in patients with panic disorder. *Journal of Anxiety Disorders, 24*, 590–595.

King, A. P., Block, S. R., Sripada, R. K., Rauch, S., Giardino, N., Favorite, T., … Liberzon, I. (2016). Altered default mode network (DMN) resting state functional connectivity following a mindfulness-based exposure therapy for posttraumatic stress disorder (PTSD) in combat veterans of Afghanistan and Iraq. *Depression and Anxiety, 33*, 289–299.

Kocovski, N. L., Fleming, J. E., & Rector, N. A. (2009). Mindfulness and acceptance-based group therapy for social anxiety disorder: An open trial. *Cognitive and Behavioral Practice, 16*, 276–289.

Kocovski, N. L., Fleming, J. E., Hawley, L. L., Huta, V., & Antony, M. M. (2013). Mindfulness and acceptance-based group therapy versus traditional cognitive behavioral group therapy for social anxiety disorder: A randomized controlled trial. *Behaviour Research and Therapy, 51*, 889–898.

Koszycki, D., Benger, M., Shlik, J., & Bradwejn, J. (2007). Randomized trial of a meditation-based stress reduction program and cognitive behavior therapy in generalized social anxiety disorder. *Behaviour Research and Therapy, 45*, 2518–2526.

Koszycki, D., Thake, J., Mavounza, C., Daoust, J. P., Taljaard, M., & Bradwejn, J. (2016). Preliminary investigation of a mindfulness-based intervention for social anxiety disorder that integrates compassion meditation and mindful exposure. *Journal of Alternative and Complementary Medicine, 22*, 363–374.

Kristeller, J. L., Baer, R. A., & Quillian-Wolever, R. (2006). Mindfulness-based approaches to eating disorders. In R. A. Baer (Ed.), *Mindfulness-based treatment approaches: Clinician's guide to evidence base and applications* (pp. 75–91). London: Academic Press.

Kuyken, W., Byford, S., Taylor, R. S., Watkins, E., Holden, E., White, K., Barrett, B., Byng, R., Evans, A., Mullan, E., & Teasdale, J. D. (2008). Mindfulness-based cognitive therapy to prevent relapse in recurrent depression. *Journal of Consulting and Clinical Psychology, 76*, 966–978.

Kuyken, W., Watkins, E., Holden, E., White, K., Taylor, R. S., Byford, S., Evans, A., Radford, S., Teasdale, J. D., & Dalgleish, T. (2010). How does mindfulness-based cognitive therapy work? *Behaviour Research and Therapy, 48*, 1105–1112.

Kuyken, W., Warren, F. C., Taylor, R. S., Whalley, B., Crane, C., Bondolfi, G., Hayes, R., Huijbers, M., Ma, H., Schweizer, S., Segal, Z., Speckens, A., Teasdale, J. D., Van Heeringen, K., Williams, M., Byford, S., Byng, R., & Dalgleish, T. (2016). Efficacy of mindfulness-based cognitive therapy in prevention of depressive relapse: An individual patient data meta-analysis from randomized trials. *JAMA Psychiatry, 73*, 565–574.

Lee, S. H., Ahn, S. C., Lee, Y. J., Choi, T. K., Yook, K. H., & Suh, S. Y. (2007). Effectiveness of a meditation-based stress management program as an adjunct to pharmacotherapy in patients with anxiety disorder. *Journal of Psychosomatic Research, 62*, 189–195.

Linehan, M. M. (1993). *Cognitive-behavioral treatment of borderline personality disorder*. New York, NY: Guilford Press.

Linehan, M. (2015). *Skills training manual for treating borderline personality disorder*. New York, NY: The Guilford Press.

Ma, S., & Teasdale, J. D. (2004). Mindfulness-based cognitive therapy for depression: Replication and exploration of differential relapse prevention effects. *Journal of Consulting and Clinical Psychology, 72*, 31–40.

MacKenzie, M. B., Abbott, K. A., & Kocovski, N. L. (2018). Mindfulness-based cognitive therapy in patients with depression: Current perspectives. *Neuropsychiatric Disease and Treatment, 14*, 1599–1605.

MacKinnon, A. L., Madsen, J. W., Giesbrecht, G. F., Campbell, T., Carlson, L. E., Dimidjian, S., Letourneau, N., Tough, S., & Tomfohr-Madsen, L. (2021). Effects of mindfulness-based cognitive therapy in pregnancy on psychological distress and gestational age: Outcomes of a randomized controlled trial. *Mindfulness*.

McEwen, B. S. (1998). Stress, adaptation, and disease: Allostasis and allostatic load. *Annals of the New York Academy of Sciences, 840*, 33–44.

McEwen, B. S. (2005). Stressed or stressed out: What is the difference? *Journal of Psychiatry & Neuroscience, 30*, 315–318.

Mennin, D. S., Fresco, D. M., Heimberg, R. G., & Ritter, M. (2015). An open trial of emotion regulation therapy for generalized anxiety disorder and co-occurring depression. *Depression and Anxiety, 32*, 614–623.

Mennin, D. S., Fresco, D. M., O'Toole, M. S., & Heimberg, R. G. (2018). A randomized controlled trial of emotion regulation therapy for generalized anxiety disorder with and without co-occurring depression. *Journal of Consulting and Clinical Psychology, 86*, 268–281.

Miller, J. J., Fletcher, K., & Kabat-Zinn, J. (1995). Three-year follow-up and clinical implications of a mindfulness meditation-based stress reduction intervention in the treatment of anxiety disorders. *General Hospital Psychiatry, 17*, 192–200.

Neff, K. D., & Germer, C. K. (2013). A pilot study and randomized controlled trial of the mindful self-compassion program. *Journal of Clinical Psychology, 69*, 28–44.

Ninomiya, A., Sado, M., Park, S., Fujisawa, D., Kosugi, T., Nakagawa, A., Shirahase, J., & Mimura, M. (2020). Effectiveness of mindfulness-based cognitive therapy in patients with anxiety disorders in secondary-care settings: A randomized controlled trial. *Psychiatry and Clinical Neurosciences, 74*, 132–139.

Ong, J. C. (2017). *Mindfulness-based therapy for insomnia*. Washington, D.C.: American Psychological Association.

Perestelo-Perez, L., Barraca, J., Peñate, W., Rivero-Santana, A., & Alvarez-Perez, Y. (2017). Mindfulness-based interventions for the treatment of depressive rumination: Systematic review and meta-analysis. *International Journal of Clinical and Health Psychology, 17*, 282–295.

Piet, J., & Hougaard, E. (2011). The effect of mindfulness-based cognitive therapy for prevention of relapse in recurrent major depressive disorder: A systematic review and meta-analysis. *Clinical Psychology Review, 31*, 1032–1040.

Piet, J., Hougaard, E., Hecksher, M. S., & Rosenberg, N. K. (2010). A randomized pilot study of mindfulness-based cognitive therapy and group cognitive-behavioral therapy for young adults with social phobia. *Scandinavian Journal of Psychology, 51*, 403–410.

Ramel, W., Goldin, P. R., Carmona, P. E., & McQuaid, J. R. (2004). The effects of mindfulness meditation on cognitive processes and affect in patients with past depression. *Cognitive Therapy and Research, 28*, 433–455.

Renna, M. E., Quintero, J. M., Soffer, A., Pino, M., Ader, L., Fresco, D. M., & Mennin, D. S. (2018). A pilot study of emotion regulation therapy for generalized anxiety and depression: Findings from a diverse sample of young adults. *Behavior Therapy, 49*, 403–418.

Renna, M. E., Fresco, D. M., & Mennin, D. S. (2020). Emotion regulation therapy and its potential role in the treatment of chronic stress-related pathology across disorders. *Chronic Stress, 4*, 1–10.

Roemer, L., Orsillo, S. M., & Salters-Pedneault, K. (2008). Efficacy of an acceptance-based behavior therapy for generalized anxiety disorder: Evaluation in a randomized controlled trial. *Journal of Consulting and Clinical Psychology, 76*, 1083–1089.

Safran, J. D., & Segal, Z. V. (1990). *Interpersonal process in cognitive therapy*. New York: Basic Books.

Schanche, E., Vøllestad, J., Binder, P.-E., Hjeltnes, A., Dundas, I., & Nielsen, G. H. (2020). Participant experiences of change in mindfulness-based stress reduction for anxiety disorders. *International Journal of Qualitative Studies on Health and Well-Being, 15*, 1776094.

Schwatrz, G. E. (1984). Psychobiology of health: A new synthesis. In B. L. Hammonds & C. J. Scheirer (Eds.), *Psychology and health: Master lecture series* (Vol. 3). Washington, DC: American Psychological Association.

Schwatrz, G. E. (1990). Psychobiology of repression and health: A systems approach. In J. Singer (Ed.), *Repression and dissociation: Implications for personality theory, psychopathology, and health*. Chicago: University of Chicago Press.

Segal, Z. V., Bieling, P., Young, T., MacQueen, G., Cooke, R., Martin, L., Bloch, R., & Levitan, R. D. (2010). Antidepressant monotherapy vs sequential pharmacotherapy and mindfulness-based cognitive therapy, or placebo, for relapse prophylaxis in recurrent depression. *Archives of General Psychiatry, 67*, 1256–1264.

Segal, Z. V., Williams, J. M. G., & Teasdale, J. D. (2013). *Mindfulness-based cognitive therapy for depression* (2nd ed.). New York: Guilford Press.

Semple, R. J., & Lee, J. (2011). *Mindfulness-based cognitive therapy for anxious children: A manual for treating childhood anxiety*. Oakland, CA: New Harbinger Publications.

Semple, R. J., Lee, J., Rosa, D., & Miller, L. F. (2010). A randomized trial of mindfulness-based cognitive therapy for children: Promoting mindful attention to enhance social-emotional resiliency in children. *Journal of Child and Family Studies, 19*, 218–229.

Shafran, R., Thordarson, D. S., & Rachman, S. (1996). Thought–action fusion in obsessive-compulsive disorder. *Journal of Anxiety Disorders, 10*, 379–391.

Shapiro, S. L., & Carlson, L. E. (2009). *The art and science of mindfulness: Integrating mindfulness into psychology and the helping professions*. Washington, DC: American Psychological Association.

Shapiro, S. L., Carlson, L. E., Astin, J. A., & Freedman, B. (2006). Mechanisms of mindfulness. *Journal of Clinical Psychology, 62*, 373–386.

Shaprio, S. L., & Schwartz, G. E. (2000a). The role of intention in self-regulation. In M. Boekaerts, P. R. Pintrich, & M. Zeidner (Eds.), *Handbook of self-regulation* (pp. 253–273). New York: Academic Press.

Shaprio, S. L., & Schwartz, G. E. (2000b). Intentional systemic mindfulness: An integrative model for self-regulation and health. *Advances in Mind-Body Medicine, 16*, 128–134.

Strauss, C., Cavanagh, K., Oliver, A., & Pettman, D. (2014). Mindfulness-based interventions for people diagnosed with a current episode of an anxiety or depressive disorder: A meta-analysis of randomised controlled trials. *PLoS One, 9*(4), e96110.

Strege, M. V., Swain, D., Bochicchio, L., Valdespino, A., & Richey, J. A. (2018). A pilot study of the effects of mindfulness-based cognitive therapy on positive affect and social anxiety symptoms. *Frontiers in Psychology, 9*, 866.

Teasdale, J. D., Segal, Z. V., Williams, J. M. G., Ridgeway, V. A., Soulsby, J. M., & Lau, M. A. (2000). Prevention of relapse/recurrence in major depression by mindfulness-based cognitive therapy. *Journal of Consulting and Clinical Psychology, 68*, 615–623.

Vøllestad, J., Sivertsen, B., & Nielsen, G. H. (2011). Mindfulness-based stress reduction for patients with anxiety disorders: Evaluation in a randomized controlled trial. *Behaviour Research and Therapy, 49*, 281–288.

Vøllestad, J., Nielsen, M. B., & Nielsen, G. H. (2012). Mindfulness- and acceptance-based interventions for anxiety disorders: A systematic review and meta-analysis. *The British Journal of Clinical Psychology, 51*, 239–260.

Witkeiwitz, K., Marlatt, G. A., & Walker, D. (2005). Mindfulness-based relapse prevention for substance use disorders. *Journal of Cognitive Psychotherapy, 19*, 211–228.

Wong, S. Y., Yip, B. H., Mak, W. W., Mercer, S., Cheung, E. Y., Ling, C. Y., Lui, W. W., Tang, W. K., Lo, H. H., Wu, J. C., Lee, T. M., Gao, T., Griffiths, S. M., Chan, P. H., & Ma, H. S. (2016). Mindfulness-based cognitive therapy v. group psychoeducation for people with generalised anxiety disorder: Randomised controlled trial. *The British Journal of Psychiatry: the Journal of Mental Science, 209*, 68–75.

Yook, K., Lee, S. H., Ryu, M., Kim, K. H., Choi, T. K., Suh, S. Y., Kim, Y. W., Kim, B., Kim, M. Y., & Kim, M. J. (2008). Usefulness of mindfulness-based cognitive therapy for treating insomnia in patients with anxiety disorders: A pilot study. *The Journal of Nervous and Mental Disease, 196*, 501–503.

Chapter 10
Neurobiology of Mindfulness-Based Interventions

Philip A. Desormeau and Norman A. S. Farb

Introduction

Mindfulness training (MT) represents a set of contemplative practices aimed at restructuring one's relationship with immediate experience to promote well-being. In recent years, MT has grown rapidly in popularity, buoyed by its increasingly validated clinical efficacy in a variety of contexts, including depression, anxiety, substance use, and chronic pain conditions (Bohlmeijer et al., 2010; Chiesa & Serretti, 2010; Goyal et al., 2014; Li & Bressington, 2019). Mindfulness also appears to have substantial benefits for stress reduction in the general population (Astin, 1997; Bowen & Marlatt, 2009; Chiesa & Serretti, 2009), and has been growing in popularity in a variety of applications ranging from more traditional meditation groups to an explosion of MT applications for smartphones and web platforms (Walsh et al., 2019). Despite the availability of MT practices for millennia, the rapid uptake and popularization of MT in contemporary Western society suggests that MT addresses a need not fully satisfied by contemporary healthcare models. What unique approaches to stress reduction does MT bring to a culture that already seems deeply invested in self-improvement and resilience? How does the neurobiology of mindfulness interventions inform our broader understanding of stress reactivity and resilience?

In this chapter, we will briefly review mindfulness theory covered in greater detail in other chapters of the book. We will discuss how multiple definitions of mindfulness correspond with types of neural systems models, and review evidence on how the most popular manualized MT interventions seem to impact brain

P. A. Desormeau
University of Toronto Scarborough, Toronto, Canada

N. A. S. Farb (✉)
University of Toronto Mississauga, Mississauga, Canada
e-mail: norman.farb@utoronto.ca

© Springer Nature Switzerland AG 2022
H. Hazlett-Stevens (ed.), *Biopsychosocial Factors of Stress, and Mindfulness for Stress Reduction*, https://doi.org/10.1007/978-3-030-81245-4_10

structure and function. The chapter will conclude with a narrative synthesis of the findings reviewed, proposing a more general neurobiological model of mindfulness and its mechanisms for reducing stress reactivity.

How Does Mindfulness Work? We begin by briefly reviewing current theory on how mindfulness works. The consensus research literature suggests that MT's efficacy stems from its ability to change maladaptive appraisal and coping habits that drive psychiatric symptoms, such as perceiving negative thoughts and feelings as objective truths rather than as transitory mental events (Farb et al., 2012; Hölzel, Lazar, et al., 2011b; Teasdale et al., 2013). Central to this theory is the idea MT promotes awareness of reactive habits that are used to cope with stress, allowing the practitioner to adjudicate between adaptive response and habitual, maladaptive responses.

For example, faced with criticism from one's boss, one might feel an urge to quit to avoid further criticism. Alternatively, one might consider reviewing the criticism as a chance to improve one's work. If avoidance is the overlearned response, one might find oneself leaving a job that one cares deeply about to protect oneself against the negative feeling of criticism. If one can recognize the impulse to quit as a natural urge to avoid criticism, one can work with that feeling and consider how else to manage it rather than making a harmful decision. Mindfulness training is precisely the practice of cultivating awareness of one's experience to limit habitual reactions and possibly replace one's most pernicious habits with alternative strategies that better align with one's personal values.

In technical terms, MT addresses the rigidity and power of overlearned perceptual and behavioral habits by leveraging awareness to de-automatize stress appraisals and their associated regulatory responses (Chong et al., 2015; Lea et al., 2015; Vago, 2014). Intentional, nonjudgmental exploration of present-moment experience serves to disrupt the habitual implementation of existing perceptual and behavioral habits. The benefits of mindfulness meditation are believed to arise from the cultivation of awareness as a buffer against "knee-jerk" reactions to stressful events, allowing for novel perceptions and appraisals to emerge. In the reflective space of mindful awareness, a practitioner is empowered to explore novel perceptions, appraisals, and responses. Practitioners are encouraged to use this freedom to explore novel perceptions and responses that align with deeply held personal values.

Given growing research establishing the efficacy of MT for addressing a range of clinical syndromes, a second wave of research has begun to identify potential mechanisms underlying MT's therapeutic effects (Alsubaie et al., 2017). Mechanistic research is critical for refining treatments to better target maladaptive processes aggravating psychological functioning (Holmes et al., 2018). To this end, neuroimaging has become an invaluable tool for the detection of putative mechanisms of MT, empowering researchers to go beyond self-report to examine both transient and lasting effects (Roffman et al., 2005). Yet despite a wealth of meta-analytic studies describing MT efficacy, we have only begun to understand MT mechanisms through a scientific lens. To address the gap between emerging mechanistic research and its integration into theory, this chapter will employ a narrative review of the neurobiological correlates of MT and their clinical significance. We will begin by briefly describing three popular paradigms in MT research, exploring mindfulness as (i) a short-term state, (ii) the product of more comprehensive clinical interventions, and (iii) a disposition or "trait". To guide discussion toward practical interpretation of function, this review will focus on large-scale, intrinsically connected brain networks rather than specific brain regions. We will review empirical studies exploring three levels of analysis: mindfulness as a transient state, the impact of MT interventions, and correlations with "trait" or dispositional mindfulness. Finally, we will discuss the implications of these findings for the study and application of contemplative practices within the clinical domain.

Conceptualizations of Mindfulness

The term "mindfulness" is multifaceted, with at least three distinct connotations and accompanying research paradigms. First, mindfulness can be thought of as a transient mental state, in which a person is induced to intentionally engage in nonjudgmental attention to the present moment (Kabat-Zinn, 1990). Experimentally, transient mindfulness can be studied as a *state induction*, examining the temporally local effects of mindfulness prompts on cognition and mood. Such research can occur within an individual, but also to evaluate whether the nature of the state changes with expertise. Second, mindfulness can be thought of as a trajectory of personal development in the capacities such as awareness or equanimity, growth that is engendered via MT, often through clinically manualized interventions (Baer, 2003). As such, MT can be studied as a form of *contemplative training*, producing lasting changes to attentional networks and well-being. This sort of research is most in line with clinical trials, where an intervention is compared against a control group, but cross-sectional research that uses training history as a covariate is also possible. Finally, a combination of natural predispositions and training could yield chronic individual differences in the tendency to be mindful (Tomlinson et al., 2018). As such, MT can be studied as a *disposition or trait*. This is most commonly the domain of cross-sectional research, correlating measures of mindfulness with other variables of interest. We will describe each of these conceptualizations to better establish the context for neuroscientific models of MT.

Mindfulness as a State

Perhaps the simplest method for studying mindfulness is through state inductions: brief tasks or meditation exercises designed to induce a state of mindfulness in participants. This is accomplished by having participants practice a specific meditation or other contemplative practice such as breath meditation and the body scan, which would normally comprise a mindfulness training or intervention program. The advantage of such an approach is the ability to explore the momentary "building blocks" of mindfulness training – to understand the immediate impact of a brief practice and to determine which of the many possible effects are most consistent and replicable. Inductions can also be used between different cohorts with varying expertise in mindfulness techniques, to examine the interplay between brief practice and more dispositional factors that will be discussed below.

While the effects of brief practice may be transient, the potential for within-study replication of effects provides an opportunity for great scientific rigor. In addition, studies that employ induction techniques may be closest to directly probing the capacity to engage in sustained mindful states, assessing a person's ability to adopt mindful awareness of the present moment given the situation at hand. Moreover, unlike more intensive manualized trainings, mindfulness induction paradigms can be adjusted in dosage (e.g., 10-min meditation vs. 40-min meditation), focus (e.g., sensations, thoughts), attitudinal stances (e.g., cognitive distancing, self-compassion), and mode of delivery (e.g., in-person vs. online).

A substantial number of induction studies now populate the research literature (Leyland et al., 2019). While the exact implementation of inductions varies substantially across studies, most studies link brief MT to improved mood, often introducing mental calmness and emotional stability (Keng et al., 2011), as well as improvements in cognitive processes such as attentional control (Gallant, 2016). Moreover, inductions also appear to improve central psychological mechanisms of MT including decentering, the ability to watch one's experience with a sense of objectivity and psychological distance (Feldman et al., 2010; Lebois et al., 2015), and experiential acceptance, the ability to tolerate and explore intense or stressful experiences (Keng et al., 2011).

However, there are some limitations to mindfulness inductions. Specifically, the preponderance of studies using mindfulness inductions rely on post-meditation self-reports to capture features of the participant's mindful state. As such, findings could be influenced by known drawbacks of self-report measures, such as social desirability bias and biases related to demand characteristics. A further limitation is that some skills may require real-life stressors, and time for reflection and repeated exposure to such stressors, before MT produces reliable alterations to emotion regulation or well-being (Garland et al., 2015). Thus, while induction techniques are attractive due to the low investment of time and resources, they may be limited in revealing the most meaningful changes associated with MT. Nevertheless, induction paradigms figure significantly in the neurobiological research literature due to their compatibility with standard experimental designs often employed in cognitive neuroscientific exploration.

Mindfulness as Contemplative Training

Ideally, inductions serve as experimental "micro-interventions" that could in theory aggregate over time to produce more impactful, lasting changes. Through practice in sustaining states of mindful awareness, interventions promote mindfulness as a regulatory skill that can be marshalled in response to life's challenges (Kabat-Zinn, 2003). In mindfulness-based interventions (MBIs), the intention is for participants to move beyond experimenting with state inductions in formal meditation to employ the mindful state in response to a stressor. The project is therefore one that extends beyond immediate symptom resolution, encouraging participants to continue their contemplative practice as a lifelong skill; indeed continued practice following intervention completion is associated with the maintenance of therapeutic gains (Crane et al., 2014). Mindfulness training is now customized for a variety of ailments through a growing cadre of manualized clinical interventions, perhaps most famously through mindfulness-based stress reduction (MBSR) (Kabat-Zinn, 1990), the first and most general form of manualized training, and mindfulness-based cognitive therapy (MBCT) (Segal et al., 2012), which was developed to address relapse and recurrence in MDD. However, MT has also been integrated as a core element of emerging cognitive-behavioral therapies such as acceptance and commitment therapy (ACT) (Hayes et al., 1999) and in dialectical behavior therapy (DBT) (Linehan et al., 1999), whose mindfulness application actually inform and precede MBCT.

Both MBSR and MBCT are group interventions that provide participants a structured 8-week protocol for practicing various mindfulness meditations and learning about their underlying principles to strengthen emotion regulation capacities and improve quality of life. These interventions are designed to stimulate metacognitive awareness and experiential acceptance of internal experiences, as it is postulated that the willingness to initiate and remain in contact with uncomfortable internal states will allow practitioners to disengage from prepotent responses (e.g., worrying, rumination) and instead deliberately engage in more adaptive regulatory strategies (e.g., problem solving, behavioral activation) in the face of stressful life events. Although MBSR and MBCT were originally tailored to target chronic pain and the prevention of depressive relapse, respectively, the growing interest in MBIs and the mounting evidence supporting their clinical efficacy have encouraged others to adapt these treatment protocols for other stress-related medical and clinical disorders.

Currently, the empirical evidence for MBSR has supported its clinical efficacy for the general reduction of stress and negative mood states (e.g., depressed mood, anxiety) and maladaptive regulatory strategies (e.g., rumination, worry) in healthy individuals (Chiesa & Serretti, 2009); patients struggling with physical conditions such as chronic pain, breast cancer, multiple sclerosis, and fibromyalgia (Fjorback et al., 2011); and primary caregivers (Lengacher et al., 2012). In contrast, MBCT has been evaluated for its prophylactic outcomes for major depressive disorder, with randomized controlled trials demonstrating that its clinical effects extend beyond treatment as usual (Ma & Teasdale, 2004; Teasdale et al., 2000), providing protection equivalent to continued antidepressant treatment (Kuyken et al., 2016). With modifications, MBIs also show for addressing even acute phases of MDD, comparable to the effects of more conventional interventions such as cognitive therapy (Bockting et al., 2015).

Mechanistic research into MBIs has revealed that intervention effects operate on both the psychological and neurobiological aspects of stress in tandem, working in partnership to reduce stress reactivity associated with affective symptom burden

(Brown et al., 2007; Chiesa & Serretti, 2009; Creswell et al., 2014). However, there are a number of limitations to the randomized controlled trials of MBSR and MBCT, as they have relied predominantly on research designs built to examine pre-post intervention effects within group or relative to waitlist controls, with relatively few studies that employ active controls (Goyal et al., 2014). As such, there is a need for more rigorous designs such as dismantling studies to better specify the MT's central mechanisms of action.

While reviewing the emergence of recent dismantling studies on psychological constructs such as decentering, acceptance, and equanimity is beyond the scope of the current chapter, neurobiological accounts provide an alternative source of data for understanding intervention mechanisms. While applying neurobiological paradigms in brief induction research helps reveal transient changes in biological processes such as brain metabolism, same paradigms hold much greater promise when applied to intervention designs. Specifically, intervention designs allow comparisons between patients and healthy controls to track the biological covariates of symptom burden and its resolution (e.g., Farb et al., 2010). When participants are instead evaluated in terms of future disorder vulnerability (e.g., Farb et al., 2011), we can learn about biomarkers of vulnerability that may not be obviously expressed as symptoms. We can therefore use contemplative training designs to learn more about how MT impacts well-being, either by normalizing brain and physiological function or by introducing novel regulatory capacities that offset enduring markers of disease vulnerability.

Mindfulness as a Disposition or Trait

Dispositional or trait mindfulness is a construct intended to capture individual differences in the tendency to engage in mindful states, representing both naturally occurring variation across the population and also the internalized "product" of efficacious MT. Measured primarily through self-report instruments, the literature contains numerous proposed structures and subprocesses comprising mindful awareness (Chiesa, 2013). Self-report measures of state mindfulness are the easiest and most popular approaches for capturing stable characteristics of mindfulness. There is disagreement regarding exactly how many components are best for defining dispositional, self-reported mindfulness, with complexity growing over time. Indeed, the most commonly used dispositional mindfulness scale 15 years ago was likely the single-factor Mindful Attention Awareness Scale (Brown & Ryan, 2003), which has now been supplanted by the current "gold standard" Five Facet Mindfulness Scale (FFMQ) (Baer et al., 2008), which incorporates much of the content of the MAAS into a more complete canvassing of mindful qualities. The FFMQ features five interrelated subfactors that contribute to a global dispositional mindfulness score: *observing* and *describing* shifting internal states, *acting with awareness* in the present moment, and approaching internal experiences with a *nonjudgmental* and *nonreactive* stance. While debate is ongoing as to the optimal number of factors by which to conduct self-assessment (Gu et al., 2016; Rudkin et al., 2018), it seems clear that dispositional mindfulness is an easily reportable and relatively stable measure of individuals' differences.

Despite their face validity, self-reports are susceptible to inaccuracies given the common methodological pitfalls of surveys and biases in retrospective recall. In addition, the subcomponents of mindfulness states measured in such questionnaires are also driven by the expectations of the questionnaire author that may miss out on canvassing central but unexpected consequences of MT. Finally, self-report may be limited in its access to subtle cognitive or physiological changes. As such, self-report questionnaires only provide limited utility in furthering our current understanding of mindfulness states, their underlying structure, and how they impact other cognitive states (e.g., awareness, attention, attitudes).

However, when integrated with neuroimaging techniques to form a multimethod approach, studying mindfulness using self-report allows researchers to investigate changes in neural processes and structures as a function of the time and effort invested in mindfulness practice. Although there is debate as to whether these questionnaires capture true changes in mindfulness levels or are driven by other nonspecific effects (Chiesa, 2013; Grossman, 2011), evidence does suggest that mindfulness measured in this fashion can differentiate between those who are practicing mindfulness and those who are not engaged in any training (Baer et al., 2008; Hölzel, Lazar, et al., 2011b). Neurobiological accounts of dispositional mindfulness therefore provide tantalizing clues around how to relate theories of stress resilience to information processing systems in the brain.

Psychological Mechanisms of Mindfulness

Currently, most research on the mechanisms of mindfulness has emphasized proposed mediators derived from the underlying theories of MBIs such as MBSR and MBCT. For this reason, investigations have been limited to psychological constructs such as decentering, ruminative thinking, and metacognition (Gu et al., 2015), which are mostly bolstered by self-report measures. Beyond these psychological constructs, other theoretical models have distinguished themselves by focusing primarily on the cognitive and affective elements associated with mindfulness meditation (Baer, 2003; Brown et al., 2007; (Baer et al., 2006; Brown et al., 2007; Hölzel, Lazar, et al., 2011b; Shapiro et al., 2006; Vago & Silbersweig, 2012). These models overlap with one another regarding the proposed subcomponents driving the effects of mindfulness and related interventions, positing self-regulation, attentional reorienting, and experiential exposure as prospective drivers of mindful awareness. However, these theoretical models have yet to explicitly establish core mediators linking mindfulness and stress reactivity and stress-related disorders.

To date, most of the research investigating the connection between mindfulness meditation and stress reactivity has been conducted with self-report measures representing the potential mechanisms of action. In addition, these studies seldom employ a strong a priori theoretical model to inform research design and the interpretation of findings, which impedes the process of building, refining, and contrasting theoretical accounts regarding intermediary factors. Though the influence of mindfulness meditation on stress outcomes has been well documented, the overreliance on self-report measures and the lack of a guiding theoretical framework have resulted in current models that are without a firm empirical foundation and only capture the subjective experience of mindfulness. Moreover, although the identification of potential psychological mediators has been invaluable for understanding the widespread effects of mindfulness meditation, these findings rely predominantly on self-report measures that presume self-awareness of such psychological states.

The complexities of mindfulness and its impact on several distinct aspects of human functioning (e.g., cognitive, biological) underscore the need for multimethod research approaches and more comprehensive theoretical accounts. For instance, assessing both phenomenological and neurobiological dimensions of stress-related conditions has potential to expand our understanding of how mindfulness operates at deeper levels of functioning to alter stress-related outcomes. Moreover, such an approach is crucial for marshalling empirical evidence at deeper levels of analysis that support the construct validity of a given psychological mediator.

Unfortunately, the mediating pathways underlying the relationship between mindfulness meditation and stress reactivity have not been fully explored. These limitations are also evident even in the popular target of relating MT to improved stress management. Although there are many studies reporting on the efficacy of MT for a variety of stress-related disorders, few studies have investigated and firmly established mechanisms of action, with even fewer adjudicating between competing mechanistic accounts in search of those that provide the greatest explanatory value (Gu et al., 2015). As the number of proposed mechanisms increases, such initiatives will become pivotal for establishing factors through which mindfulness meditation and related interventions function, especially for improving the design and delivery of such interventions. For instance, the identification of the major mediating players would allow for the augmentation of therapeutic mechanisms likely to produce the greatest amount of symptom alleviation while also differentiating between the unique contributions of MBIs and those that are attributable to common treatment features (Kazdin, 2007).

Given the limitations of relying on self-report measures and focusing primarily on the phenomenological experience of mindfulness, neuroscience methods have burgeoned in interest and appeal due to their methodological features that overcome such pitfalls. Neuroimaging and other biological measures are capable of noninvasively isolating and monitoring psychological processes that characterize pathological conditions and the influence of mindfulness at a deeper level of analysis, without requiring participant self-report and insight into their subjective experience. This is especially helpful when studying treatment effects in psychiatric disorders that are associated with lower levels of insight and psychological mindedness than that which is observed in the general population (e.g., schizophrenia). Ultimately, applying neuroscientific methods measuring the biological underpinnings of stress reactivity and related conditions following a stressor allows researchers to formulate more realistic models of vulnerability and evaluate predicted therapeutic processes underlying mindfulness meditation. When integrated into a single methodological approach, such studies can effectively identify the basic mechanisms of action and how they interact to produce the phenomenological experience of mindful awareness. Recent investigations of the neural correlates of mindfulness meditation and MBIs have greatly expanded our current understanding of the major brain networks operating during a mindful state. These investigations have largely focused on examining the effects of mindfulness meditation using two approaches of neurological analysis, *functional* and *structural analysis*, which together have the potential to produce a more comprehensive neural account of mindfulness. *Functional analyses* are designed to detect differences in brain activation between experimental conditions (e.g., mindfulness meditation vs. control) or time points (e.g., pre- vs. post-treatment), as well as correlations of activation patterns between brain regions. For instance, in *regional functional analysis*, participants perform an experimental task (e.g., focusing on the breath) during brain scanning to detect task-specific neural activations that underlie a specified mental state. In addition, given its capacity to reveal activation patterns unique to an experimental condition (e.g., reward task, emotion provocation), researchers have also implemented functional analyses to measure the impact of contemplative training and therapeutic interventions on cognitive and affective processes (e.g., reward processing, emotion regulation). Similarly, *resting-state functional analysis* is also applied to uncover activation patterns across brain regions or networks while the participant is at rest and unengaged in an explicit task. During the imposed resting states, participants are often automatically engaged in internally focused processes such as memory retrieval and perspective taking, allowing researchers to further measure the generalizability of mindfulness meditation and MBIs to self-referential processes. In contrast, *structural analyses* detect changes in the anatomy of neural structures (e.g., amygdala) or pathways (e.g., corpus callosum) across time points. Within the context of mindfulness meditation, structural analyses are generally applied to ascertain the depth of the impact of contemplative training and related interventions on brain structures implicated in well-being (e.g., amygdala) or putative cognitive processes related to mindfulness (e.g., attentional neural hubs).

Proposed Neural Networks

Early on, neuroscientists traditionally studied the neural correlates of cognitive and affective states via the detection of activation rates and structural changes in isolated brain regions. The findings from these analyses catalyzed the production of modular accounts of neural functioning wherein the interconnectivity between brain regions was surmised, expanding the rudimentary understanding of various neural structures. However, over the past decade, research objectives have gradually transitioned to modeling circuits within and between brain networks rather than relying predominantly on theoretical accounts, as more methodological advancements were made available to empirically validate such claims. This shift in methodological approach was accompanied by a growing acknowledgment for the synergistic interactions among neural networks for producing complex cognitions and behaviors.

In the case of mindfulness meditation, several distinct advantages were obtained from network analyses of brain function. For instance, whereas the modular accounts of mindfulness meditation provide a snapshot of the neural structures involved, network accounts are better suited for modeling the sequential processing of information as it travels and changes from one network to the next. In addition, researchers can model bidirectional connectivity among brain networks and the mechanisms that determine the flow of processing across such networks given their appropriateness for the task at hand. Altogether, network accounts move beyond modular accounts by capturing the dynamic interactions of neural networks in the brain.

In the subsequent section, we will provide a brief introduction of the major neural networks due to their relevance to the current empirical literature and findings on mindfulness meditation and related interventions. We will describe three crucial networks: the default mode network (DMN), the salience network (SLN), and the central executive network (CEN).

Default Mode Network The default mode network (DMN) was first discovered in a neuroimaging study by Raichle et al. (2001), wherein participants underwent brain scanning during task-positive periods – periods wherein the participant is engaged in goal-directed behaviors – and task-negative periods, also known as a rest period. After examining brain activity during task-negative periods, it was revealed that some neural structures were activated during these rest periods in comparison to task-positive periods (Raichle et al., 2001). These findings led Raichle and colleagues to conclude that the synchronous activations during task-negative periods are indicative of a non-goal-oriented intrinsic brain network comprised of functionally interconnected set of regions.

Since then, our understanding of the DMN has grown exponentially. Subsequent studies have extended the aforementioned findings and led to newer theoretical accounts that have refined original conceptualizations and linked the DMN to the cognitive processing of internal stimuli, including information about the self, mind-wandering, and cognitive elaborations in general (Buckner & DiNicola, 2019). In addition, researchers have identified core brain regions implicated in the DMN, three of which include the posterior cingulate cortex (PCC) and the precuneus, the medial prefrontal cortex (MPFC), and the angular gyrus. Individually, these brain regions have been implicated in numerous cognitive processes that contribute to the overall function of the DMN. For instance, the PCC is implicated in integrating incoming information with memory and perceptions to allow for such cognitive processes as memory recall, future-oriented thinking, and *mentalization*, which refers to understanding the mental states of others (Leech & Sharp, 2014). The precuneus and angular gyrus are also implicated in memory recall in addition to visuo-spatial imagery and navigation and attentional deployment (Cavanna & Trimble, 2006), whereas the MPFC represents self-referential schemas and supports decision-making affecting oneself in the future or in situations concerning social partners (Lieberman et al., 2019). In combination with other neural regions such as the temporoparietal junction (TPJ) and temporal poles, the DMN is capable of producing a continuous sense of self, as well as the capacity to understand and predict others (Qin & Northoff, 2011).

Within the context of mindful awareness, researchers have also found associations between DMN activations and mind-wandering, which typically occurs during task-negative periods, involves thinking about the self or others, and is generally categorized as a non-mindful state (Fox et al., 2015). Given that mindfulness meditation has also been shown to influence self-reported mind-wandering and self-focused cognitions (Mrazek et al., 2012; Rahl et al., 2017), it has been suggested that contemplative practices fostering mindful awareness could be targeting these cognitive processes by tuning DMN connectivity (Doll et al., 2015; Taylor et al., 2013). Moreover, it is widely believed that mindfulness meditation also has implications for the treatment of stress-related conditions that are purportedly driven by dysfunctions in the DMN (Marchetti et al., 2016). According to review (Buckner et al., 2008), DMN dysfunction is related to numerous clinical disorders, including major depressive disorder (MDD), posttraumatic stress disorder (PTSD), and schizophrenia. For instance, individuals diagnosed with MDD show greater functional connectivity among neural hubs in the DMN (Mulders et al., 2015), and the stronger interconnectivity within the DMN is also associated with ruminative thinking (Zhou et al., 2020), which is linked to the onset and maintenance of depressive symptoms. Altogether, if mindfulness meditation is capable of shifting the neural activation of the DMN, it could potentially correct maladaptive activation patterns in the DMN of individuals susceptible to psychiatric disorders, in turn predicting less self-referential thinking and a more consistent sense of self.

Central Executive Network Whereas the DMN is associated with task-negative periods, the central executive network (CEN) is most activated during task-positive periods (Fox & Raichle, 2007). The CEN represents the neural underpinnings' executive functions including attentional control, working memory, and planning (Collette & Van der Linden, 2002; Salmon et al., 1996). These hubs include the dorsolateral prefrontal cortex (DLPFC), the dorsal medial prefrontal cortex (DMPFC), the inferior parietal lobule (IPL), and the orbitofrontal cortex (OFC). Each of these brain regions are lynchpins of the network, as they integrate incoming information from other brain structures to perform functions such as planning and problem solving, impulse control, self-monitoring, and social behaviors (Baumgartner et al., 2011). As such, the CEN is seen as critical for cognitive forms of emotion regulation (Goldin et al., 2008). Generally, the CEN is activated when planning a response to novel situations that cannot be managed with learned, habitual responses associated with the DMN (Ballard et al., 2011). This is accomplished primarily by overriding DMN-driven prepotent responses that would typically be prompted by external cues and instead engaging in novel goal-directed actions and intentions to optimally handle the situation at hand (Koshino, 2017).

In addition to general executive functions, the CEN has also been implicated in mindfulness meditation and related practices, especially since mindful awareness is necessarily about orienting and sustaining attention toward internal cues, inhibiting prepotent responses that might follow from such cues, and adopting actions that best meet the individual's goals in that moment (Farb et al., 2012). The CEN is most associated with controlled behavioral process, which puts it into direct opposition to the DMN, and indeed a CEN hub region appears to inhibit DMN activity when the CEN is activated (Chen et al., 2013). However, while the DMN is generally inhibited as the CEN works to overcome habitual responses, the two networks may become positively connected when the CEN is applied to monitoring internal processes and prepotent urges (Christoff et al., 2009).

Given the apparent association between the CEN and mindful awareness, mindfulness meditation could also have implications for the treatment of stress-related conditions associated with CEN dysfunction (Alfonso et al., 2011; Crowe & McKay, 2016). For instance, the CEN is found to have fewer connections in children diagnosed with PTSD, while the DMN is hyper-connected, possibly reflecting their inability to disentangle themselves from ruminative thinking processes (Suo et al., 2015). Similarly, individuals with anxiety disorders tend to exhibit hypo-connectivity within the CEN but increased connectivity between the CEN and orbitofrontal cortex, potentially reflecting the influence of worrying on their behavioral actions for coping with environmental stressors (Geiger et al., 2016). Improving attentional control through MT could therefore be expressed as a reconfiguration of CEN connectivity, increasing local connections with the CEN while re-tuning its connections to the DMN other brain areas.

Salience Network To adjudicate between effortful processing and reliance on habit, the salience network (SLN) is purported to detect events that prompt switching between the DMN and CEN, or in other words, from task-negative to task-positive functioning (Menon & Uddin, 2010). Ultimately, the SLN represents the motivational relevance of internal and external cues by filtering incoming information and sending forward the most pertinent for goal-directed behavior (Craig, 2009). The primary hubs of the SLN are the anterior insula (AI) and the dorsal anterior cingulate cortex (DACC), both of which are linked via a specialized tract of Von Economo neurons (Allman et al., 2011). The SLN is associated with the detection and integration of emotional and sensory stimuli, as well as in modulating the switch between the internally directed cognition of the default mode network and the externally directed cognition of the central executive network (Seeley et al., 2007). More specifically, the AI has been primarily implicated in sensory monitoring, whereas the DACC has been implicated in motor monitoring (Medford & Critchley, 2010).

In addition to these core hubs, the SLN also includes brain regions such as the amygdala, putamen, and ventral striatum, which are broadly involved in arousal and emotional responses (Reynolds, 2005). As such, the SLN is the most "emotional" of the three brain networks described here, and it is also associated with visceral feelings rather than "conceptual thought" (Craig, 2002). Given its role in determining the motivational relevance of situational cues, the SLN is heavily recruited by stressors (Hermans et al., 2014), serving to allocate necessary resources to meet the current needs of the situation at hand, by tapping into the DMN and CEN forwarding information for processing as deemed necessary (Sridharan et al., 2008). As such, the SLN may support transitions out of self-referential thinking to support feelings of agency by integrating momentary sensation and available motor responses.

As for its relation to mindfulness meditation, the SLN might be implicated in mindfulness given that mindfulness meditation has been found to increase noticing of the somatic and visceral sensations, and catching oneself when the mind wanders (Price & Hooven, 2018). As such, it could very well be that mindfulness results in lower DMN activation through the SLN switching between the DMN and CEN (Sridharan et al., 2008). By operating on these networks, MT and related interventions could also have implications for the treatment of stress-related conditions driven by SLN dysfunction (Farb et al., 2012). For instance, the AI node of the salience network has been observed to be hyperactive in anxiety disorders, which is thought to reflect predictions of aversive bodily states leading to worrisome thoughts and anxious behaviors (Pannekoek et al., 2013). Increasing awareness of worrisome thoughts may allow the CEN to offload the activity of the SLN, reducing anxious feelings in the face of greater perceived agency and intentionality of responding.

Narrative Review of Neuroscientific Findings

This section will review neuroimaging findings around MT, focusing on regional activations, resting-state activations, and structural changes. The summary of findings will canvas the three levels of analysis discussed above: transient states, training effects, and trait-like or dispositional individual differences. However, before proceeding, it should be noted that the described findings are to be interpreted with caution. A common methodological drawback among most brain imaging studies is the small number of participants recruited and assigned to each experimental condition, and the neural investigation of mindfulness is no different. Although there are numerous practical explanations for the limited sample sizes, we nonetheless recommend that all findings and interpretations be treated as preliminary rather than conclusive.

Mindfulness as a State

Emotion Processing Neuroscientific investigations of mindfulness customarily seek to address two questions: what neural networks support the mindful state, and are mindful brain states impacted by MT? To address these questions, researchers commonly recruit both novice and experienced meditators and have them engage in mindfulness to discern the impact of mindfulness practice on brain function. By comparing MT to a non-meditative control condition, researchers can detect more immediate neural activations associated with the mindful state, whereas the comparison of novice and experienced meditators would speak to the role of expertise in entering this state. To date, studies in this vein have identified significant widespread changes in neural hubs that together constitute large-scale brain networks such as the DMN and SLN.

For example, one study instructed novice and experienced meditators to view positive and negative images designed to elicit emotional reactivity while either mindfully observing the presented stimuli or viewing them passively without any attentional modifications (Taylor et al., 2011). Compared to the passive-viewing condition, mindful viewing of both positive and negative images attenuated the subjective emotional intensity reported by novice *and* experienced meditators. However, although MT reduced emotional reactivity equally across both experienced and novice meditators, the neural structures underlying increased emotional stability varied as a function of meditation experience. Compared to experienced meditators, novice meditators lower activation of the amygdala, an SLN hub region associated with emotional processing and reactivity. In contrast, compared to novice meditators, experienced meditators exhibited reduced activation in neural hubs comprising the DMN, including the MPFC and the posterior cingulate cortex (PCC). In this way, the mindful state in beginners may differ from experts: while the average person may experience mindfulness as being less emotionally reactive to events, more experienced meditators may experience the state as one more broadly free from habitual ways of responding to events rather than blunting emotional salience. A second recent study replicated the finding that task-related deactivations of DMN cortical midline structures were more pronounced in experienced meditators when compared to novice meditators (Lutz et al., 2016).

A third study corroborates the finding of DMN deactivation during mindful states, even in novice meditators with only brief exposure to MT (Doll et al., 2016). Researchers recruited novice meditators for a 2-week attention-to-breath mindfulness training. During brain scanning, participants were instructed to attend to aversive images either by i) mindfully attending to the breath or ii) passively viewing the images. Following 2 weeks of MT, the novice meditators increased deactivation of the amygdala during mindful viewing versus passive viewing of aversive stimuli, supported by greater connectivity between the CEN and SLN. Furthermore, these novice meditators increased deactivation of the dorsal MPFC, a DMN hub, during mindful viewing regardless of whether the aversive cue was present or not, potentially signaling a greater capacity for disengaging from cognitive elaborations in general. A fourth study by yet another independent group supports characterizing the mindful state as low DMN activation, but high CEN and SLN recruitment (Tomasino & Fabbro, 2016). After an 8-week mindfulness training program, novice meditators showed increased activation in CEN and SLN regions involved in attention regulation (i.e., dorsolateral PFC, caudate/anterior insula) and decreased activation of DMN hubs (e.g., rostral PFC) at post-training.

Thus, after even short-term MT, practitioners begin disengaging from cognitive elaborations by downregulating DMN processing and relying more on modulating attention. However, expertise seems to dictate to what extent DMN deactivation is also coupled with SLN deactivation, blunting the affective salience of events. Specifically, high-expertise meditators seem able to deactivate the DMN without also blunting activation of the SLN, including the amygdala, which acts as a detector of emotionally salient events (Cunningham et al., 2008). In addition, repeated finding of increased amygdala-PFC connectivity may suggest that, with increased trait mindfulness through continued training, meditators are better able to regulate their emotions by integrating CEN-driven cognitive control with SLN-driven momentary emotional experience.

The inclusion of both novice and expert samples adds further nuance to the neural depiction of the mindful state. The reviewed studies suggest that early MT may reduce the emotional impact of events by exerting control over affective processing, which may also indirectly reduce habitual patterns of emotional reactivity to evocative cues, as indexed by attenuated DMN recruitment. However, with more meditation experience, emotional stability might then be sustained by primarily acting upon these habitual ways of responding via a focus on reducing DMN recruitment rather than SLN inhibition, disengaging from cognitive elaboration (e.g., self-referential thinking) that heightens emotional reactivity without needing to control affective processing directly. In effect, with more experience, MT practitioners may be able to experience greater emotional clarity without responding, whereas early practitioners may need to inhibit emotional responses themselves to get a handle on reactive habits.

Pain Although neuroscientific research on MT has largely focused on experimental conditions incorporating emotionally evocative cues, a growing number of studies suggest that the benefits of mindfulness could also extend to pain perception and regulation (Hilton et al., 2017). The neural mechanisms underlying these benefits are now becoming clearer: reductions in pain intensity are often associated with increased activation of SLN hubs (i.e., ACC, anterior insula), reflecting changes in saliency processing and attentional monitoring of painful somatosensory cues. Across a variety of experimental paradigms, similar results were also found for pain processing and anxiety in naïve and experienced meditators, with increases in SLN regions relating to reductions in pain anticipation and unpleasantness and improved anxiety relief (Gard, 2014; Lutz et al., 2013; Zeidan & Vago, 2016). Intriguingly, the shift from DMN- to SLN-oriented processing during mindfulness of painful stimulation *elevates* activity of sensory regions involved in processing pain. For instance, reductions in pain unpleasantness were associated with increased activation of the orbitofrontal cortex (OFC), a region involved in the contextual evaluation of aversive cues, and decreased activation of the thalamus, the sensorimotor switchboard of the brain (Zeidan et al., 2011). As such, although both experienced and naïve meditators demonstrate higher activity in the somatosensory cortex while mindfully attending to emotions and body sensations (e.g., Lutz et al., 2016), experienced meditators shift more of their cognitive resources to interoceptive awareness and basic sensory functioning while downregulating executive processes and cognitive elaboration in self-regulation.

Reward Beyond impacting the processing of evocative emotional or painful stimuli, MT also appears to produce meaningful effects on the reward processing of positive cues. Although behavioral differences were not detected between participant groups, experienced meditators showed greater activation of reward-related hubs (e.g., caudate, putamen) and sensory hubs (e.g., posterior insula) and an attenuated connection between the caudate and anterior insula compared to non-meditators during reward anticipation (Kirk et al., 2015; Kirk & Montague, 2015). Compared to novice meditators, experienced meditators showed greater deactivation of cortical hubs of the DMN (e.g., MPFC) during the receipt of reward. The regional activations underlying formal mindfulness practices might therefore produce downstream neural changes to non-meditative states and mindful states relying on different sensory anchors (e.g., breath, body sensation), which supports the notion that mindfulness is generalizable and not strictly task specific. Moreover, experienced meditators may downregulate the saliency and valence of aversive stimuli *and* rewarding events, becoming less reactive to both intrinsic and extrinsic cues as they instead attend to incoming sensory information in a nonreactive manner.

Structural Changes   Numerous studies have investigated the structural brain changes associated with various meditation traditions, but to the authors, it is implausible that these structural changes could occur quickly enough to characterize a transient mindful states opposed to more intensive changes over longer periods of time. To our knowledge, the shortest training interval required to detect changes in brain morphology (as opposed to activity) is only an hour in the context of learning a complex balancing task (Taubert et al., 2016). However, no mindfulness have employed a comparable design to our knowledge. Structural changes will therefore be covered in more detail in the following sections on longitudinal training and mindfulness as an enduring disposition or trait.

Mindfulness as Contemplative Training

Enhanced Sensory Processing The neuroscientific evidence for MBIs has grown rapidly over the past decade, with most studies investigating the neural correlates of mindful states and other cognitive processes ameliorated by such treatments. As these interventions were designed to promote psychological well-being, such studies have aimed to uncover the neural changes underlying MBIs and the active treatment mechanisms that lead to improved symptom burden and quality of life in both healthy and psychiatric populations. To date, studies have connected MBIs to a range of cognitive processes for a range of psychiatric conditions, most of which emphasize shifting away from self-referential processing and toward heightened body awareness. For instance, our group compared MBSR participants and untrained novices on a brain-imaging task, wherein they were instructed to focus on their moment-to-moment experience (experiential focus) or on self-referential thinking (narrative focus) while reading evocative trait descriptive words (Farb et al., 2007).

Compared to untrained novices, MBSR participants showed a more pronounced reduction in activation of the cortical midline structures of the DMN (e.g., MPFC) while adopting an experiential focus. Furthermore, MT completers showed additional activation of neural structures involved in somatic and interoceptive awareness (e.g., insula, secondary somatosensory cortex, IPL), in accordance with characterization of the mindful state as one of increased sensory representation. Finally, MT completers showed greater connectivity between sensory regions and the CEN, and reduced connectivity with more ventral PFC hubs of the DMN, in keeping with less habitual and evaluative attention toward sensation. As MBSR participants become better able to disengage from self-referential thinking supported by the DMN, they were able to increase direct representation of momentary body sensations, shifting the criteria for self-reference.

Our program of research joins other studies in showing that MT has powerful effects on not only the upregulation of sensory representation but also altered communication between sensory regions and the prefrontal cortex, moving from ventral hubs that are part of the DMN to more dorsal hubs that are typical of CEN activity. For example, we recently showed that, relative to visual attention, attention to the breath resulted in greater activation of the body representation regions in the posterior insula and somatosensory cortex (Farb et al., 2013a). When MBSR completers were compared to a waitlisted control group using this paradigm, MT was associated with greater local connectivity within the posterior insula and middle insula, and increased connectivity between the posterior insula and the dorsal PFC associated with CEN processing (Farb et al., 2013b). Despite increased posterior insula connectivity, the anterior insula, the hub of the SLN, was predominantly deactivated during internal attention relative to external attention. Thus, greater SLN activation is not a general end-point of MT, even if greater SLN activity connectivity is often shown during mindful states.

This finding is corroborated by an independent study (Ives-Deliperi et al., 2011), which reported on training-related changes in brain activity following MBSR-based MT while participants meditated in the brain scanner. Although MBSR resulted in deactivated midline cortical structures belonging to the DMN, MBSR was also associated with greater deactivations of SLN hubs (e.g., anterior insula, ventral ACC). The findings suggest that MBIs do not unilaterally enhance emotional salience, but rather such activation may be context dependent – when exposed to external stimulation, greater SLN activation may represent a sense of engagement with momentary experience and be enhanced by MT. Conversely, when engaged in internal reflection, greater expertise is linked with more pronounced SLN deactivation, as participants aim to approach internal experience with equanimity and nonattachment. From this perspective, MT is more about tuning sensory salience (SLN), cognitive elaboration (CEN), and habitual perception and appraisal (DMN) than it is about promoting dominance of any one brain system over another. Such flexibility is particularly important when one considers that many affective disorders such as major depressive disorder are characterized by abnormally elevated recruitment of all three of these brain networks simultaneously (Kaiser et al., 2015; Sheline et al., 2010), belying the notion that "more activation means better mental health." Instead, more selective recruitment of networks, and better communication particularly with the CEN and SLN and/or sensory cortices, seem to suggest the type of increased clarity and focus qualitatively associated with MT.

Emotion Processing Beyond the neural changes observed in mindful states, MBIs are also expected to cause downstream effects on the neural profiles of other cognitive-affective states such as emotion processing and regulation. A large (N = 158) study of emotion processing compared experienced practitioners, MBSR participants, and participants enrolled in a health enhancement program (HEP) on an emotional face processing task (Kral et al., 2018). No group differences were reported when processing negative faces, though experienced practitioners showed greater deactivation of the amygdala when viewing positive pictures. However, this pattern was not observed in MBSR participants, who instead exhibited increased functional connectivity from pre- to post-treatment between the amygdala and ventromedial prefrontal cortex, SLN, and DMN regions respectively implicated in emotion processing and regulation. These findings suggest that, like novice meditators in the early stages of mindfulness practice, MBIs might be promoting efforts directed at self-regulation rather than the nonjudgmental, present-moment openness that is observed in more experienced practitioners.

Encouragingly, studies have found comparable effects of MT on emotion processing for individuals diagnosed with psychiatric conditions. For instance, individuals diagnosed with social anxiety disorder (SAD) who underwent MBSR were asked to complete a regulation task during brain scanning (Goldin & Gross, 2010). Participants were asked to regulate their emotions in response to presented negative self-beliefs by either attending to the breath or distracting themselves. Compared to distraction, attuning to the breath resulted in deactivations of SLN regions and greater activation of visual attention hubs in the occipital and parietal cortices. This finding parallels studies of advanced practitioners discussed above (Farb et al., 2013b; Kilpatrick et al., 2011), in which greater engagement sensory representation, and less engagement with affective salience, might be a relief for participants in conditions where the impact of emotional information is itself a contributor to symptom burden.

However, increasing sensory processing is not always a boon; we have discussed above how depressive disorders are characterized not by hyperactivity in the SLN, but rather a conflation of prefrontal network activity, to the exclusion of nonevaluative forms of processing (Sheline et al., 2010). Our group examined depressive symptom burden in a community sample, comparing MBSR completers to a waitlist control group (Farb et al., 2010). We observed that it was the suppression of middle and posterior insula activity rather than prefrontal hyperactivity that best predicted symptom burden, as though the prefrontal activation was starving processing of momentary sensation that might interfere with overarching negative schema. Completion of MBSR was associated with a restoration of insula activity, linked to processing of the body's internal state (Craig, 2002), and a commensurate reduction of activation in the posterior cingulate posterior hub of the DMN, associated with conceptual self-reference (Farb et al., 2007; Yang et al., 2019). Indeed, learning to better recruit even visual attention regions may serve as a sign of protection against depression in addition to reduction of prefrontal reactivity (Farb et al., 2011). Thus, in the context of psychopathology, simply increasing affective salience may be insufficient to reduce symptom burden – instead, a return to sensory processing may be needed to disentangle maladaptive neurobiological responses to stress. Even though a shift from conceptual (DMN) to affective salience (SLN) may be sufficient in healthy individuals, the conflation of affective salience with habitual, maladaptive appraisals in disorders like depression may require a more drastic move away from DMN/SLN processing altogether, supported primary though direct sensory access via relatively non"emotional" or "selfish" dorsal CEN structures.

Yet as more clinical intervention studies emerge, it becomes increasingly important to remember that some mechanisms of treatment are likely disorder-specific, and mindfulness' benefits may involve a skillful application of CEN integration into other forms of representation that extends beyond any "one-size-fits-all" pattern of connectivity. Unlike social anxiety disorder, in which treatment response led to reduced SLN activity, and depression, in which pathology treatment was associated with recovered sensory activation in the insula and other sensory cortices, posttraumatic stress disorder (PTSD) research suggests a third distinct mechanism of action. A study of combat veterans diagnosed with PTSD who underwent Mindfulness-Based Exposure Therapy (MBET) or Person-Centered Group Therapy found evidence consistent with this CEN-sensory connectivity account of symptom resolution (King et al., 2016). Both treatment groups showed decreased PTSD symptoms from pre- to post-treatment, and these improvements were associated with increased connectivity between the DMN and both the SLN and CEN. Thus, greater CEN engagement in both momentary salience attribution and cognitive elaborative habits was a sign of increased capacity to engage in the volitional shifting of attention, similar to the literature on MT states in health participants. Finally, a study of participants with bipolar disorder undergoing MBCT found yet another profile of results (Ives-Deliperi et al., 2013), in which training was associated with increased activation in both DMN and CEN hubs, commensurate with a need to better integrate self-knowledge into momentary decision-making and self-regulation.

Pain Extending research on pain processing in mindful states, it seems that even brief training can alter pain processing habits. In a seminal study by Zeidan et al., novice meditators underwent 4 days of MT training and complete a pain stimulation task during brain imaging before and after training (Zeidan et al., 2011). During the task, the participants were asked to attend to their breath while receiving noxious thermal stimuli to the skin. While engaged in mindful breathing during pain stimulation, participants at post-training compared to pre-training reported increased reductions in pain unpleasantness and intensity, as well as a dampening of pain-related activation of the somatosensory cortex, which processes somatic sensations. Moreover, reductions in pain intensity were associated with increased activation of SLN hubs (i.e., ACC, anterior insula), reflecting changes in saliency processing and attentional monitoring of painful somatosensory cues.

Reward Mindfulness may also allow practitioners to improve the feeling of pleasure or meaning associated with relatively mundane or normal events. The goal of many substance use recovery programs, such as Mindfulness-Oriented Recovery Enhancement (MORE), is to restructure reward processing away from the conditioned substance and more toward regular life events, thus rebalancing the award system away from focused drug craving and toward savoring of other potentially meaningful or pleasurable aspects of daily life (Garland et al., 2014). Accordingly, pilot work on participants enrolled in MORE for smoking cessation demonstrates reduced activation to substance-related cues within the ventral striatum, the putative neural hub of reward processing. Concurrently, the same participants demonstrated increased ventral striatum responses to nondrug emotionally evocative cues, supporting the stated intentions of the program (Froeliger et al., 2017). Thus, the same principle of increasing sensory representations may generalize to reward processing to help expand and normalize the brain's selection of rewarding experiences in the face of addiction.

Reviewing examples of research across distinct disorders, it seems as though the implication of greater cognitive control by the CEN seems to be ubiquitous in the MT literature, but the application of such control seems to vary in observing the skillful resolution of different mental health conditions. Additional research may clarify the reliability of these disorder-specific response patterns, as in many conditions research is being driven only by a handful of labs, making it hard to determine how replicable and generalizable these finds truly are – the authors' lab included! Nevertheless, increased CEN connectivity seems to be a more reliable, transdiagnostic marker of intervention-related MT effects on brain networks.

Structural Changes With the use of more comprehensive interventions, the possible changes to brain structure seem more feasible than examinations of transient state changes. Although limited, a few studies have investigated structural changes linked to MBIs from pre- to post-treatment. Participants completing MBSR demonstrated increased gray matter density in the right insula and somatosensory cortex, with greater increases predicting decreased alexithymia symptoms (Santarnecchi et al., 2014). Hölzel et al. (2010) recruited participants who self-reported high levels of stress and underwent MBSR training and showed that compared to pre-treatment, MBSR participants at post-treatment reported lower levels of stress, and the stress reduction was associated with decreased gray matter density in the amygdala (Hölzel et al., 2010), suggesting that short-term mindfulness practice can result in neuroplastic changes to network hubs underlying emotion processing that underlie attenuated stress responses. These effects may play out in white matter tracts, both in connecting the two hemispheres of the brain (Luders et al., 2012), and also in connecting memory circuits in the hippocampus with the anterior insula, the sensory hub of the SLN (Britta K. Hölzel et al., 2016). Subsequent connectivity analyses suggested that this uncinate tract may help support extinction learning of fear-inducing stimuli, improving stress resilience (Sevinc et al., 2019).

Mindfulness as a Disposition or Trait

Resting-State Imaging The discovery of intrinsically connected brain networks has led to a rapid expansion of "resting-state" fMRI, which examines brain connectivity in the absence of task demands. In this way, resting-state studies may be close analogues to psychometric studies of dispositional mindfulness, examining the "default" state of a participant rather than focusing on a transiently induced mindfulness state. Focusing on the connectivity between brain regions in the absence of an explicit task, resting-state activations can be used to ascertain whether MT influences habitual patterns of neural representation and computation. Accordingly, investigations linking MT with resting-state activity seek to determine the extent to which habitual modes of mind can be altered by mindfulness interventions.

Researchers investigating the resting-state effects of MT aim to uncover the degree to which mindfulness can alter connectivity within and between brain networks while the individual is at rest and not focused on a given task. In pursuit of this aim, researchers commonly recruit both novice and experienced meditators and compare their resting-state activations to evaluate these questions, although comparisons of participants before and after training, and in theory even before and after state inductions, could occur.

Compared to the number of neuroscientific investigations into regional activations, fewer studies have examined the impact of MT on resting-state activations. Nevertheless, these studies correspond with previous findings for regional activations, such that the DMN is downregulated while other networks such as the SLN are upregulated. In one pioneering study, experienced meditators reported on the spontaneous noticing of mind-wandering in a "task-free" neuroimaging paradigm (Hasenkamp & Barsalou, 2012). From this single key press, four cognitive processes related to focused-attention meditation, including mind-wandering (which preceded the button press), awareness of mind-wandering (the key press itself), refocusing of attention (immediately after the key press), and sustained attention (the period leading up to the next mind-wandering period).

Ultimately, more experienced meditators exhibited greater trait-like connectivity between the CEN (i.e., dorsolateral PFC) and SLN (i.e., right insula) and decreased functional connectivity between the DMN and SLN while sustaining their attention to the breath (Hasenkamp & Barsalou, 2012). The fact that Hasenkamp and Barsalou (2012) found increased connectivity between the CEN and SLN suggests that experienced meditators could be relying on less habitual processing as they increasingly cultivate present-moment awareness and acceptance of their internal experiences. These findings are supported by an independent study (Berkovich-Ohana et al., 2016), which reports that experienced meditators (relative to novice meditators) showed lower spontaneous fluctuations of neural responses in DMN hubs while exhibiting greater fluctuations in the visual cortex.

Given the putative role of the right anterior insula in switching between brain networks, the resting-state findings suggest that MT engenders a greater capacity for neural and cognitive shifts from the DMN and states of cognitive elaboration to the CEN and states of focused attention. As the neural hubs of the SLN and CEN become functionally connected, meditators become more apt at de-automatization, that is, breaking free from habitual, automatic processing for other cognitive states that might be of more utility in the moment. Finally, local prefrontal effects may further be reinforced by greater functional connectivity with other regions of the brain involved in sensory integration, emotion processing, and other important cognitive-affective functions (Kilpatrick et al., 2011; Tang et al., 2017).

Nevertheless, some studies have suggested that the current understanding of the resting-state effects of mindfulness practice might not be capturing more nuanced neural changes *within* brain networks. One study (Taylor et al., 2013) examined experienced meditators and novice practitioners who underwent a weeklong meditation training, later comparing them on resting-state functional connectivity. Compared to the novice meditators, the experienced meditators exhibited weaker functional connectivity between cortical midline hubs within the DMN, areas associated with conceptual self-elaboration. However, relative to novice meditators, experienced meditators also showed greater connectivity between other clusters of DMN that are associated with perspective taking and theory of mind. A complementary set of findings demonstrated that untrained participants who engaged in 2 weeks of daily mindfulness meditations showed a link between trait mindfulness and reduced functional connectivity between the DMN and SLN, indicating perhaps less habitual reactivity to emotionally salient events (Doll et al., 2015). This finding was replicated in a study of university students who had undergone an MBSR-inspired treatment, in which the meditative state was associated with weaker DMN and SLN connectivity, and better connectivity within the SLN (Yang et al., 2016). Together, these results emphasize that mindfulness mechanisms should not be simplified to a general weakening or strengthening of a neural network; instead, the DMN, which supports cognitive habits, can be tuned to shape habits through training rather than seeking to eliminate reliance on habits. These findings support an account where habits turn less toward reactive self-evaluation and more toward perspective taking. Thus, we must be careful not to demonize any one brain network in understanding MT, but instead ask how that network is being tuned by training to afford more adaptive perceptions and responses.

Accordingly, it is possible that the beneficial effects of MT stem not from weakening the DMN, but rather from rewiring in connectivity allowing practitioners to engage in more adaptive cognitive elaborations and self-referential processes. For instance, participants enrolled in a 4-day meditation intervention and underwent brain imaging at pre- and post-training to measure the impact of MT on resilience compared to relaxation training (Kwak et al., 2019). Participants from the MT and relaxation training exhibited increased self-reported mindfulness and resiliency from pre- to post-training, and the meditation group sustained these gains at a 3-month follow-up. Interestingly, the meditation group also exhibited strengthened resting-state connectivity between neural hubs of the DMN (i.e., rostral ACC) and CEN (e.g., dMPFC) compared to the relaxation group at post-training. Moreover, the strength of the DMN-CEN connectivity across time points mediated the association between self-reported mindfulness and resilience at post-training and positively predicted resilience at follow-up. These findings suggest that through MT, practitioners learned to bring awareness to their self-referential cognitions and other internal experiences, which in turn enhanced their resilience in the face of adversity.

Together, these findings support the notion that mindfulness practices can cultivate an orientation of open awareness toward internal experiences that can also translate to nonspecific states. Transfer may extend beyond simple reductions in self-referential processing and other forms of cognitive elaboration, increasing awareness of mental habits while reducing reactive responses to salient events; in neural terms, transfer may reduce reliance on the SLN in favor of attentional monitoring supported by the CEN.

The role of the SLN also appears to be modulated by MT; rather than broadly reducing SLN activity and blunting emotional responses to motivationally salient events, MT may upregulate exploratory, sensory responses to such events in place of cognitive judgments supported by the DMN. For example, when compared to untrained participants, participants completing MBSR showed increased connectivity within the auditory and visual networks and between these networks and the SLN (Kilpatrick et al., 2011). The auditory and visual networks also showed greater anticorrelations more strongly negatively associated with MBSR participants, suggesting a better capability of inhibiting irrelevant sensory information for mindfulness practitioners. Training has also been linked to greater interconnectivity between sensory and motor hubs of the SLN, as participants who suffered from moderate to severe pain showed greater coherence between the anterior insula and dorsal anterior cingulate cortex following MBSR training compared to healthy controls (Su et al., 2016). Intriguingly, this greater connectivity was linked to reduced pain scores, indicating that stronger communication with the SLN may afford greater regulation rather than just serving to alert an individual to a painful or distressing sensation. A recent systematic review bolstered such interpretations (Gotink et al., 2016), showing that on average neuroimaging reduced activity but greater connectivity between the amygdala, an SLN hub, and the prefrontal cortex, indicating better communication but less reactivity to emotional events. While it may be too early to speak definitively, the "dorsal shift" in connectivity between the SLN and the prefrontal cortex may be a characteristic of mindful states, leveraging CEN rather than DMN processing to response to motivationally relevant information (Farb et al., 2007).

These results may extend beyond just SLN-CEN connectivity to also show DMN activity becoming coupled with CEN control. Patients with trauma-related symptoms who underwent mindfulness-based exposure therapy (MBET) showed increased connectivity between the CEN and both the DMN and SLN (King et al., 2016), and effect that seems to be supported by recent meta-analysis (Boyd et al., 2018). Similarly, unemployed participants who underwent a 3-day intensive residential MT program showed increased connectivity between the DMN and CEN (Creswell et al., 2016). These findings were further supported by a large (n = 14) study aimed at replicating the functional enhanced connectivity findings between the DMN and CEN (Kral et al., 2019). Untrained participants were recruited and assigned to MBSR or an active or waitlist control condition, and each participant underwent brain scanning at three time points: baseline, post-treatment, and follow-up approximately 5 months post-treatment. MBSR participants exhibited increased PCC-dlPFC connectivity from pre- to post-treatment relative to controls, but these differences were not sustained at follow-up. Nevertheless, the increased PCC-dlPFC connectivity was more strongly related to the number of days spent practicing therapy skills in the MBSR group compared to the active control group, as well as strengthened white matter connectivity between these same regions. Again, these findings support the notion that the connection between the CEN and DMN may be strengthened through mindfulness practice and may be related to an improved capacity for attuning to sensory cues while inhibiting or containing cognitive elaborations.

Together, these findings suggest that dispositional mindfulness may be indicated by a normative profile of brain connectivity which features higher-than-normal connectivity between the CEN and the SLN/DMN, and perhaps greater connectivity within the DMN and SLN as well. Typically, high dispositional mindfulness is to greater network connectivity between prefrontal networks and sensory representation regions, and increased DMN or SLN connectivity strength may be a sign of successful MT to the extent that these networks more richly correspond to activation in sensory regions.

Structural Changes Perhaps the most obvious level of analysis in which to detect brain structure changes is at the level of stable differences in mindfulness as a disposition or trait. Accordingly, one of the first papers leverages lifetime MT experience as a predictor of gray matter volume, demonstrating that meditation practice seemed to reduce the rate of cortical atrophy that naturally occurs with human aging (Lazar et al., 2005). In a cross-sectional analysis of cortical thickness, meditators showed preserved gray matter density in a variety of brain regions, including all three of the major networks (SLN/CEN/DMN) discussed in this chapter. The implication of this first study was that meditation practice helps preserve brain health over the life span, as MT-related differences were particularly pronounced in older participants included in the sample. These general protective effects were recently replicated in a large ($n = 100$) cross-sectional study (Luders et al., 2015).

Since the Lazar et al. (2005) paper, many other investigators have explored the effects of MT on brain structure, employing both cross-sectional and intervention studies. Generally, results have been inconsistent and appear biased toward findings of increased gray matter, as to our knowledge almost no gray matter reductions have been linked to MT. The lone exception is a study relating dispositional mindfulness to brain structure, in which higher levels of trait mindfulness were linked to less gray matter in the amygdala and caudate, elements of the SLN (Taren et al., 2013), putatively indicating less "practice" in evaluating stimuli for their emotional importance, although the cross-sectional nature of the study leaves the possibility that those with smaller amygdala are also more likely to meditate.

On the other hand, many studies have linked greater levels of gray matter with MT, and somewhat surprisingly these reports tend to include aspects of the DMN. One cross-sectional investigation found greater gray matter in meditators than controls in the left inferior temporal gyrus and right hippocampus in long-term meditators, aspects of the DMN memory system (Holzel et al., 2008). A within-participant MT intervention study by the same group found increased gray matter in DMN hubs such as the cingulate cortex and temporoparietal junction (Hölzel, Carmody, et al., 2011a). Another small ($n = 6$) within-participant longitudinal study of MT over 6 weeks found precuneus (a posterior DMN hub) increases in older adults (Kurth et al., 2014).

What can we make of preserved DMN and (possibly) reduced SLN gray matter associated with mindfulness? The interpretation of gray matter density is not so straightforward as greater gray matter indicating greater use or function of a region, because pruning (reduction) of brain structures is a critical aspect of healthy brain development (Casey et al., 2008; Sowell et al., 2001). Greater DMN volume may therefore not necessarily indicate a greater reliance on habit, but rather a restoration/preserving of plasticity in these regions, whereas lower gray matter regions may already have been "pruned" to a support more efficient but more static population of neurons.

Effects within the DMN aside, other investigations have focused more on increases in gray matter in brainstem regions associated with the production of neurotransmitters and supporting the cranial nerves. Studies have found increased matter in these regions (Vestergaard-Poulsen et al., 2009), and subsequent research has linked this increased gray matter to greater levels of personal well-being (Singleton et al., 2014).

Evidently, more research is needed to test the consistency of such effects and the relationship between MT and function. As a broad principle, MT appears to help preserve gray and white matter from age-related atrophy, and greater gray matter in the DMN specifically may reflect greater plasticity in habitual cognition rather than a greater reliance on any one particular set of cognitive habits.

Neural Models of Mindfulness

From this review, it is evident that the prefrontal brain networks (CEN/DMN/SLN) can help to explain mindful awareness and are potentially linked to clinical outcomes in the therapeutic context. Beyond these empirical investigations, the sheer range of findings across various dimensions of brain structure and functioning prompted the development of theoretical models to integrate said findings into a cohesive and comprehensive account of mindfulness.

One of the most notable models comes from Holzel et al. (2011b), who proposed that the effects of mindfulness meditation are primarily driven by four internal processes. These processes include *attention regulation*, the capacity to sustain attention on a given sensory cue and return one's attention to said cue when distracted; *body awareness*, the capacity to attend and bring awareness to sensory and visceral sensations; *emotion regulation*, particularly the ability to adopt a nonjudgmental attitude toward emotional experiences and contact unpleasant internal cues in a nonreactive manner; and *change in self-perspective*, the ability to refrain from over-identifying with inner narratives regarding self-identity (Hölzel, Lazar, et al., 2011b). Moreover, Holzel et al. (2011b) also proposed that the four internal processes were associated with neural activation of the ACC (attention regulation), the insular cortex (body awareness), the dorsolateral and ventromedial PFC and amygdala (emotion regulation), and the PCC and cortical midline structures (self-perspective). In essence, it is argued that mindfulness meditation and related interventions exert its therapy effects by correcting maladaptive changes in brain function and structure associated with a given clinical disorder (e.g., attention regulation deficits in bipolar disorder, elevated emotion reactivity in major depression and anxiety disorders) (Holzel et al., 2011b), which could explain the divergent findings regarding the effects of mindfulness training when studying different psychiatric populations.

Another noteworthy model comes from Vago and Silbersweig (2012), who proposed that mindfulness practice influences self-processing through the strengthening of three internal processes: *self-awareness*, the ability to focus and maintain awareness on internal states; *self-regulation*, the ability to inhibit and modulate pre-potent responses; and *self-transcendence*, the ability to transcend self-focused needs and adopt more prosocial characteristics. Similar to Hozel et al. (2011b), this model proposes that changes in these internal processes reflect functional and structural alterations in brain networks broadly associated with attention and emotion regulation, as well as motivation and prosociality (Vago & Silbersweig, 2012). The authors propose that clinical disorders stem from negative self-beliefs that are reinforced by attentional biases toward emotionally evocative cues, which subsequently aggravate and are reinforced by psychiatric symptoms. Through this continuous cycle, negative self-beliefs can become deeply entrenched into internal schemas of the self, others, and the world, leading one to habitually engage in maladaptive cognitive processes (e.g., rumination) and behaviors (e.g., situational avoidance). Moreover, these maladaptive cognitions and behaviors can also interfere with learning from present-moment contingencies and engaging in novel behaviors or adopting more functional interpretations of the self and evocative situations. However, through the practice mindfulness meditation, psychiatric populations can minimize the deleterious influence of negative self-beliefs by assuming an experiential and transcendent sense of self. More specifically, mindfulness training modulates attention regulation and executive monitoring via alterations in brain activation and structural changes to integrate the evaluative self with experiential and transcendent senses of self and in turn correcting affect-biased attention. Similar to Holzel et al. (2011b), many of the neural structures proposed to underlie these varying senses of self are associated with the DMN (e.g., ventromedial PFC, PCC), SLN (e.g., ACC, ventral striatum), and CEN (e.g., dMPFC). Interestingly, many of the neural structures described by Holzel et al. (2011a,b) and Vago and Silbersweig (2012) serve as neural hubs within the DMN, SLN, and CEN, suggesting that apparent deficits in one of the aforementioned internal processes and senses of self could reflect broader maladaptive functioning and structural changes in the network subsuming that process.

Concluding Remarks

The literature reviewed suggests an emerging theory of mindfulness-based stress reduction, enhaning attentional control while reducing habitual evaluation. We must consider such mechanistic inferences cautiously, as few studies have been replicated in the research literature. At the same time, several themes have become apparent in our narrative review.

First, no single brain network can explain or track mindfulness, as mindfulness is a multifaceted construct with many components, and likely requires the coordination of multiple brain systems to alter existing mental habits and eventually replace them with more adaptive cognitive patterns for perceiving and responding to life events. The central elements of mindfulness are themselves under investigation, and a new generation of dismantling studies that seek to hone in on the minimum essential ingredients of mindfulness are currently underway, for example, examining the role of homework practices in MBCT (Williams et al., 2013), or comparing training in focused attention against training in open monitoring (Chin et al., 2019). As these dismantling studies mature, they will inform neurobiological paradigms by allowing for more specific forms of induction and training. Over time, this may allow for a closer correspondence between specific mechanisms of MT and the underlying neural dynamics that best support efficacious contemplative training.

Second, there is room for optimism in that mindfulness seems associated with multiple levels of analysis with a "dorsal" shift, i.e., the more dorsal CEN seems to become increasingly empowered to activate and communicate with other brain regions through MT, and offers an alternative to evaluation and self-referencing in the face of stressful experience. Learning to "decenter" or see things from a wider, more objective perspective is a central goal of MT, and is very consistent with a shift from DMN/SLN connectivity with sensory cortices to greater connectivity with the CEN.

Third, there is evidence from multiple sources that MT involves not just enhanced CEN connectivity, but also greater representation and integration of sensory information to the prefrontal cortex. This is often reported via CEN connectivity, but occasionally through SLN or DMN connectivity, especially in clinical studies. The integration of sensory information represents a marked shift away from the brain's tendency to prune out sensory signals in favor of more rarefied conceptual and response distinctions in the prefrontal cortex. As such, mindfulness may represent a radical shift toward neural plasticity as sensory information is prioritized, likely provoking downstream changes in how we represent and respond to perceived events. This move to become "sensory learners" may also account for the widespread findings of cortical growth or reduced atrophy across the life span in meditation practitioners.

Fourth, the cultivation of mindfulness may be expressed differently depending on the expertise and mental health of the practitioner. This means that no single neural configuration is going to accurately represent an ideal mindful state or proof of training across diverse populations that include novices, experts, and those with clinical conditions. For example, MT in healthy individuals seems to involve reducing DMN activity and increasing sensory connectivity, but expert meditators and people recovering from depression both demonstrate patterns of sensory activation through a combination of SLN and CEN connectivity to sensory cortices. So, we must urge caution in interpreting any one study as providing normative data for all practitioners.

In conclusion, neurobiological accounts of mindfulness are growing rapidly as mechanistic research becomes increasingly justified in the face of a growing clinical and public consensus that MT provides benefits in a variety of populations and conditions. The neuroimaging literature largely supports accounts of nonjudgmental attention that integrates sensory representations in an exploratory way, which may eventually be reintegrated into a set of perceptual and behavioral habits. The coming decades of research will hopefully clarify the critical components of efficacious MT and their underlying neurobiological mechanisms, to the benefit of all those interested in the emerging field of contemplative science.

References

Alfonso, J. P., Caracuel, A., Delgado-Pastor, L. C., & Verdejo-García, A. (2011). Combined goal management training and mindfulness meditation improve executive functions and decision-making performance in abstinent polysubstance abusers. *Drug and Alcohol Dependence, 117*(1), 78–81. https://doi.org/10.1016/j.drugalcdep.2010.12.025

Allman, J. M., Tetreault, N. A., Hakeem, A. Y., Manaye, K. F., Semendeferi, K., Erwin, J. M., Park, S., Goubert, V., & Hof, P. R. (2011). The von Economo neurons in the frontoinsular and anterior cingulate cortex: Allman et al. *Annals of the New York Academy of Sciences, 1225*(1), 59–71. https://doi.org/10.1111/j.1749-6632.2011.06011.x

Alsubaie, M., Abbott, R., Dunn, B., Dickens, C., Keil, T. F., Henley, W., & Kuyken, W. (2017). Mechanisms of action in mindfulness-based cognitive therapy (MBCT) and mindfulness-based stress reduction (MBSR) in people with physical and/or psychological conditions: A systematic review. *Clinical Psychology Review, 55*, 74–91. https://doi.org/10.1016/j.cpr.2017.04.008

Astin, J. A. (1997). Stress reduction through mindfulness meditation. Effects on psychological symptomatology, sense of control, and spiritual experiences. *Psychotherapy and Psychosomatics, 66*(2), 97–106.

Baer, R. A. (2003). Mindfulness training as a clinical intervention: A conceptual and empirical review. *Clinical Psychology: Science and Practice, 10*(2), 125–143.

Baer, R. A., Smith, G. T., Hopkins, J., Krietemeyer, J., & Toney, L. (2006). Using self-report assessment methods to explore facets of mindfulness. *Assessment, 13*(1), 27–45. https://doi.org/10.1177/1073191105283504

Baer, R. A., Smith, G. T., Lykins, E., Button, D., Krietemeyer, J., Sauer, S., Walsh, E., Duggan, D., & Williams, J. M. (2008). Construct validity of the five facet mindfulness questionnaire in meditating and nonmeditating samples. *Assessment, 15*(3), 329–342. https://doi.org/10.1177/1073191107313003

Ballard, I. C., Murty, V. P., Carter, R. M., MacInnes, J. J., Huettel, S. A., & Adcock, R. A. (2011). Dorsolateral prefrontal cortex drives mesolimbic dopaminergic regions to initiate motivated behavior. *The Journal of Neuroscience, 31*(28), 10340–10346. https://doi.org/10.1523/JNEUROSCI.0895-11.2011

Baumgartner, T., Knoch, D., Hotz, P., Eisenegger, C., & Fehr, E. (2011). Dorsolateral and ventromedial prefrontal cortex orchestrate normative choice. *Nature Neuroscience.* https://doi.org/10.1038/nn.2933

Berkovich-Ohana, A., Harel, M., Hahamy, A., Arieli, A., & Malach, R. (2016). Alterations in task-induced activity and resting-state fluctuations in visual and DMN areas revealed in long-term meditators. *NeuroImage, 135,* 125–134. https://doi.org/10.1016/j.neuroimage.2016.04.024

Bockting, C. L., Hollon, S. D., Jarrett, R. B., Kuyken, W., & Dobson, K. (2015). A lifetime approach to major depressive disorder: The contributions of psychological interventions in preventing relapse and recurrence. *Clinical Psychology Review, 41,* 16–26. https://doi.org/10.1016/j.cpr.2015.02.003

Bohlmeijer, E., Prenger, R., Taal, E., & Cuijpers, P. (2010). The effects of mindfulness-based stress reduction therapy on mental health of adults with a chronic medical disease: A meta-analysis. *Journal of Psychosomatic Research, 68*(6), 539–544. https://doi.org/10.1016/j.jpsychores.2009.10.005

Bowen, S., & Marlatt, A. (2009). Surfing the urge: Brief mindfulness-based intervention for college student smokers. *Psychology of Addictive Behaviors, 23*(4), 666–671. https://doi.org/10.1037/a0017127

Boyd, J. E., Lanius, R. A., & McKinnon, M. C. (2018). Mindfulness-based treatments for posttraumatic stress disorder: A review of the treatment literature and neurobiological evidence. *Journal of Psychiatry & Neuroscience, 43*(1), 7–25. https://doi.org/10.1503/jpn.170021

Brown, K. W., & Ryan, R. M. (2003). The benefits of being present: Mindfulness and its role in psychological Well-being. *Journal of Personality and Social Psychology, 84*(4), 822–848.

Brown, K. W., Ryan, R. M., & Creswell, J. D. (2007). Mindfulness: Theoretical foundations and evidence for salutary effects. *Psychological Inquiry, 18*(4), 211–237.

Buckner, R. L., & DiNicola, L. M. (2019). The brain's default network: Updated anatomy, physiology and evolving insights. *Nature Reviews Neuroscience, 20*(10), 593–608. https://doi.org/10.1038/s41583-019-0212-7

Buckner, R. L., Andrews-Hanna, J. R., & Schacter, D. L. (2008). The brain's default network: Anatomy, function, and relevance to disease. *Annals of the New York Academy of Sciences, 1124,* 1–38. https://doi.org/10.1196/annals.1440.011

Casey, B. J., Getz, S., & Galvan, A. (2008). The adolescent brain. *Developmental Review, 28*(1), 62–77. https://doi.org/10.1016/j.dr.2007.08.003

Cavanna, A. E., & Trimble, M. R. (2006). The precuneus: A review of its functional anatomy and behavioural correlates. *Brain, 129*(3), 564–583. https://doi.org/10.1093/brain/awl004

Chen, A. C., Oathes, D. J., Chang, C., Bradley, T., Zhou, Z.-W., Williams, L. M., Glover, G. H., Deisseroth, K., & Etkin, A. (2013). Causal interactions between fronto-parietal central executive and default-mode networks in humans. *Proceedings of the National Academy of Sciences, 110*(49), 19944–19949. https://doi.org/10.1073/pnas.1311772110

Chiesa, A. (2013). The difficulty of defining mindfulness: Current thought and critical issues. *Mindfulness, 4*(3), 255–268. https://doi.org/10.1007/s12671-012-0123-4

Chiesa, A., & Serretti, A. (2009). Mindfulness-based stress reduction for stress management in healthy people: A review and meta-analysis. *Journal of Alternative and Complementary Medicine, 15*(5), 593–600. https://doi.org/10.1089/acm.2008.0495

Chiesa, A., & Serretti, A. (2010). A systematic review of neurobiological and clinical features of mindfulness meditations. *Psychological Medicine, 40*(8), 1239–1252. https://doi.org/10.1017/S0033291709991747

Chin, B., Lindsay, E. K., Greco, C. M., Brown, K. W., Smyth, J. M., Wright, A. G. C., & Creswell, J. D. (2019). Psychological mechanisms driving stress resilience in mindfulness training: A randomized controlled trial. *Health Psychology, 38*(8), 759–768. https://doi.org/10.1037/hea0000763

Chong, Y. W., Kee, Y. H., & Chaturvedi, I. (2015). Effects of brief mindfulness induction on weakening habits: Evidence from a computer mouse control task. *Mindfulness, 6*(3), 582–588. https://doi.org/10.1007/s12671-014-0293-3

Christoff, K., Gordon, A. M., Smallwood, J., Smith, R., & Schooler, J. W. (2009). Experience sampling during fMRI reveals default network and executive system contributions to mind wandering. *Proceedings of the National Academy of Sciences of the United States of America, 106*(21), 8719–8724. https://doi.org/10.1073/pnas.0900234106

Collette, F., & Van der Linden, M. (2002). Brain imaging of the central executive component of working memory. *Neuroscience & Biobehavioral Reviews, 26*(2), 105–125. https://doi.org/10.1016/S0149-7634(01)00063-X

Craig, A. D. (2002). How do you feel? Interoception: The sense of the physiological condition of the body. *Nature Reviews. Neuroscience, 3*(8), 655–666. https://doi.org/10.1038/nrn894

Craig, A. D. (2009). How do you feel—Now? The anterior insula and human awareness. *Nature Reviews. Neuroscience, 10*(1), 59–70. https://doi.org/10.1038/nrn2555

Crane, C., Crane, R. S., Eames, C., Fennell, M. J. V., Silverton, S., Williams, J. M. G., & Barnhofer, T. (2014). The effects of amount of home meditation practice in mindfulness based cognitive therapy on hazard of relapse to depression in the staying well after depression trial. *Behaviour Research and Therapy, 63*, 17–24. https://doi.org/10.1016/j.brat.2014.08.015

Creswell, J. D., Pacilio, L. E., Lindsay, E. K., & Brown, K. W. (2014). Brief mindfulness meditation training alters psychological and neuroendocrine responses to social evaluative stress. *Psychoneuroendocrinology, 44*, 1–12. http://dx.doi.org.myaccess.library.utoronto.ca/10.1016/j.psyneuen.2014.02.007

Creswell, J. D., Taren, A. A., Lindsay, E. K., Greco, C. M., Gianaros, P. J., Fairgrieve, A., Marsland, A. L., Brown, K. W., Way, B. M., Rosen, R. K., & Ferris, J. L. (2016). Alterations in resting-state functional connectivity link mindfulness meditation with reduced Interleukin-6: A randomized controlled trial. *Biological Psychiatry, 80*(1), 53–61. https://doi.org/10.1016/j.biopsych.2016.01.008

Crowe, K., & McKay, D. (2016). Mindfulness, obsessive–compulsive symptoms, and executive dysfunction. *Cognitive Therapy and Research, 40*(5), 627–644. https://doi.org/10.1007/s10608-016-9777-x

Cunningham, W. A., Van Bavel, J. J., & Johnsen, I. R. (2008). Affective flexibility: Evaluative processing goals shape amygdala activity. *Psychological Science, 19*(2), 152–160. https://doi.org/10.1111/j.1467-9280.2008.02061.x

Department of Preventive Medicine, USC School of Medicine, Los Angeles, CA 90032, USA, Black, D. S., Kurth, F., Department of Neurology, UCLA School of Medicine, Los Angeles, CA 90095, USA, Luders, E., Department of Neurology, UCLA School of Medicine, Los Angeles, CA 90095, USA, Wu, B., & USC School of Medicine, Los Angeles, CA 90032, USA. (2014). Brain Gray matter changes associated with mindfulness meditation in older adults: An exploratory pilot study using Voxelbased morphometry. Neuro – Open Journal, 1(1), 23–26. https://doi.org/https://doi.org/10.17140/NOJ-1-106.

Doll, A., Hölzel, B. K., Boucard, C. C., Wohlschläger, A. M., & Sorg, C. (2015). Mindfulness is associated with intrinsic functional connectivity between default mode and salience networks. *Frontiers in Human Neuroscience, 9*, 461. https://doi.org/10.3389/fnhum.2015.00461

Doll, A., Hölzel, B. K., Mulej Bratec, S., Boucard, C. C., Xie, X., Wohlschläger, A. M., & Sorg, C. (2016). Mindful attention to breath regulates emotions via increased amygdala–prefrontal cortex connectivity. *NeuroImage, 134*, 305–313. https://doi.org/10.1016/j.neuroimage.2016.03.041

Farb, N. A., Segal, Z. V., Mayberg, H., Bean, J., McKeon, D., Fatima, Z., & Anderson, A. K. (2007). Attending to the present: Mindfulness meditation reveals distinct neural modes of self-reference. *Social Cognitive and Affective Neuroscience, 2*(4), 313–322. https://doi.org/10.1093/scan/nsm030

Farb, N. A., Anderson, A. K., Mayberg, H., Bean, J., McKeon, D., & Segal, Z. V. (2010). Minding one's emotions: Mindfulness training alters the neural expression of sadness. *Emotion, 10*(1), 25–33. https://doi.org/10.1037/a0017151

Farb, N. A., Anderson, A. K., Bloch, R. T., & Segal, Z. V. (2011). Mood-linked responses in medial prefrontal cortex predict relapse in patients with recurrent unipolar depression. *Biological Psychiatry, 70*(4), 366–372. https://doi.org/10.1016/j.biopsych.2011.03.009

Farb, N. A., Anderson, A. K., & Segal, Z. V. (2012). The mindful brain and emotion regulation in mood disorders. *Canadian Journal of Psychiatry, 57*(2), 70–77.

Farb, N. A., Segal, Z. V., & Anderson, A. K. (2013a). Attentional modulation of primary interoceptive and exteroceptive cortices. *Cerebral Cortex, 23*(1), 114–126. https://doi.org/10.1093/cercor/bhr385

Farb, N. A. S., Segal, Z. V., & Anderson, A. K. (2013b). Mindfulness meditation training alters cortical representations of interoceptive attention. *Social Cognitive and Affective Neuroscience, 8*(1), 15–26.

Feldman, G., Greeson, J., & Senville, J. (2010). Differential effects of mindful breathing, progressive muscle relaxation, and loving-kindness meditation on decentering and negative reactions to repetitive thoughts. *Behaviour Research and Therapy, 48*(10), 1002–1011. https://doi.org/10.1016/j.brat.2010.06.006

Fjorback, L. O., Arendt, M., Ørnbøl, E., Fink, P., & Walach, H. (2011). Mindfulness-based stress reduction and mindfulness-based cognitive therapy – A systematic review of randomized controlled trials. *Acta Psychiatrica Scandinavica, 124*(2), 102–119. https://doi.org/10.1111/j.1600-0447.2011.01704.x

Fox, M. D., & Raichle, M. E. (2007). Spontaneous fluctuations in brain activity observed with functional magnetic resonance imaging. *Nature Reviews. Neuroscience, 8*(9), 700–711. https://doi.org/10.1038/nrn2201

Fox, K. C. R., Spreng, R. N., Ellamil, M., Andrews-Hanna, J. R., & Christoff, K. (2015). The wandering brain: Meta-analysis of functional neuroimaging studies of mind-wandering and related spontaneous thought processes. *NeuroImage, 111*, 611–621. https://doi.org/10.1016/j.neuroimage.2015.02.039

Froeliger, B., Mathew, A. R., McConnell, P. A., Eichberg, C., Saladin, M. E., Carpenter, M. J., & Garland, E. L. (2017). Restructuring reward mechanisms in nicotine addiction: A pilot fMRI study of mindfulness-oriented recovery enhancement for cigarette smokers. *Evidence-based Complementary and Alternative Medicine, 2017*, 1–10. https://doi.org/10.1155/2017/7018014

Gallant, S. N. (2016). Mindfulness meditation practice and executive functioning: Breaking down the benefit. *Consciousness and Cognition, 40*, 116–130. https://doi.org/10.1016/j.concog.2016.01.005

Gard, T. (2014). Different neural correlates of facing pain with mindfulness: Contributions of strategy and skill: Comment on "facing the experience of pain: A neuropsychological perspective" by Fabbro and Crescentini. *Physics of Life Reviews, 11*(3).

Garland, E. L., Froeliger, B., & Howard, M. O. (2014). Effects of mindfulness-oriented recovery enhancement on reward responsiveness and opioid cue-reactivity. *Psychopharmacology*, 1–10.

Garland, E. L., Farb, N. A., Goldin, P., & Fredrickson, B. L. (2015). Mindfulness broadens awareness and builds Eudaimonic meaning: A process model of mindful positive emotion regulation. *Psychological Inquiry, 26*(4), 293–314. https://doi.org/10.1080/1047840X.2015.1064294

Geiger, M. J., Domschke, K., Ipser, J., Hattingh, C., Baldwin, D. S., Lochner, C., & Stein, D. J. (2016). Altered executive control network resting-state connectivity in social anxiety disorder. *The World Journal of Biological Psychiatry, 17*(1), 47–57. https://doi.org/10.3109/15622975.2015.1083613

Goldin, P. R., & Gross, J. J. (2010). Effects of mindfulness-based stress reduction (MBSR) on emotion regulation in social anxiety disorder. *Emotion, 10*(1), 83–91. https://doi.org/10.1037/a0018441

Goldin, P. R., McRae, K., Ramel, W., & Gross, J. J. (2008). The neural bases of emotion regulation: Reappraisal and suppression of negative emotion. *Biological Psychiatry, 63*(6), 577–586. https://doi.org/10.1016/j.biopsych.2007.05.031

Gotink, R. A., Meijboom, R., Vernooij, M. W., Smits, M., & Hunink, M. G. M. (2016). 8-week mindfulness based stress reduction induces brain changes similar to traditional long-term meditation practice – A systematic review. *Brain and Cognition, 108*, 32–41. https://doi.org/10.1016/j.bandc.2016.07.001

Goyal, M., Singh, S., Sibinga, E. M., Gould, N. F., Rowland-Seymour, A., Sharma, R., Berger, Z., Sleicher, D., Maron, D. D., & Shihab, H. M. (2014). Meditation programs for psychological stress and Well-being: A systematic review and meta-analysis. *JAMA Internal Medicine, 174*(3), 357–368.

Grossman, P. (2011). Defining mindfulness by how poorly I think I pay attention during everyday awareness and other intractable problems for psychology's (re) invention of mindfulness: Comment on Brown et al. (2011). *Psychological Assessment, 23*(4), 1034–1040.; discussion 1041-6. https://doi.org/10.1037/a0022713

Gu, J., Strauss, C., Bond, R., & Cavanagh, K. (2015). How do mindfulness-based cognitive therapy and mindfulness-based stress reduction improve mental health and wellbeing? A systematic review and meta-analysis of mediation studies. *Clinical Psychology Review, 37*, 1–12. https://doi.org/10.1016/j.cpr.2015.01.006

Gu, J., Strauss, C., Crane, C., Barnhofer, T., Karl, A., Cavanagh, K., & Kuyken, W. (2016). Examining the factor structure of the 39-item and 15-item versions of the five facet mindfulness questionnaire before and after mindfulness-based cognitive therapy for people with recurrent depression. *Psychological Assessment, 28*(7), 791–802. https://doi.org/10.1037/pas0000263

Hasenkamp, W., & Barsalou, L. W. (2012). Effects of meditation experience on functional connectivity of distributed brain networks. *Frontiers in Human Neuroscience*, 6. https://doi.org/10.3389/fnhum.2012.00038

Hayes, S. C., Strosahl, K., & Wilson, K. G. (1999). *Acceptance and commitment therapy: An experiential approach to behavior chagne*. Guilford Press.

Hermans, E. J., Henckens, M. J. A. G., Joëls, M., & Fernández, G. (2014). Dynamic adaptation of large-scale brain networks in response to acute stressors. *Trends in Neurosciences, 37*(6), 304–314. https://doi.org/10.1016/j.tins.2014.03.006

Hilton, L., Hempel, S., Ewing, B. A., Apaydin, E., Xenakis, L., Newberry, S., Colaiaco, B., Maher, A. R., Shanman, R. M., Sorbero, M. E., & Maglione, M. A. (2017). Mindfulness meditation for chronic pain: Systematic review and meta-analysis. *Annals of Behavioral Medicine, 51*(2), 199–213. https://doi.org/10.1007/s12160-016-9844-2

Holmes, E. A., Ghaderi, A., Harmer, C. J., Ramchandani, P. G., Cuijpers, P., Morrison, A. P., Roiser, J. P., Bockting, C. L. H., O'Connor, R. C., Shafran, R., Moulds, M. L., & Craske, M. G. (2018). The lancet psychiatry commission on psychological treatments research in tomorrow's science. *The Lancet. Psychiatry, 5*(3), 237–286. https://doi.org/10.1016/S2215-0366(17)30513-8

Holzel, B. K., Ott, U., Gard, T., Hempel, H., Weygandt, M., Morgen, K., & Vaitl, D. (2008). Investigation of mindfulness meditation practitioners with voxel-based morphometry. *Social Cognitive and Affective Neuroscience, 3*(1), 55–61. https://doi.org/10.1093/scan/nsm038

Hölzel, B. K., Carmody, J., Vangel, M., Congleton, C., Yerramsetti, S. M., Gard, T., & Lazar, S. W. (2011). Mindfulness practice leads to increases in regional brain gray matter density. Psychiatry research: neuroimaging, 191(1), 36–43

Hölzel, B. K., Carmody, J., Evans, K. C., Hoge, E. A., Dusek, J. A., Morgan, L., Pitman, R. K., & Lazar, S. W. (2010). Stress reduction correlates with structural changes in the amygdala. *Social Cognitive and Affective Neuroscience, 5*(1), 11–17. https://doi.org/10.1093/scan/nsp034

Hölzel, B. K., Carmody, J., Vangel, M., Congleton, C., Yerramsetti, S. M., Gard, T., & Lazar, S. W. (2011a). Mindfulness practice leads to increases in regional brain gray matter density. *Psychiatry Research, 191*(1), 36–43. https://doi.org/10.1016/j.pscychresns.2010.08.006

Hölzel, B. K., Lazar, S. W., Gard, T., Schuman-Olivier, Z., Vago, D. R., & Ott, U. (2011b). How does mindfulness meditation work? Proposing mechanisms of action from a conceptual and neutral perspective. *Perspectives on Psychological Science, 6*(6), 537–559.

Hölzel, B. K., Brunsch, V., Gard, T., Greve, D. N., Koch, K., Sorg, C., Lazar, S. W., & Milad, M. R. (2016). Mindfulness-based stress reduction, fear conditioning, and the Uncinate fasciculus: A pilot study. *Frontiers in Behavioral Neuroscience, 10*. https://doi.org/10.3389/fnbeh.2016.00124

Ives-Deliperi, V. L., Solms, M., & Meintjes, E. M. (2011). The neural substrates of mindfulness: An fMRI investigation. *Social Neuroscience, 6*(3), 231–242. https://doi.org/10.1080/1747091 9.2010.513495

Ives-Deliperi, V. L., Howells, F., Stein, D. J., Meintjes, E. M., & Horn, N. (2013). The effects of mindfulness-based cognitive therapy in patients with bipolar disorder: A controlled functional MRI investigation. *Journal of Affective Disorders, 150*(3), 1152–1157. https://doi.org/10.1016/j.jad.2013.05.074

Kabat-Zinn, J. (1990). *Full catastrophe living: Using the wisdom of your body and mind to face stress, pain and illness.* Delacorte.

Kabat-Zinn, J. (2003). Mindfulness-based interventions in context: Past, present, and future. *Clinical Psychology: Science and Practice, 10*(2), 144–156.

Kaiser, R. H., Andrews-Hanna, J. R., Wager, T. D., & Pizzagalli, D. A. (2015). Large-scale network dysfunction in major depressive disorder: A meta-analysis of resting-state functional connectivity. *JAMA Psychiatry, 72*(6), 603–611. https://doi.org/10.1001/jamapsychiatry.2015.0071

Kazdin, A. E. (2007). Mediators and mechanisms of change in psychotherapy research. *Annual Review of Clinical Psychology, 3*(1), 1–27. https://doi.org/10.1146/annurev.clinpsy.3.022806.091432

Keng, S. L., Smoski, M. J., & Robins, C. J. (2011). Effects of mindfulness on psychological health: A review of empirical studies. *Clinical Psychology Review, 31*(6), 1041–1056. https://doi.org/10.1016/j.cpr.2011.04.006

Kilpatrick, L. A., Suyenobu, B. Y., Smith, S. R., Bueller, J. A., Goodman, T., Creswell, J. D., Tillisch, K., Mayer, E. A., & Naliboff, B. D. (2011). Impact of mindfulness-based stress reduction training on intrinsic brain connectivity. *NeuroImage, 56*(1), 290–298. https://doi.org/10.1016/j.neuroimage.2011.02.034

King, A. P., Block, S. R., Sripada, R. K., Rauch, S., Giardino, N., Favorite, T., Angstadt, M., Kessler, D., Welsh, R., & Liberzon, I. (2016). Altered default mode network (DMN) resting state functional connectivity following a mindfulness-based exposure therapy for posttraumatic stress disorder (PTSD) in combat veterans of Afghanistan and Iraq. *Depression and Anxiety, 33*(4), 289–299. https://doi.org/10.1002/da.22481

Kirk, U., & Montague, P. R. (2015). Mindfulness meditation modulates reward prediction errors in a passive conditioning task. *Frontiers in Psychology, 6.* https://doi.org/10.3389/fpsyg.2015.00090

Kirk, U., Brown, K. W., & Downar, J. (2015). Adaptive neural reward processing during anticipation and receipt of monetary rewards in mindfulness meditators. *Social Cognitive and Affective Neuroscience, 10*(5), 752–759. https://doi.org/10.1093/scan/nsu112

Koshino, H. (2017). Coactivation of default mode network and executive network regions in the human brain. In M. Watanabe (Ed.), *The prefrontal cortex as an executive, emotional, and social brain* (pp. 247–276). Springer. https://doi.org/10.1007/978-4-431-56508-6_13

Kral, T. R. A., Schuyler, B. S., Mumford, J. A., Rosenkranz, M. A., Lutz, A., & Davidson, R. J. (2018). Impact of short- and long-term mindfulness meditation training on amygdala reactivity to emotional stimuli. *NeuroImage, 181*, 301–313. https://doi.org/10.1016/j.neuroimage.2018.07.013

Kral, T. R. A., Imhoff-Smith, T., Dean, D. C., Grupe, D., Adluru, N., Patsenko, E., Mumford, J. A., Goldman, R., Rosenkranz, M. A., & Davidson, R. J. (2019). Mindfulness-based stress reduction-related changes in posterior cingulate resting brain connectivity. *Social Cognitive and Affective Neuroscience, 14*(7), 777–787. https://doi.org/10.1093/scan/nsz050

Kurth, F., Luders, E., Wu, B., & Black, D. S. (2014). Brain gray matter changes associated with mindfulness meditation in older adults: an exploratory pilot study using voxel-based morphometry. Neuro: open journal, 1(1), 23

Kuyken, W., Warren, F. C., Taylor, R. S., Whalley, B., Crane, C., Bondolfi, G., Hayes, R., Huijbers, M., Ma, H., Schweizer, S., Segal, Z., Speckens, A., Teasdale, J. D., Van Heeringen, K., Williams, M., Byford, S., Byng, R., & Dalgleish, T. (2016). Efficacy of mindfulness-based cognitive therapy in prevention of depressive relapse: An individual patient data meta-analysis from randomized trials. *JAMA Psychiatry, 73*(6), 565–574. https://doi.org/10.1001/jamapsychiatry.2016.0076

Kwak, S., Lee, T. Y., Jung, W. H., Hur, J.-W., Bae, D., Hwang, W. J., Cho, K. I. K., Lim, K.-O., Kim, S.-Y., Park, H. Y., & Kwon, J. S. (2019). The immediate and sustained positive effects of meditation on resilience are mediated by changes in the resting brain. *Frontiers in Human Neuroscience, 13,* 101. https://doi.org/10.3389/fnhum.2019.00101

Lazar, S. W., Kerr, C. E., Wasserman, R. H., Gray, J. R., Greve, D. N., Treadway, M. T., McGarvey, M., Quinn, B. T., Dusek, J. A., Benson, H., Rauch, S. L., Moore, C. I., & Fischl, B. (2005). Meditation experience is associated with increased cortical thickness. Neuroreport, 16 (17), 1893–1897. https://doi.org/10.1097/01.wnr.0000186598.66243.19

Lea, J., Cadman, L., & Philo, C. (2015). Changing the habits of a lifetime? Mindfulness meditation and habitual geographies. *Cultural Geographies, 22*(1), 49–65. https://doi.org/10.1177/1474474014536519

Lebois, L. A. M., Papies, E. K., Gopinath, K., Cabanban, R., Quigley, K. S., Krishnamurthy, V., Barrett, L. F., & Barsalou, L. W. (2015). A shift in perspective: Decentering through mindful attention to imagined stressful events. *Neuropsychologia, 75,* 505–524. https://doi.org/10.1016/j.neuropsychologia.2015.05.030

Leech, R., & Sharp, D. J. (2014). The role of the posterior cingulate cortex in cognition and disease. *Brain: A Journal of Neurology, 137*(Pt 1), 12–32. https://doi.org/10.1093/brain/awt162

Lengacher, C. A., Kip, K. E., Barta, M. K., Post-White, J., Jacobsen, P., Groer, M., Lehman, B., Moscoso, M. S., Kadel, R., Le, N., Loftus, L., Stevens, C., Malafa, M., & Shelton, M. M. (2012). A pilot study evaluating the effect of mindfulness-based stress reduction on psychological status, physical status, salivary cortisol, and Interleukin-6 among advanced-stage Cancer patients and their caregivers. *Journal of Holistic Nursing.* https://doi.org/10.1177/0898010111435949

Leyland, A., Rowse, G., & Emerson, L.-M. (2019). Experimental effects of mindfulness inductions on self-regulation: Systematic review and meta-analysis. *Emotion, 19*(1), 108–122. https://doi.org/10.1037/emo0000425

Li, S. Y. H., & Bressington, D. (2019). The effects of mindfulness-based stress reduction on depression, anxiety, and stress in older adults: A systematic review and meta-analysis. *International Journal of Mental Health Nursing, 28*(3), 635–656. https://doi.org/10.1111/inm.12568

Lieberman, M. D., Straccia, M. A., Meyer, M. L., Du, M., & Tan, K. M. (2019). Social, self, (situational), and affective processes in medial prefrontal cortex (MPFC): Causal, multivariate, and reverse inference evidence. *Neuroscience & Biobehavioral Reviews, 99,* 311–328. https://doi.org/10.1016/j.neubiorev.2018.12.021

Linehan, M. M., Schmidt, H., Dimeff, L. A., Craft, J. C., Kanter, J., & Comtois, K. A. (1999). Dialectical behavior therapy for patients with borderline personality disorder and drug-dependence. *The American Journal on Addictions, 8*(4), 279–292.

Luders, E., Phillips, O. R., Clark, K., Kurth, F., Toga, A. W., & Narr, K. L. (2012). Bridging the hemispheres in meditation: Thicker callosal regions and enhanced fractional anisotropy (FA) in long-term practitioners. *NeuroImage, 61*(1), 181–187. https://doi.org/10.1016/j.neuroimage.2012.02.026

Luders, E., Cherbuin, N., & Kurth, F. (2015). Forever young (er): Potential age-defying effects of long-term meditation on gray matter atrophy. *Frontiers in Psychology, 5.* https://doi.org/10.3389/fpsyg.2014.01551

Lutz, A., McFarlin, D. R., Perlman, D. M., Salomons, T. V., & Davidson, R. J. (2013). Altered anterior insula activation during anticipation and experience of painful stimuli in expert meditators. *NeuroImage, 64,* 538–546. https://doi.org/10.1016/j.neuroimage.2012.09.030

Lutz, J., Brühl, A. B., Scheerer, H., Jäncke, L., & Herwig, U. (2016). Neural correlates of mindful self-awareness in mindfulness meditators and meditation-naïve subjects revisited. *Biological Psychology, 119,* 21–30. https://doi.org/10.1016/j.biopsycho.2016.06.010

Ma, S. H., & Teasdale, J. D. (2004). Mindfulness-based cognitive therapy for depression: Replication and exploration of differential relapse prevention effects. *Journal of Consulting and Clinical Psychology, 72*(1), 31–40. https://doi.org/10.1037/0022-006X.72.1.31

Marchetti, I., Koster, E. H. W., Klinger, E., & Alloy, L. B. (2016). Spontaneous thought and vulnerability to mood disorders: The dark side of the wandering mind. *Clinical Psychological Science: A Journal of the Association for Psychological Science, 4*(5), 835–857. https://doi.org/10.1177/2167702615622383

Medford, N., & Critchley, H. D. (2010). Conjoint activity of anterior insular and anterior cingulate cortex: Awareness and response. *Brain Structure and Function, 214*(5–6), 535–549. https://doi.org/10.1007/s00429-010-0265-x

Menon, V., & Uddin, L. Q. (2010). Saliency, switching, attention and control: A network model of insula function. *Brain Structure & Function, 214*(5–6), 655–667. https://doi.org/10.1007/s00429-010-0262-0

Mrazek, M. D., Smallwood, J., & Schooler, J. W. (2012). Mindfulness and mind-wandering: Finding convergence through opposing constructs. *Emotion, 12*(3), 442–448. Scopus. https://doi.org/10.1037/a0026678

Mulders, P. C., van Eijndhoven, P. F., Schene, A. H., Beckmann, C. F., & Tendolkar, I. (2015). Resting-state functional connectivity in major depressive disorder: A review. *Neuroscience & Biobehavioral Reviews, 56*, 330–344. https://doi.org/10.1016/j.neubiorev.2015.07.014

Pannekoek, J. N., Veer, I. M., van Tol, M.-J., van der Werff, S. J. A., Demenescu, L. R., Aleman, A., Veltman, D. J., Zitman, F. G., Rombouts, S. A. R. B., & van der Wee, N. J. A. (2013). Resting-state functional connectivity abnormalities in limbic and salience networks in social anxiety disorder without comorbidity. *European Neuropsychopharmacology, 23*(3), 186–195. https://doi.org/10.1016/j.euroneuro.2012.04.018

Price, C. J., & Hooven, C. (2018). Interoceptive awareness skills for emotion regulation: Theory and approach of mindful awareness in body-oriented therapy (MABT). *Frontiers in Psychology, 9*. https://doi.org/10.3389/fpsyg.2018.00798

Qin, P., & Northoff, G. (2011). How is our self related to midline regions and the default-mode network? *NeuroImage, 57*(3), 1221–1233. https://doi.org/10.1016/j.neuroimage.2011.05.028

Rahl, H. A., Lindsay, E. K., Pacilio, L. E., Brown, K. W., & Creswell, J. D. (2017). Brief mindfulness meditation training reduces mind wandering: The critical role of acceptance. *Emotion, 17*(2), 224–230. https://doi.org/10.1037/emo0000250

Raichle, M. E., MacLeod, A. M., Snyder, A. Z., Powers, W. J., Gusnard, D. A., & Shulman, G. L. (2001). A default mode of brain function. *Proceedings of the National Academy of Sciences of the United States of America, 98*(2), 676–682. https://doi.org/10.1073/pnas.98.2.676

Reynolds, S. M. (2005). Specificity in the projections of prefrontal and insular cortex to ventral Striatopallidum and the extended amygdala. *Journal of Neuroscience, 25*(50), 11757–11767. https://doi.org/10.1523/JNEUROSCI.3432-05.2005

Roffman, J. L., Marci, C. D., Glick, D. M., Dougherty, D. D., & Rauch, S. L. (2005). Neuroimaging and the functional neuroanatomy of psychotherapy. *Psychological Medicine, 35*(10), 1385–1398. https://doi.org/10.1017/S0033291705005064

Rudkin, E., Medvedev, O. N., & Siegert, R. J. (2018). The five-facet mindfulness questionnaire: Why the observing subscale does not predict psychological symptoms. *Mindfulness, 9*(1), 230–242. https://doi.org/10.1007/s12671-017-0766-2

Salmon, E., Van der Linden, M., Collette, F., Delfiore, G., Maquet, P., Degueldre, C., Luxen, A., & Franck, G. (1996). Regional brain activity during working memory tasks. *Brain, 119*(5), 1617–1625. https://doi.org/10.1093/brain/119.5.1617

Santarnecchi, E., D'Arista, S., Egiziano, E., Gardi, C., Petrosino, R., Vatti, G., Reda, M., & Rossi, A. (2014). Interaction between neuroanatomical and psychological changes after mindfulness-based training. *PLoS One, 9*(10), e108359. https://doi.org/10.1371/journal.pone.0108359

Seeley, W. W., Menon, V., Schatzberg, A. F., Keller, J., Glover, G. H., Kenna, H., Reiss, A. L., & Greicius, M. D. (2007). Dissociable intrinsic connectivity networks for salience processing and executive control. *The Journal of Neuroscience, 27*(9), 2349–2356. https://doi.org/10.1523/JNEUROSCI.5587-06.2007

Segal, Z. V., Williams, J. M. G., & Teasdale, J. D. (2012). *Mindfulness-based cognitive therapy for depression.* Guilford Press.

Sevinc, G., Hölzel, B. K., Greenberg, J., Gard, T., Brunsch, V., Hashmi, J. A., Vangel, M., Orr, S. P., Milad, M. R., & Lazar, S. W. (2019). Strengthened hippocampal circuits underlie enhanced retrieval of extinguished fear memories following mindfulness training. *Biological Psychiatry, 86*(9), 693–702. https://doi.org/10.1016/j.biopsych.2019.05.017

Shapiro, S. L., Carlson, L. E., Astin, J. A., & Freedman, B. (2006). Mechanisms of mindfulness. *Journal of Clinical Psychology, 62*(3), 373–386. https://doi.org/10.1002/jclp.20237

Sheline, Y. I., Price, J. L., Yan, Z., & Mintun, M. A. (2010). Resting-state functional MRI in depression unmasks increased connectivity between networks via the dorsal nexus. *Proceedings of the National Academy of Sciences of the United States of America, 107*(24), 11020–11025. https://doi.org/10.1073/pnas.1000446107

Singleton, O., Hölzel, B. K., Vangel, M., Brach, N., Carmody, J., & Lazar, S. W. (2014). Change in brainstem Gray matter concentration following a mindfulness-based intervention is correlated with improvement in psychological Well-being. *Frontiers in Human Neuroscience, 8.* https://doi.org/10.3389/fnhum.2014.00033

Sowell, E. R., Thompson, P. M., Tessner, K. D., & Toga, A. W. (2001). Mapping continued brain growth and Gray matter density reduction in dorsal frontal cortex: Inverse relationships during Postadolescent brain maturation. *The Journal of Neuroscience, 21*(22), 8819–8829. https://doi.org/10.1523/JNEUROSCI.21-22-08819.2001

Sridharan, D., Levitin, D. J., & Menon, V. (2008). A critical role for the right fronto-insular cortex in switching between central-executive and default-mode networks. *Proceedings of the National Academy of Sciences of the United States of America, 105*(34), 12569–12574. https://doi.org/10.1073/pnas.0800005105

Su, I.-W., Wu, F.-W., Liang, K.-C., Cheng, K.-Y., Hsieh, S.-T., Sun, W.-Z., & Chou, T.-L. (2016). Pain perception can be modulated by mindfulness training: A resting-state fMRI study. *Frontiers in Human Neuroscience, 10.* https://doi.org/10.3389/fnhum.2016.00570

Suo, X., Lei, D., Li, K., Chen, F., Li, F., Li, L., Huang, X., Lui, S., Li, L., Kemp, G. J., & Gong, Q. (2015). Disrupted brain network topology in pediatric posttraumatic stress disorder: A resting-state fMRI study: Suo et al. *Human Brain Mapping, 36*(9), 3677–3686. https://doi.org/10.1002/hbm.22871

Tang, Y.-Y., Tang, Y., Tang, R., & Lewis-Peacock, J. A. (2017). Brief mental training reorganizes large-scale brain networks. *Frontiers in Systems Neuroscience, 11.* https://doi.org/10.3389/fnsys.2017.00006

Taren, A. A., Creswell, J. D., & Gianaros, P. J. (2013). Dispositional mindfulness co-varies with smaller amygdala and caudate volumes in community adults. *PLoS One, 8*(5), e64574. https://doi.org/10.1371/journal.pone.0064574

Taubert, M., Mehnert, J., Pleger, B., & Villringer, A. (2016). Rapid and specific gray matter changes in M1 induced by balance training. *NeuroImage, 133,* 399–407. https://doi.org/10.1016/j.neuroimage.2016.03.017

Taylor, V. A., Grant, J., Daneault, V., Scavone, G., Breton, E., Roffe-Vidal, S., Courtemanche, J., Lavarenne, A. S., & Beauregard, M. (2011). Impact of mindfulness on the neural responses to emotional pictures in experienced and beginner meditators. *NeuroImage, 57*(4), 1524–1533. https://doi.org/10.1016/j.neuroimage.2011.06.001

Taylor, V. A., Daneault, V., Grant, J., Scavone, G., Breton, E., Roffe-Vidal, S., Courtemanche, J., Lavarenne, A. S., Marrelec, G., Benali, H., & Beauregard, M. (2013). Impact of meditation training on the default mode network during a restful state. *Social Cognitive and Affective Neuroscience, 8*(1), 4–14. https://doi.org/10.1093/scan/nsr087

Teasdale, J. D., Segal, Z. V., Williams, J. M., Ridgeway, V. A., Soulsby, J. M., & Lau, M. A. (2000). Prevention of relapse/recurrence in major depression by mindfulness-based cognitive therapy. *Journal of Consulting and Clinical Psychology, 68*(4), 615–623.

Teasdale, J. D., Williams, J. M. G., & Segal, Z. V. (2013). *The mindful Way workbook: An 8-week program to free yourself from depression and emotional distress.* Guilford Publications.

Tomasino, B., & Fabbro, F. (2016). Increases in the right dorsolateral prefrontal cortex and decreases the rostral prefrontal cortex activation after-8 weeks of focused attention based mindfulness meditation. *Brain and Cognition, 102,* 46–54. https://doi.org/10.1016/j.bandc.2015.12.004

Tomlinson, E. R., Yousaf, O., Vittersø, A. D., & Jones, L. (2018). Dispositional mindfulness and psychological health: A systematic review. *Mindfulness, 9*(1), 23–43. https://doi.org/10.1007/s12671-017-0762-6

Vago, D. R. (2014). Mapping modalities of self-awareness in mindfulness practice: A potential mechanism for clarifying habits of mind: Clarifying habits of mind. *Annals of the New York Academy of Sciences, 1307*(1), 28–42. https://doi.org/10.1111/nyas.12270

Vago, D. R., & Silbersweig, D. A. (2012). Self-awareness, self-regulation, and self-transcendence (S-ART): A framework for understanding the neurobiological mechanisms of mindfulness. *Frontiers in Human Neuroscience, 6.*

Vestergaard-Poulsen, P., van Beek, M., Skewes, J., Bjarkam, C. R., Stubberup, M., Bertelsen, J., & Roepstorff, A. (2009). Long-term meditation is associated with increased gray matter density in the brain stem. *Neuro Report, 20*(2), 170–174. https://doi.org/10.1097/WNR.0b013e328320012a

Walsh, K. M., Saab, B. J., & Farb, N. A. (2019). Effects of a mindfulness meditation app on subjective Well-being: Active randomized controlled trial and experience sampling study. *JMIR Mental Health, 6*(1), e10844. https://doi.org/10.2196/10844

Williams, J. M., Crane, C., Barnhofer, T., Brennan, K., Duggan, D. S., Fennell, M. J., Hackmann, A., Krusche, A., Muse, K., Von Rohr, I. R., Shah, D., Crane, R. S., Eames, C., Jones, M., Radford, S., Silverton, S., Sun, Y., Weatherley-Jones, E., Whitaker, C. J., … Russell, I. T. (2013). Mindfulness-based cognitive therapy for preventing relapse in recurrent depression: A randomized dismantling trial. *Journal of Consulting and Clinical Psychology.* https://doi.org/10.1037/a0035036

Yang, C.-C., Barrós-Loscertales, A., Pinazo, D., Ventura-Campos, N., Borchardt, V., Bustamante, J.-C., Rodríguez-Pujadas, A., Fuentes-Claramonte, P., Balaguer, R., Ávila, C., & Walter, M. (2016). State and training effects of mindfulness meditation on brain networks reflect neuronal mechanisms of its antidepressant effect. *Neural Plasticity, 2016*, 1–14. https://doi.org/10.1155/2016/9504642

Yang, C.-C., Barrós-Loscertales, A., Li, M., Pinazo, D., Borchardt, V., Ávila, C., & Walter, M. (2019). Alterations in brain structure and amplitude of low-frequency after 8 weeks of mindfulness meditation training in meditation-Naïve subjects. *Scientific Reports, 9*(1), 10977. https://doi.org/10.1038/s41598-019-47470-4

Zeidan, F., & Vago, D. R. (2016). Mindfulness meditation-based pain relief: A mechanistic account. *Annals of the New York Academy of Sciences, 1373*(1), 114–127. https://doi.org/10.1111/nyas.13153

Zeidan, F., Martucci, K. T., Kraft, R. A., Gordon, N. S., McHaffie, J. G., & Coghill, R. C. (2011). Brain mechanisms supporting the modulation of pain by mindfulness meditation. *Journal of Neuroscience, 31*(14), 5540–5548. https://doi.org/10.1523/JNEUROSCI.5791-10.2011

Zhou, H.-X., Chen, X., Shen, Y.-Q., Li, L., Chen, N.-X., Zhu, Z.-C., Castellanos, F. X., & Yan, C.-G. (2020). Rumination and the default mode network: Meta-analysis of brain imaging studies and implications for depression. *NeuroImage, 206*, 116287. https://doi.org/10.1016/j.neuroimage.2019.116287

Index

Lightning Source UK Ltd.
Milton Keynes UK
UKHW020606130922
408788UK00001B/16